Traffic and Pavement
Engineering

Solved Practical Problems in Transportation Engineering

Volume I: Traffic and Pavement Engineering

Volume II: Highway Planning, Survey, and Design

Traffic and Pavement Engineering

Ghazi G. Al-Khateeb

University of Sharjah, UAE and Jordan University of Science and Technology, Jordan

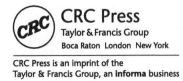

CRC Press
Taylor & Francis Group
Boca Raton London New York

CRC Press is an imprint of the
Taylor & Francis Group, an **informa** business

CRC Press
Taylor & Francis Group
6000 Broken Sound Parkway NW, Suite 300
Boca Raton, FL 33487-2742

© 2021 by Taylor & Francis Group, LLC
CRC Press is an imprint of Taylor & Francis Group, an Informa business

No claim to original U.S. Government works

Printed on acid-free paper

International Standard Book Number-13: 978-0-367-14983-3 (Hardback)
International Standard Book Number-13: 978-0-367-50010-8 (Paperback)

Visit the Taylor & Francis Web site at
http://www.taylorandfrancis.com

and the CRC Press Web site at
http://www.crcpress.com

This book is dedicated to my parents;

to the soul of my mother, and

my father, Gaseem;

to my beloved family;

my wife, Nuha; my sons, Khalid, Amr, and Ayham; and my daughter, Dana;

to

my sisters, Shareefah and Suhad;

and my brothers, Ja'fer, Mohammed, Faisal, Hashem, and Tareq;

and to

all my teachers, supervisors, colleagues, friends, and students
in every place I studied, taught, or researched …

Without the science, knowledge, support, and encouragement I received
over the years and the tremendous time and persistence devoted
in writing this book, this huge work would not be possible.

Contents

PART I   Traffic Engineering

PART II   Pavement Materials, Analysis, and Design

Preface

This book, comprising of two volumes and encompassing a total of five parts, is the outcome of a great deal of effort and time spent over the years in studying and teaching at outstanding academic institutes like Jordan University of Science and Technology, Applied Science University, University of Illinois at Urbana-Champaign, the American University of Sharjah, and the University of Sharjah, and in conducting advanced scientific research at some of the highly recognized research centers worldwide like the Advanced Transportation Research and Engineering Laboratory (ATREL) at the University of Illinois in Urbana-Champaign, the Federal Highway Administration's Turner-Fairbank Highway Research Center (TFHRC), advanced research labs at Jordan University of Science and Technology and the Advanced Pavement Research Lab at the University of Sharjah.

The book implements a unique kind of approach and categorizes transportation engineering topics into five major key areas, as shown below:

- Volume I: Traffic and Pavement Engineering
  - Part I "Traffic Engineering" deals with the functional part of transportation systems and introduces engineering techniques, practices, and models that are applied to design traffic systems, control traffic flow and movement, and construct proper roads and highways to achieve safe and efficient movement of people and traffic on roadways.
  - Part II "Pavement Materials, Analysis, and Design" deals with both the structural and functional parts of transportation facilities and introduces engineering techniques and principles of the uses of high-quality and sustainable materials that are employed to design, maintain, and construct asphalt-surfaced road pavements and concrete rigid pavements. The ultimate goal of pavement engineering is to provide a pavement structure that is safe, durable, sustainable, and capable of carrying the predicted traffic loads under prevailing climatic conditions. Proper structural design of pavements is one that takes into consideration the mechanistic analysis of pavements for stresses and strains that can predict the performance of the pavement with time. This section fulfills this goal by presenting the subject in a unique manner.

- Volume II: Highway Planning, Survey, and Design

  - Part I "Urban Transportation Planning" presents a process that involves a multi-modal approach and comprehensive planning steps and models to design and evaluate a variety of alternatives for transportation systems and facilities, predict travel demand and future needs, and manage the facilities and services for the different modes of transportation to finally achieve a safe, efficient, and sustainable system for the movement of people and goods.
  - Part II "Highway Survey" presents the basic concepts and standard procedures necessary to make precise and accurate distance, angle, and level measurements for highway alignment, cross-sections, and earth quantities used in the design of highways.
  - Part III "Geometric Design of Highways" deals with engineering design techniques, standards, and models that control the three main elements of highway geometric design: horizontal and vertical alignments, profile, and cross-section to achieve the primary objectives of geometric design: safety, efficiency, and sustainability.

The book is designed to benefit students in engineering programs at academic institutes where courses in pavement engineering, highway engineering, transportation engineering, traffic engineering, urban transportation planning, and survey are offered. The book is intended to be used as a state-of-the-art textbook for engineering students at the undergraduate and graduate level as well as professionals and technologists in the civil engineering field.

The main goals of this book are:

(1) To serve as a textbook in traffic and pavement engineering as well as highway and transportation engineering at the undergraduate level.
(2) To serve as a reference book in advanced courses or special topics that deal with contemporary subjects at the graduate level.
(3) To serve as a reliable professional reference for academic professors, practitioners, professional engineers, professional/licenser exams, site engineers, researchers, lab managers, quality control/quality assurance (QC/QA) engineers, and technologists in the field of civil engineering.

The distinctiveness of this book emanates from the plentiful number of problems on each topic and the broad range of ideas and practical problems that are included in all areas of the book. Furthermore, the problems cover theory, concepts, practice, and applications. The solution of each problem in the book follows a step-by-step procedure that includes the theory and the derivation of the formulas in some cases and the computations. Besides this, almost all problems in the five parts of the book include detailed calculations that are solved using MS Excel worksheets where mathematical, trigonometric, statistical, and logical formulas are used by inserting the correct function in the worksheet to perform the computations more rapidly and efficiently. The MS Excel Solver tool is at times used for solving complex equations in several problems in the book. Additionally, numerical methods, linear algebraic methods, and least squares regression techniques are utilized in some problems to assist in solving the problem and make the solution much easier. The advantage of these MS Excel worksheets and computations is that each one can be used to solve other practical problems with similar type of inputs by just changing the input values to obtain the outputs or the results. The book is supplemented by a CD that includes all the MS Excel worksheets for the computational problems of the book.

In summary, the book is designed to be informative and filled with an abundance of solutions to problems in the engineering science of transportation. It is hoped that this book will enrich the knowledge and science in transportation engineering, thereby elevating the civil engineering profession in general and the transportation engineering practice in particular as well as advancing the transportation engineering field to the best levels possible. It is also hoped that the targeted domain including students, academic professors, and professionals will benefit considerably from this book.

Author's Bio

Dr. Ghazi G. Al-Khateeb received a Bachelor of Science (B.S.) and Master of Science (M.S.) in Civil Engineering/ Transportation from Jordan University of Science and Technology (JUST) in 1991 and 1994, respectively, and the doctoral (Ph.D.) degree in Civil Engineering/ Transportation from the University of Illinois at Urbana-Champaign, USA in 2001.

Dr. Al-Khateeb is currently a professor at the University of Sharjah (UOS) in the United Arab Emirates (September 1, 2015 to present). He also served as a visiting professor at the American University of Sharjah, UAE (September 1, 2014 to August 31, 2015). He is on currently leave from Jordan University of Science and Technology. He has been on the academic staff of JUST since September of 2006. During his work at JUST, Dr. Al-Khateeb held the position of Vice Dean of Engineering for two years (September 2012 to September 2014) and the Vice Director for the Consultative Center for Science and Technology (September 2009 to September 2010).

Previously, he worked as a senior research scientist at the Turner-Fairbank Highway Research Center (TFHRC) of the Federal Highway Administration (FHWA) in Virginia, USA for 6 years (November 2000 to September 2006). Dr. Khateeb's research is in the area of pavement and transportation engineering. He has published more than 90 papers in international scientific refereed journals and conferences as well as book chapters. Dr. Al-Khateeb teaches undergraduate as well as graduate courses in civil engineering, transportation engineering, and pavement engineering, and has supervised many senior design projects for undergraduate students and thesis work for graduate students.

Dr. Al-Khateeb served as a member of the American Society of Civil Engineers (ASCE), the Association of Asphalt Paving Technologies (AAPT), Jordan Engineers Association (JEA), Jordan Society for Scientific Research (JSSR), and Jordan Road Accidents Prevention (JRAP).

Dr. Al-Khateeb is listed in *Who's Who* in Engineering Academia. In addition, he serves as an editorial / advisory board member for several international journals and publishers such as the *Materials Analysis and Characterization* at Cambridge Scholars Publishing, *Science Progress* in the Materials Science and Engineering Section with Sage Publishing, the *International Journal of Recent Development in Civil and Environmental Engineering*. He also served as a lead guest editor for a special issue on "Innovative Materials, New Design Methods, and Advanced Characterization Techniques for Sustainable Asphalt Pavements," for the *International Journal of Advances in Materials Science and Engineering*. In addition, he is an active reviewer for more than twenty international indexed and refereed scientific journals.

Dr. Al-Khateeb has been the principal investigator and co-investigator for many funded research projects in his field. He has also received several honor awards for his study, teaching, and research accomplishments during his career.

Dr. Al-Khateeb has served on several technical national and international committees such as the Technical Committee for the Road Master Plan of Jordan, the Asphalt and Roads Technical Committee and Superpave Technical Committee for the Jordan's Ministry of Public Works and Housing, and the Technical Traffic Committee for Irbid City. At the universities where he worked, he has served on many technical committees at the university level, college level, and department level.

Acknowledgments

Special thanks to the publisher of this book, Taylor & Francis Group/CRC Press, particularly to Joseph Clements, senior publisher; Lisa Wilford, editorial assistant; Joette Lynch, production editor; and Bryan Moloney, project manager at Deanta Global, for their support and help.

Part I

Traffic Engineering

1 Terminology, Concepts, and Theory

Chapter 1 includes questions and problems that cover the terminology, concepts, and theory used in traffic engineering. A traffic engineering practitioner should have the minimum understanding and grasp of the terms used in traffic engineering as well as knowledge and perception of the concepts and theory behind this type of engineering. The questions will shed light on this aspect, which will comprise the four major themes of traffic engineering in this book. The questions and the answers are available at the end of Part I.

2 Characteristics of Driver, Vehicle, and Roadway

Chapter 2 includes practical problems and questions related to the characteristics of driver, vehicle, and roadway. The characteristics of driver (road user), vehicle (moving object on the road), and the road itself are the three major components that compose the whole traffic system. Hence, safety and efficiency of the system is dependent on these three components. The ideal characteristics of these modules will lead to idea conditions and operation of the traffic system. However, variability, inconsistency, and unpredictability commonly prevail in any traffic system, which result in some complications. Consequently, there is a need to understand these characteristics and to compute motion parameters and travel measures in order to predict the operation of the system. The following practical problems and questions will focus on these aspects of traffic engineering.

2.1 The design speed on a highway is 100 km/h (62.1 mph), and the highway has an upgrade of 3%. In this case and by assuming the perception–reaction time=2.5 s and the average deceleration rate of vehicles is 3.4 m/s² (11.2 ft/s²), estimate the minimum stopping sight-distance on this highway.

Solution:

The following two formulas are used; the first one is in SI units and the second one is in US customary units:

$$SSD = 0.278 \, ut + \frac{u^2}{254\left(\dfrac{a}{g} \pm G\right)} \quad \text{(SI units)} \tag{2.1}$$

$$SSD = 1.47 \, ut + \frac{u^2}{30\left(\dfrac{a}{g} \pm G\right)} \quad \text{(US customary units)} \tag{2.2}$$

Since the highway has an upgrade of 3%, a +ve sign will be used for the G in the denominator of the formula:

$$SSD = 0.278(100)(2.5) + \frac{(100)^2}{254\left(\dfrac{3.4}{9.81} + 0.03\right)}$$

$$SSD = 174.0 \text{ m (570.3 ft)}$$

The MS Excel worksheet used for rapid and efficient solution is shown in Figure 2.1.

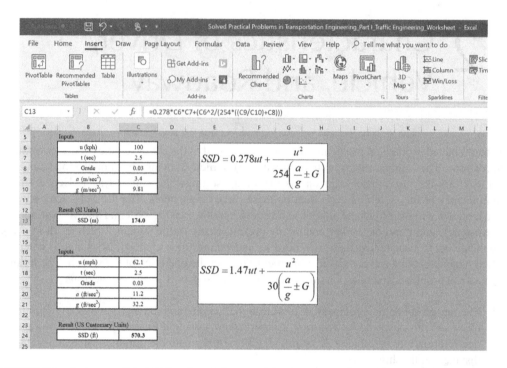

FIGURE 2.1 An image of the MS Excel worksheet used for the computations of Problem 2.1.

2.2 In a study to verify the design speed on a roadway with a downgrade of 4%, the estimated stopping sight distance for vehicles is 200 m (656.2 ft), what would the recommended design speed be on this roadway assuming that the average perception–reaction time of drivers is 2.5 s and the average deceleration rate of vehicles is 3.4 m/s² (11.2 ft/s²).

Solution:

The same two formulas as above will be used:

$$SSD = 0.278\,ut + \frac{u^2}{254\left(\dfrac{a}{g} \pm G\right)} \quad \text{(SI units)}$$

$$SSD = 1.47\,ut + \frac{u^2}{30\left(\dfrac{a}{g} \pm G\right)} \quad \text{(US customary units)}$$

Since the highway has a downgrade of 4%, a −ve sign will be used for the G in the denominator of the formula:

$$200 = 0.278(2.5)u + \frac{u^2}{254\left(\dfrac{3.4}{9.81} - 0.04\right)}$$

By simplifying and reformulating the above equation, the following equation is obtained:

$$0.012791\,u^2 + 0.695\,u - 200 = 0 \tag{2.3}$$

Following the solution of a quadratic equation, the speed can be obtained as follows:

$$u = -0.695 \mp \frac{\sqrt{(0.695)^2 - 4(0.012791)(-200)}}{2(0.012791)}$$

Two values for the speed are obtained; one is negative (not possible) and the other one is positive (practical).

$$u = 100.6 \text{ kph (62.6 mph)}.$$

The MS Excel Solver tool can be used to solve this problem as well.

Excel Solver Tool Procedure

If you do not have it: in MS Office 2003-2000, from "File" in the top menu, select "Options", "Add-Ins", "Solver Add-In", and hit "Go", Select "Solver Add-In" and hit "OK". Now, you will find it under "Data" in the top menu of the Excel worksheet.

Set up the Solver worksheet before using the tool:

The left side of the equation in this case = 200.

The right side is a function of the speed and the other known variables. Start with an initial value for the speed, the right side will have a computed value now based on the speed initial value and the other known parameters. Define the error, which basically represents the deviation between the left side and the right side of the equation. It is defined as the square of the difference between the two.

Now, go to "Data", "Solver", you will see a window like the one below.

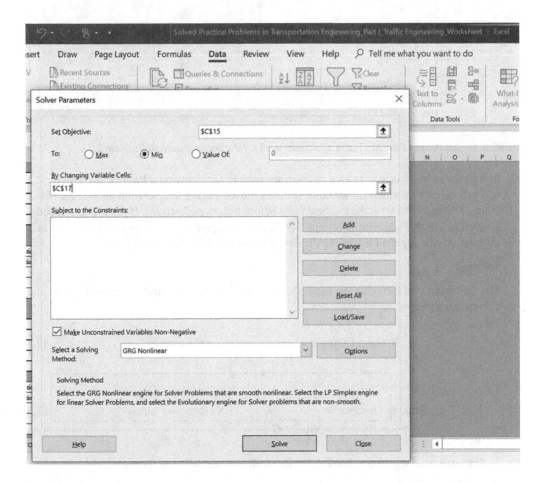

FIGURE 2.2 An image of the MS Excel Solver used for determining the design speed for Problem 2.2.

Use "Set Objective" to set the error to a value of zero or minimum; choose the cell of the error in the Excel worksheet, and use "By Changing Variable Cells" to select the cell of the "speed" value. Afterwards, hit "Solve". The Excel Solver will keep doing iterations until the error is close to zero by changing the speed value. The results obtained using the Excel Solver optimization are shown in Figure 2.2.

The MS Excel worksheet is shown in Figure 2.3.

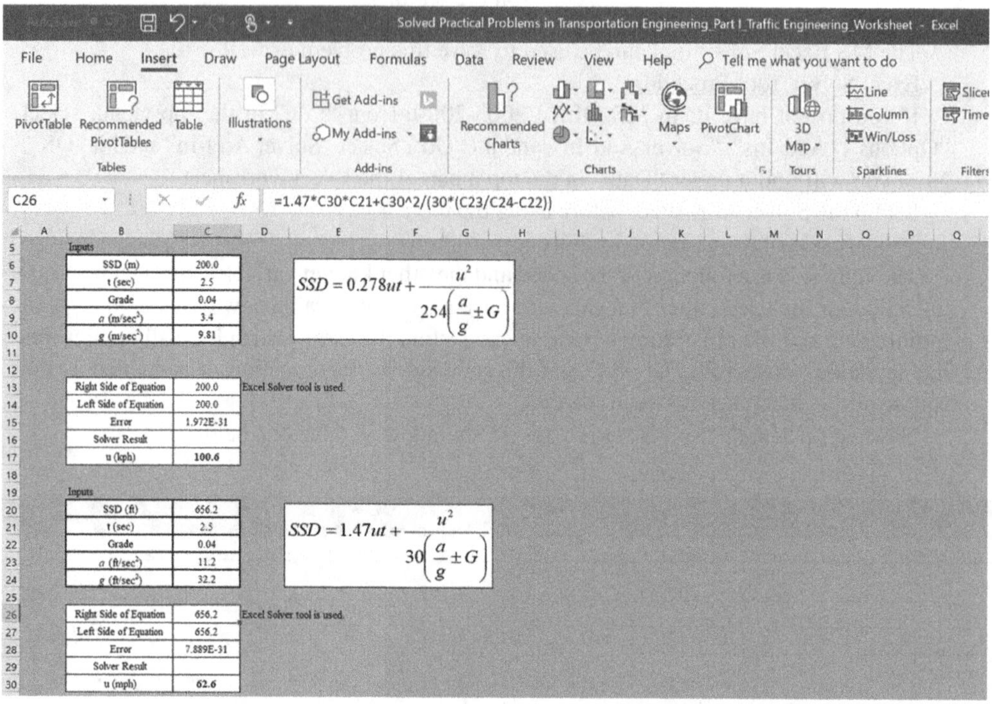

FIGURE 2.3 An image of the MS Excel worksheet used for the computations of Problem 2.2.

2.3 In a traffic safety study to reduce pedestrian–vehicle accidents on a level road segment, it was estimated by a video camera that vehicles moved during deceleration an average distance of 60 m (590.6 ft) at the moment right before hitting pedestrians. If the average speed of vehicles on this segment is 80 kph (49.7 mph), compute the <u>maximum safest speed</u> of vehicles on this segment assuming the deceleration rate of vehicles is 3.4 m/s²:

Solution:

The following two formulas are used; the first one is in SI units and the second one is in US customary units:

$$\text{Braking Distance} = \frac{u^2}{254\left(\dfrac{a}{g} \pm G\right)} \qquad \text{(SI units)} \qquad (2.4)$$

$$\text{Braking Distance} = \frac{u^2}{30\left(\dfrac{a}{g} \pm G\right)} \qquad \text{(US customary units)} \qquad (2.5)$$

Since the road segment is level, $G = 0$ in the denominator of the formula:

$$60 = \frac{u^2}{254\left(\dfrac{3.5}{9.81}\right)}$$

$\Rightarrow$

$$u = \sqrt{60(254)\left(\frac{3.4}{9.81}\right)}$$

$$u = 72.7 \text{ kph } (45.2 \text{ mph})$$

The MS Excel worksheet is shown in Figure 2.4.

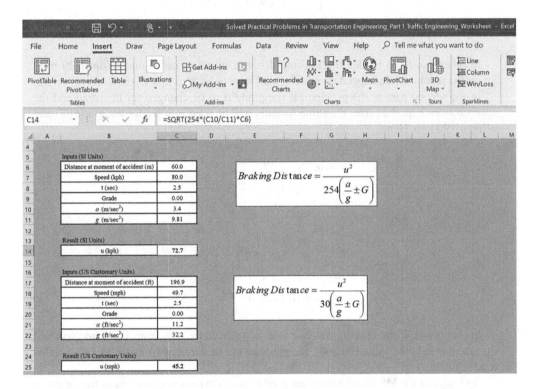

FIGURE 2.4   An image of the MS Excel worksheet used for the computations of Problem 2.3.

2.4 The average speed of vehicles on a road with a downgrade of 4% is 90 kph. A traffic study was conducted to improve the braking distance (reduce the braking distance) due to a high risk of accident occurrence on this road during the night. If the existing braking distance is 120 m, determine what would be the <u>recommended speed limit</u> based on a reduction in braking distance of 20%. If the grade of the road is to be reduced to achieve this reduction in braking distance, what would the <u>new grade</u> of the road be?

(Assume the deceleration rate of vehicles is 3.4 m/s²):

Solution:

The recommended speed is computed using the formula shown below, and at an improved braking distance equal to 80 m:

$$\text{Braking Distance} = \frac{u^2}{254\left(\dfrac{a}{g} \pm G\right)}$$

$$96 = \frac{u^2}{254\left(\dfrac{3.5}{9.81} - 0.04\right)}$$

$\Rightarrow$

$$u = \sqrt{96(254)\left(\frac{3.4}{9.81} - 0.04\right)}$$

$$u = 86.5 \text{ kph } (53.8 \text{ mph})$$

In other words: to reduce the braking distance on this road segment from 120 m to 96 m, the recommended speed of vehicles will have to be 86.5 kph (53.8 mph).

In the second part of the problem: if the grade is to be changed for this road and the speed is to be kept the same, the same formula will be used to solve for the grade (G) and using u = 90 kph. Hence:

$$96 = \frac{(90)^2}{254\left(\dfrac{3.4}{9.81} - G\right)}$$

$\Rightarrow$

$$G = \left(\frac{3.4}{9.81} - \frac{(90)^2}{96 \times 254}\right)$$

$$G = 0.014 \text{ or } 1.4\%$$

In conclusion, if the braking distance is to be reduced by 20% from 120 m to 96 m and the speed is kept the same (90 kph), the grade has to be changed from 4% to 1.4% on this road segment.

An image of the MS Excel worksheet used to perform the computations of this problem is shown in Figures 2.5 and 2.6.

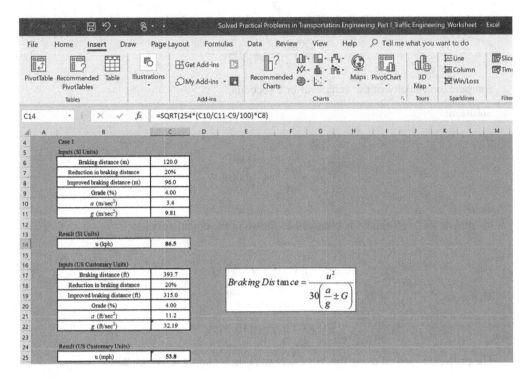

FIGURE 2.5 An image of the MS Excel worksheet used for the computations of Problem 2.4.

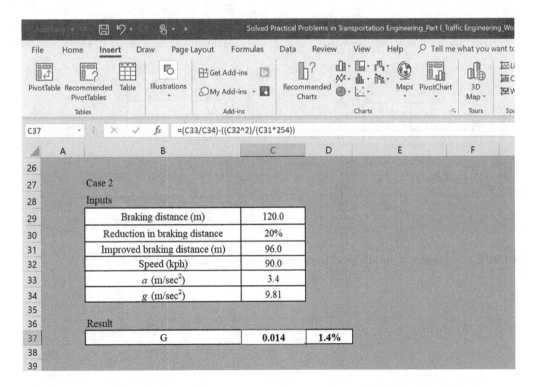

FIGURE 2.6 An image of the MS Excel worksheet used for the computations of Problem 2.4.

2.5 A vehicle is traveling at 80 kph. If the acceleration of the vehicle is given by the equation:

$$a = \frac{du}{dt} = 4 - 0.06u \qquad (2.6)$$

compute the velocity of the vehicle after 5 seconds:

Solution:

If

$$a = \frac{du}{dt} = 4 - 0.06u$$

then:

by integration, the velocity is obtained as shown in the equation below:

$$u = \frac{\alpha}{\beta}\left(1 - e^{-\beta t}\right) + u_0 e^{-\beta t} \qquad (2.7)$$

Therefore,

$$u = 33.7 \text{ m/s or } 110.7 \text{ ft/s}$$

The problem is solved using the MS Excel worksheet shown in Figure 2.7.

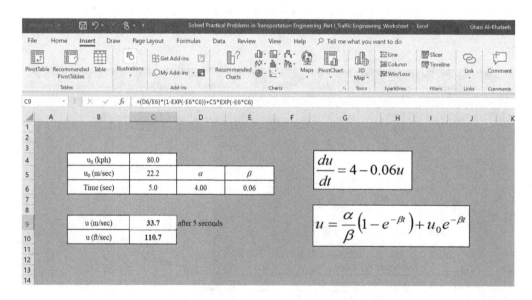

FIGURE 2.7   An image of the MS Excel worksheet used for the computations of Problem 2.5.

2.6 In Problem 2.5 above, compute the acceleration of the vehicle after 8 seconds.

Solution:

After 8 seconds, the vehicle will have a velocity of:

$$u = \frac{\alpha}{\beta}\left(1 - e^{-\beta t}\right) + u_0 e^{-\beta t}$$

$$u = 39.2 \text{ m/s or } 128.5 \text{ ft/s}$$

And therefore, the acceleration is computed using the formula given in Problem 2.5:

$$u = \frac{\alpha}{\beta}\left(1 - e^{-\beta t}\right) + u_0 e^{-\beta t}$$

$$a = \frac{du}{dt} = 4 - 0.06u$$

$$a = 1.65 \text{ m/s}^2 \text{ or } 5.41 \text{ ft/s}^2.$$

The problem is solved using the MS Excel worksheet shown in Figure 2.8.

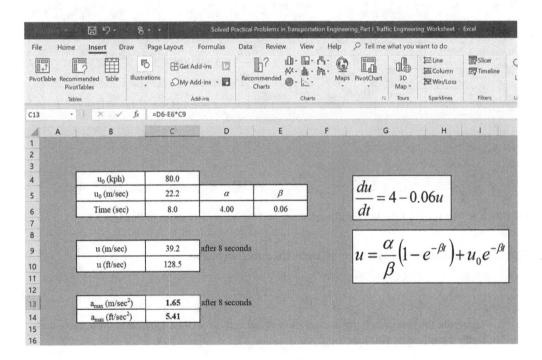

FIGURE 2.8 An image of the MS Excel worksheet used for the computations of Problem 2.6.

2.7   In the same Problem, 2.5, what is the distance travelled by the vehicle after 10 seconds?

Solution:

By integrating the velocity given in the formula in Problem 2.5, the following expression is obtained for the distance:

$$x = \frac{\alpha}{\beta}t - \frac{\alpha}{\beta^2}\left(1 - e^{-\beta t}\right) + \frac{u_0}{\beta}\left(1 - e^{-\beta t}\right) \tag{2.8}$$

And hence,

$$x = 332.5 \text{ m or } 1090.7 \text{ ft.}$$

The problem is solved using the MS Excel worksheet shown in Figure 2.9.

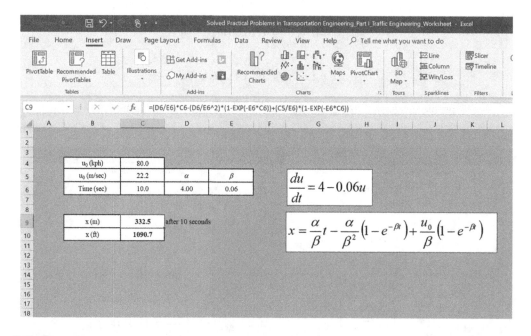

FIGURE 2.9   An image of the MS Excel worksheet used for the computations of Problem 2.7.

2.8   The acceleration of a vehicle is given by the equation below:

$$\frac{du}{dt} = 2 - 0.04\,u \tag{2.9}$$

Compute the time when the acceleration of the vehicle will be 1.0 m/s² if the vehicle was initially travelling at 100 kph:

Solution:

The acceleration of the vehicle is given by:

$$a = \frac{du}{dt} = 2 - 0.04\,u$$

When a = 1.0 m/s², u = 25.0 m/s.

By integrating the acceleration, the velocity is obtained as below:

$$u = \frac{\alpha}{\beta}\left(1 - e^{-\beta t}\right) + u_0 e^{-\beta t} \tag{2.10}$$

The time is obtained at $u = 25.0$ m/s, and given the parameters $\alpha = 2$ and $\beta = 0.04$ and the initial speed, $u_0 = 80$ kph $= 22.2$ m/s:

$$25 = \frac{2}{0.04}\left(1 - e^{-0.04t}\right) + 22.2 e^{-0.04t}$$

Solving the above equation for t provides:
The Excel Solver tool is used to obtain t:

$$t = 2.635 \text{ seconds} \cong 2.64 \text{ seconds.}$$

The MS Excel worksheet along with the Excel Solver tool are used to obtain the time in this problem as shown in Figure 2.10.

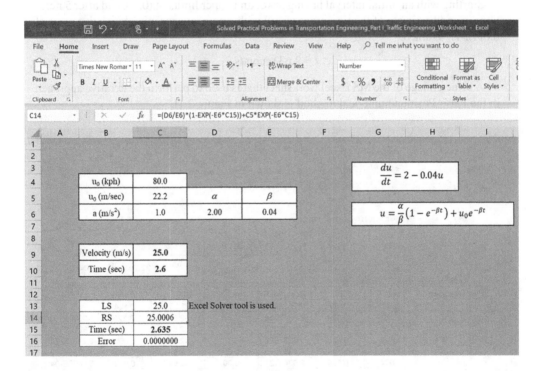

FIGURE 2.10 An image of the MS Excel worksheet used for the computations of Problem 2.8.

The False-Position numerical method or the Newton–Rapshon method (*Numerical Methods for Engineers* by Chapra and Canale, 2015) can also be used to solve the above equation as shown below:

$$f(t) = \frac{2}{0.04}\left(1 - e^{-0.04t}\right) + 22.2e^{-0.04t} - 25 = 0 \tag{2.11}$$

According to the False-Position numerical method, the root (time) at any iteration is obtained using the formula:

$$\text{Root} = \text{UL} - \frac{f(\text{UL})(\text{LL} - \text{UL})}{f(\text{LL}) - f(\text{UL})} \tag{2.12}$$

Where:

UL = upper limit of the selected interval in the previous iteration
LL = lower interval of the selected interval in the previous iteration
f(LL) = the function value at the selected upper limit, and
f(LL) = the function value at the selected lower limit

Starting with an initial interval having lower and upper limits of (0, 5), and after 5 iterations, a good estimate of the time (root of f(t)) with a percent approximate relative error of 0.011% is obtained:

$$t = 2.63 \text{ seconds.}$$

A screen image of the MS Excel worksheet that is used to perform the iterative approach in the False-Position numerical method is shown in Figure 2.11.

$$f(t) = \frac{2}{0.04}(1 - e^{-0.04t}) + 22.2e^{-0.04t} - 25 = 0$$

False-Position Numerical Method:

Iteration	LL	UL	Root (Tme)	f(LL)	f(UL)	f(Root)	ε_a (%)
1	0.00	5.00	2.7583278	-2.78	2.26	0.124	
2	0.000	0.124	2.5062054	-2.78	-2.64	-0.128	1.0060E+01
3	2.506	0.124	2.6277095	-0.13	-2.64	-0.006	4.6240E+00
4	2.628	0.124	2.6337018	-0.01	-2.64	0.000	2.2752E-01
5	2.634	0.124	2.6339975	0.00	-2.64	0.000	1.1229E-02
6	2.634	0.124	2.6340121	0.00	-2.64	0.000	5.5426E-04
7	2.634	0.124	2.6340129	0.00	-2.64	0.000	2.7358E-05
8	2.634	0.124	2.6340129	0.00	-2.64	0.000	1.3504E-06
9	2.634	0.124	2.6340129	0.00	-2.64	0.000	6.6656E-08
10	2.634	0.124	2.6340129	0.00	-2.64	0.000	3.2904E-09

FIGURE 2.11 An image of the MS Excel worksheet used for performing the False-Position numerical method for Problem 2.8.

Using the Newton–Raphson method, the root (time) at any iteration is obtained using the formula:

$$t_{i+1} = t_i - \frac{f(t_i)}{f'(t_i)}$$ (2.13)

Where:

t_{i+1} = the root after iteration i+1
t_i = the root after iteration i
$f(t_i)$ = the function value at t_i, and
$f'(t_i)$ = the derivative value at t_i

Using an initial value $t_0 = 0$, the estimate of the root after five iterations is equal to 2.654 with a percent relative error of 0.000%.
Therefore,

$$t = 2.65 \text{ seconds.}$$

A screen image of the MS Excel worksheet that is used to perform the iterative approach in the Newton–Raphson numerical method is shown in Figure 2.12.

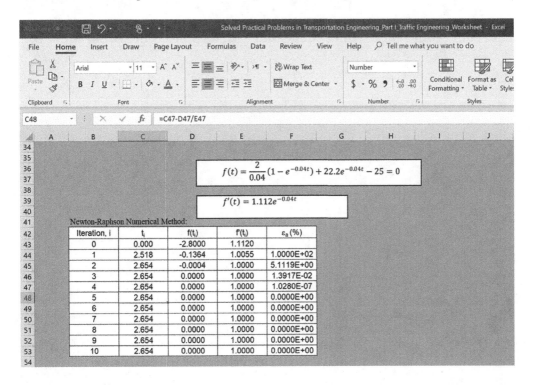

FIGURE 2.12 An image of the MS Excel worksheet used for performing the Newton–Raphson numerical method for Problem 2.8.

3 Traffic Flow Theory and Models

Chapter 3 covers the theory that controls traffic flow on highways. The fundamental relationships between flow, speed, and density of vehicles on highways are also presented. Different macroscopic models are presented in this part, which describe the relationship between speed and density on the highway. Uncongested conditions as well as congested condition will also be discussed. Bottleneck conditions that accompany a sudden reduction in the capacity of the highway as a result of urgent (up normal) situations on the highway such as accidents, construction on one or more lanes of the highway, etc., will be discussed as well. And finally, the concept of gap and gap acceptance in traffic streams is introduced. The methods of determining the critical gap for merging vehicles are also discussed. The practical problems presented in the following sections will focus on the aforementioned topics.

3.1 If the traffic flow on a highway segment is estimated to be 1800 vph, compute the average time headway on the highway segment.

Solution:

The average space headway (d) is related to the density of vehicles (k) through the following equation:

$$\bar{d} = \frac{1}{k} \tag{3.1}$$

And also, the space headway (d) is related to the time headway (h) and the space mean speed (u_s) through the following formula:

$$\bar{d} = u_s \bar{h} \tag{3.2}$$

And therefore, the time headway (h) can be

$$\bar{h} = \frac{1}{ku_s} = \frac{1}{q} \tag{3.3}$$

As the flow is given in the units of vph, the units of the time headway obtained from the above formula will be h/veh. To convert that into sec/veh, the answer will be multiplied by 3600:

$$\bar{h} = \frac{3600}{q} = \frac{3600}{1800} = 2 \text{ sec/veh}$$

A screen image of the MS Excel worksheet used to perform the computations of this problem is shown in Figure 3.1.

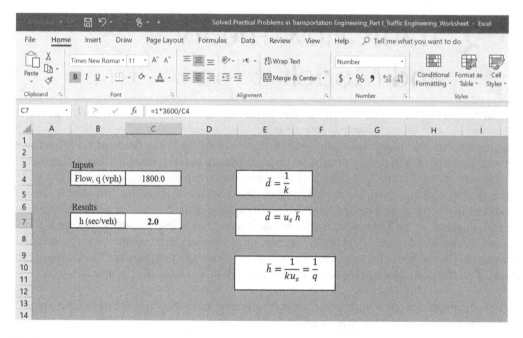

FIGURE 3.1 A screen image of the MS Excel worksheet used for the computations of Problem 3.1.

3.2 The space mean speed (u_s) on a highway segment is 60 mph (96.6 kph) and the average time headway is 3 sec/veh. Estimate the density and the flow on this highway segment.

Solution:

Using Equation (3.2), the space headway (d) can be computed as shown below:

$$\bar{d} = u_s \bar{h}$$

But before using this equation, the units must be consistent; in other words, the unit of the time headway should be in h/veh. Therefore:

Time headway = 3 sec/veh = 3/3600 h/veh.

$$\bar{d} = 60 \times \frac{3}{3600}$$

$$\bar{d} = 0.05 \text{ mi/veh} \left(0.08 \text{ km/veh} \right)$$

And since:

$$\bar{d} = \frac{1}{k}$$

$$k = \frac{1}{0.05} = 20 \text{ vpm} \left(12.4 \text{ veh/km} \right)$$

Since the space mean speed was given and the density was computed, the flow can be computed using the equation given below:

$$q = u_s k \tag{3.4}$$

$$q = 60 \times 20 = 1200 \text{ vph}$$

The MS Excel worksheet used to solve this problem is shown in Figure 3.2.

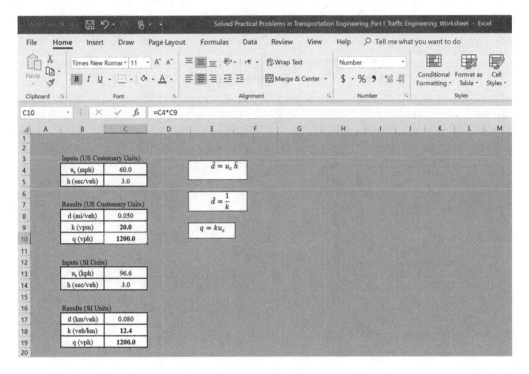

FIGURE 3.2   A screen image of the MS Excel worksheet used for the computations of Problem 3.2.

3.3   If the traffic flow and the average space headway on a highway segment are 1000 vph and 240 ft/veh, respectively, determine the space mean speed and the density on this highway segment.

Solution:

First the units of the time headway should be converted from ft/veh into mi/veh, the following is obtained:

$$\bar{d} = \frac{240}{5280} \text{ mi/veh}$$

Using Equation (3.1), the density can be obtained:

$$\bar{d} = \frac{1}{k}$$

$\Rightarrow$

$$k = \frac{5280}{240} = 22 \text{ vpm} \left(13.7 \text{ veh/km}\right)$$

Since:

$$q = u_s k$$

$$u_s = \frac{q}{k} = \frac{1000}{22} = 45.5 \text{ mph} \left(73.2 \frac{\text{km}}{\text{h}} \right)$$

The MS Excel worksheet shown in Figure 3.3 illustrates the computed results of this problem.

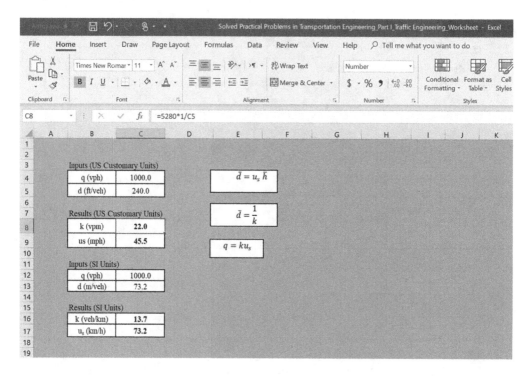

FIGURE 3.3   A screen image of the MS Excel worksheet used for the computations of Problem 3.3.

3.4 The number of vehicles passing a point on a highway segment was counted to be 500 vehicles during a time interval of 15 minutes. Determine the equivalent hourly flow rate on the highway segment.

Solution:

The hourly flow rate is computed using the formula:

$$q = \frac{N \times 3600}{T} \tag{3.5}$$

Where:
   q = equivalent hourly flow rate
   N = number of vehicles
   T = time period (seconds)

Alternatively:

$$q = \frac{N \times 60}{T} \tag{3.6}$$

Where:

T = time period (minutes)

Therefore,

$$q = \frac{500 \times 4}{15 * 60} = 2000 \text{ vph}$$

Or:

$$q = \frac{500 \times 60}{15} = 2000 \text{ vph}$$

For this problem, the MS Excel worksheet used to compute the flow is shown in Figure 3.4.

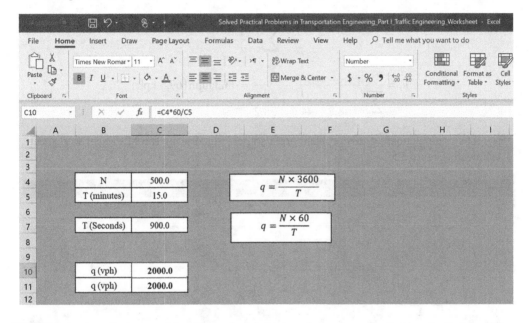

FIGURE 3.4 A screen image of the MS Excel worksheet used to compute the flow rate for Problem 3.4.

3.5 At a particular time on a highway, the speeds of three vehicles were 48.2, 44.6, and 38.2 mph (77.6, 71.8, and 61.5 kph). Compute the time mean speed and the space mean speed of the vehicles.

Solution:

The time mean speed (u_t) represents the arithmetic average of the speed of vehicles. Hence, it is computed using the following formula:

$$u_t = \frac{\sum_{i=1}^{n} u_i}{n} \qquad (3.7)$$

Where:

u_i = speed of vehicle i

n = number of vehicles

Therefore,

$$u_t = \frac{\sum_{i=1}^{3} u_i}{3}$$

$$u_t = \frac{48.2 + 44.6 + 38.2}{3} = 43.7 \text{ mph} (70.3 \text{ kph})$$

The space mean speed (u_s) is the harmonic mean of the speeds of vehicles. In other words, it is estimate d using the following formula:

$$u_s = \frac{n}{\sum_{i=1}^{n} \frac{1}{u_i}} \tag{3.8}$$

Therefore,

$$u_s = \frac{3}{\sum_{i=1}^{3} \frac{1}{u_i}}$$

$$u_s = \frac{3}{\sum_{i=1}^{3} \frac{1}{48.2} + \frac{1}{44.6} + \frac{1}{38.2}} = 43.3 \text{ mph} (69.6 \text{ kph})$$

The computations of this problem are also performed using the MS Excel worksheet shown in Figure 3.5.

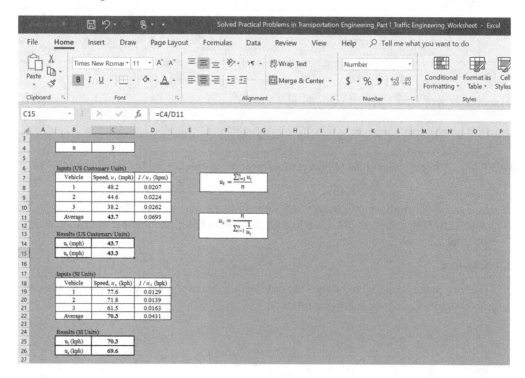

FIGURE 3.5 A screen image of the MS Excel worksheet used to compute the time mean speed and the space mean speed for Problem 3.5.

3.6 If the space mean speed for three vehicles on a highway segment is 40.3 mph (64.9 kph), and the individual speeds for two vehicles are 45.0 and 40.4 mph (72.4 and 65.0 kph), then what is the speed for the third vehicle?

Solution:

$$u_s = \frac{n}{\sum_{i=1}^{n} \frac{1}{u_i}}$$

Therefore,

$$u_s = \frac{3}{\sum_{i=1}^{3} \frac{1}{u_i}}$$

$$u_s = \frac{3}{\sum_{i=1}^{3} \frac{1}{45.0} + \frac{1}{40.4} + \frac{1}{u_3}} = 40.3$$

$$\Rightarrow$$

$$\frac{1}{u_3} = 0.0275$$

$$\Rightarrow$$

$$u_3 = 36.4 \text{ mph} (58.6 \text{ kph})$$

The MS Excel worksheet shown in Figure 3.6 is used to compute the required results in this problem.

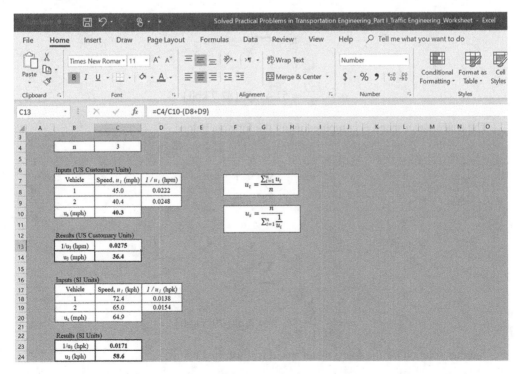

FIGURE 3.6 A screen image of the MS Excel worksheet used for the computations of Problem 3.6.

3.7 Five vehicles pass a 1000-ft (304.8-m) highway segment in time periods of 10, 14, 18, 15, 12 seconds, respectively. Determine the time mean speed and the space mean speed of the vehicles.

Solution:

The speeds of the five vehicles are computed using the following formula:

$$u_i = \frac{L}{t_i} \tag{3.9}$$

Where:
   u_i = speed of vehicle i
   L = length of segment
   t = time for vehicle i to pass the segment

The length is divided by 5280 to convert from ft to mile, and the time is divided by 3600 to convert from seconds to hr.

The time mean speed is then computed using the following formula as the arithmetic mean of the five speeds:

$$u_t = \frac{\sum_{i=1}^{n} u_i}{n}$$

$\Rightarrow$

$$u_t = \frac{\sum_{i=1}^{5} u_i}{5}$$

$$u_t = \frac{\sum_{i=1}^{5} u_i}{5}$$

$$u_t = \frac{257}{5} = 51.4 \text{ mph} (82.7 \text{ kph})$$

The space mean speed is computed using the formula below:

$$u_s = \frac{n}{\sum_{i=1}^{n} \frac{1}{u_i}}$$

$\Rightarrow$

$$u_s = \frac{5}{\sum_{i=1}^{5} \frac{1}{u_i}}$$

$$u_s = \frac{5}{0.1012} = 49.4 \text{ mph} (79.5 \text{ kph})$$

Another solution for the space mean speed is illustrated below:

$$u_s = \frac{n}{\sum_{i=1}^{n} \frac{1}{u_i}}$$

But:

$$u_i = \frac{L}{t_i}$$

Therefore,

$$u_s = \frac{n}{\sum_{i=1}^{n} \frac{t_i}{L}}$$

Or:

$$u_s = \frac{nL}{\sum_{i=1}^{n} t_i} \qquad (3.10)$$

$\Rightarrow$

$$u_s = \frac{5(1000)}{69} = 72.5 \text{ ft/s} = 49.4 \text{ mph} \left(79.5 \text{ kph}\right)$$

The screen images of the MS Excel worksheets used to perform the computations in this problem to determine the time mean speed and space mean speed are shown in Figures 3.7 and 3.8.

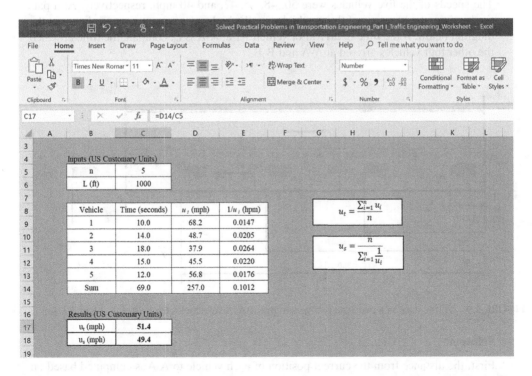

FIGURE 3.7   A screen image of the MS Excel worksheet used to compute the time mean speed and the space mean speed for Problem 3.7 (US customary units).

					Solved Practical Problems in Transportation Engineering_Part I_Traffic Engineering_Worksheet - Excel	

File   Home   Insert   Draw   Page Layout   Formulas   Data   Review   View   Help   Tell me what you want to do

C34   fx   =C21/E30

	A	B	C	D	E	F	G	H	I	J	K	L
19												
20		Inputs (SI Units)										
21		n	5									
22		L (m)	304.8									
23												
24		Vehicle	Time (seconds)		u_i (kph)	$1/u_i$ (hpm)						
25		1	10.0		109.7	0.0091						
26		2	14.0		78.4	0.0128						
27		3	18.0		61.0	0.0164						
28		4	15.0		73.2	0.0137						
29		5	12.0		91.4	0.0109						
30		Sum	69.0		413.7	0.0629						
31												
32		Results (SI Units)										
33		u_t (mph)	82.7									
34		u_s (mph)	79.5									
35												
36												

FIGURE 3.8   A screen image of the MS Excel worksheet used to compute the time mean speed and the space mean speed for Problem 3.7 (SI Units).

3.8   Traffic data were collected on a two-lane two-way highway segment between A-A and B-B 400-ft (121.9-m) apart as shown in Figure 3.9. Five vehicles passed A-A at different times. The speeds of the five vehicles were 50, 48, 45, 42, and 40 mph, respectively. At a particular time, the locations of these vehicles on the highway segment were as shown in the diagram below. If the time of arrival of vehicle #1 at A-A is t_0, compute the time at which the other four vehicles arrived at A-A and B-B.

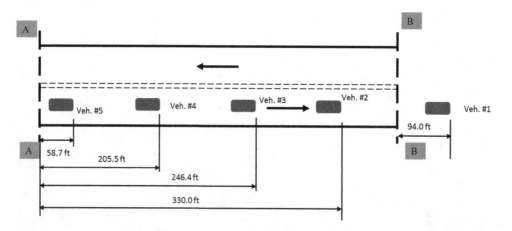

FIGURE 3.9   Traffic data on a two-lane two-way highway for five vehicles for Problems 3.8 and 3.9.

Solution:

First, the distance from the current position of each vehicle to A-A is computed based on the distances given in the diagram as follows:
For vehicle #1: distance = 400 + 94 = 494.0 ft.

The other distances for the other vehicles are simply given in the diagram.

Second, the travel time (seconds) from A-A to reach the current position for each vehicle is calculated using the formula given below taking into consideration the conversions of the units:

$$t_i = \frac{\text{Distance}}{u_i} \tag{3.11}$$

For vehicle #1:

$$t_1 = \frac{494}{50 \times 5280 / 3600} = 6.74 \text{ seconds}$$

The other results are shown in Table 3.1.

Third, if the time of arrival of vehicle #1 at A-A is t_0, then the time of arrival of the other vehicles at A-A can be calculated by determining the difference in time between vehicle #1 and each of the other four vehicles from A-A to reach the current position.

Vehicle #1 took 6.74 seconds to reach its current position in the diagram. Since vehicle #2 took 4.69 seconds to reach its current position and vehicle #1 reached A-A at t_0, then vehicle #2 arrived at A-A at $t_0+(6.74-4.69)=t_0+2.05$. In a similar manner, the time of arrival at A-A for the other vehicles is computed as shown in Table 3.1.

Fourth, the time of arrival at B-B can be calculated by adding the time of arrival at A-A and the travel time required to pass from A-A to B-B for each vehicle.

Travel time from A-A to B-B (length of segment=L=400 ft) is determined as:

$$t_i = \frac{L}{u_i}$$

Therefore, for vehicle #1:

$$t_i = \frac{400}{50 \times 5280 / 3600} = 5.45 \text{ seconds}$$

Hence, the arrival time at B-B$=t_0+5.45$

In a similar manner, the time of arrival at B-B for the other four vehicles is computed as shown in Table 3.1. The results for the five vehicles are summarized in Table 3.1.

TABLE 3.1
The Results for the Arrival Times at A-A and B-B for Problem 3.8

Vehicle #	Speed (mph)	Distance from A-A (ft)	Travel Time from A-A to reach current position (sec)	Difference in Time between vehicle #1 and the others (sec)	Time of Arrival at A-A (sec)	Travel Time from A-A to B-B (sec)	Time of Arrival at B-B (sec)
1	50.0	494.0	6.74	0.00	t_0	5.45	$t_0+5.45$
2	48.0	330.0	4.69	2.05	$t_0+2.05$	5.68	$t_0+7.73$
3	45.0	246.4	3.73	3.00	$t_0+3.00$	6.06	$t_0+9.06$
4	42.0	205.5	3.34	3.40	$t_0+3.40$	6.49	$t_0+9.89$
5	40.0	58.7	1.00	5.74	$t_0+5.74$	6.82	$t_0+12.56$

The screen images of the MS Excel worksheet used to perform the computations of this problem are shown in Figures 3.10 and 3.11.

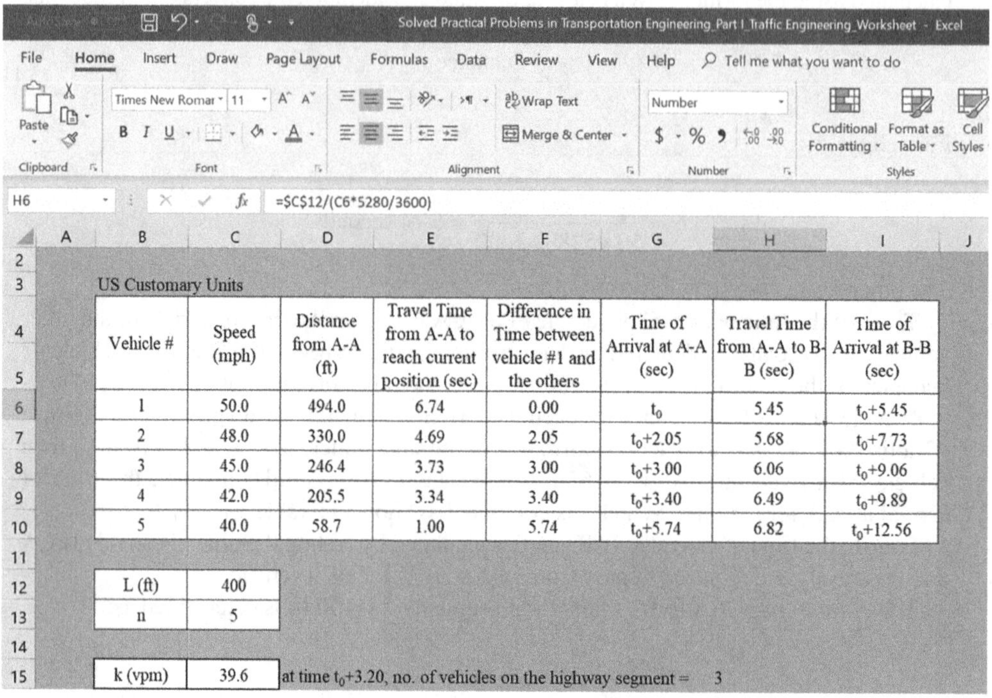

FIGURE 3.10 Screen images of the MS Excel worksheet for the computations of Problem 3.8 (US customary units).

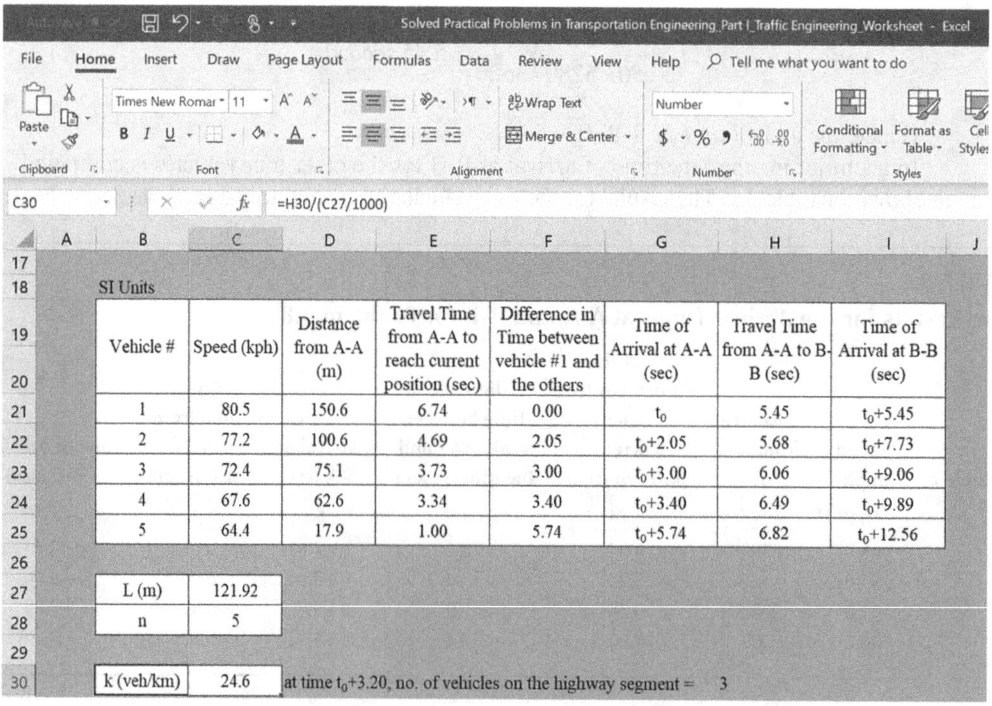

FIGURE 3.11 Screen images of the MS Excel worksheet for the computations of Problem 3.8 (SI units).

3.9   In Problem 3.8, determine the density of the highway segment at $t_0 + 3.20$ sec.

Solution:

At time $t_0 + 3.20$ sec, the number of vehicles on the highway segment are determined. Watching the times of arrival of each vehicle at A-A and B-B, it can be determined which vehicle existed on the segment at that time. For instance, vehicle #1 arrived at A-A at t_0 and reached B-B at $t_0 + 5.45$, that means at time $t_0 + 3.20$, the vehicle was still on the highway segment. In the same way, vehicle #2 arrived at A-A at $t_0 + 2.05$ and reached B-B at $t_0 + 7.73$, which means again that at time $t_0 + 3.2$, the vehicle was on the highway segment. Vehicle #3 also existed on the highway segment at $t_0 + 3.2$. The other two vehicles (#4 and #5) arrived at A-A at times $t_0 + 3.40$ and $t_0 + 5.74$, respectively; which means that at time $t_0 + 3.20$, both vehicles were out of the segment (specifically before A-A; i.e., they didn't reach A-A yet). Therefore, the number of vehicles that existed on the highway segment at time $t_0 + 3.2$ was three vehicles. The density is computed as shown below:

$$k = \frac{N}{L} \qquad (3.12)$$

Where:
    N = number of vehicles on the highway segment
    L = length of the highway segment

$\Rightarrow$

$$k = \frac{3}{400 / 5280} = 39.6 \text{ vpm} \left(24.6 \text{ veh/km} \right)$$

The 5280 is a conversion factor to convert the unit of ft to mile.

The computation of the density is also shown in the screen images of the MS Excel worksheet of the previous problem (Problem 3.8).

3.10 Figure 3.12 shows the time–space diagram for four vehicles. Determine the following:
    (a) Space headway between vehicles #3 and #4 at time = 6 second
    (b) Speed of vehicle #3 at time = 3 seconds
    (c) Speed of vehicle #1

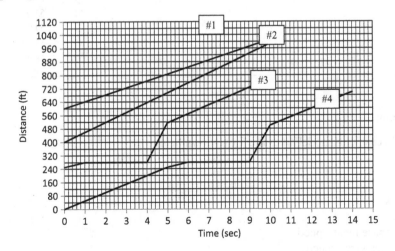

FIGURE 3.12   Time–space diagram for four vehicles for Problem 3.10.

Solution:

(a) From the diagram, at time=6 seconds, vehicle #3 and vehicle #4 traveled distances of 560 ft and 280 ft, respectively. Therefore, the space headway between the two vehicles is equal to:

$$d = 560 - 280 = 280 \text{ ft} (85.3 \text{ m})$$

(b) At time=3 seconds, the location of vehicle #3 is 280 ft, and the vehicle is stopping since the distance is constant. Therefore, speed=0.

(c) The time-distance relationship for vehicle #1 is linear as seen from this diagram and does not change all the way. The slope of this line is simply the speed. Therefore:

$$\text{Speed of vehicle } \#1 = \frac{\text{Distance traveled}}{\text{Time Interval}} \tag{3.13}$$

$$\text{Speed of vehicle } \#1 = \frac{(1010 - 600)}{(10 - 0)} = 41.0 \text{ ft/sec} = 28.0 \text{ mph} (45 \text{ kph})$$

3.11 The traffic density and speed data shown in Table 3.2 were obtained on a highway segment. Use linear regression analysis to fit the data to Greenshields model.

TABLE 3.2
Traffic Density and Speed Data on a Highway Segment for Problem 3.11

Density, k (vpm)	Speed, u_s (mph)
37.5	30.0
22.5	35.0
18.0	40.0
15.0	44.0
12.0	47.0
9.0	50.0
7.5	52.0
6.0	54.0
4.5	55.0
3.0	58.0

Solution:

The Greenshields model takes the form shown below:

$$u_s = u_f - \frac{u_f}{k_j} k \tag{3.14}$$

Where:
   u_s = space mean speed
   u_f = mean free speed
   k_j = jam density
   k = density

For a linear equation with the form: $y = a_0 + a_1 x$, the following linear system (with matrix notation) is obtained using linear squares regression techniques and numerical methods:

$$\begin{bmatrix} n & \sum x \\ \sum x & \sum x^2 \end{bmatrix} \begin{bmatrix} a_0 \\ a_1 \end{bmatrix} = \begin{bmatrix} \sum y \\ \sum xy \end{bmatrix} \tag{3.15}$$

This system represents the two normal equations that will be used to solve for the coefficients a_0 and a_1 in the linear equation.

The solution of the above linear system is:

$$a_1 = \frac{\sum_{i=1}^{n} x_i y_i - \frac{1}{n} \left(\sum_{i=1}^{n} x_i \right) \left(\sum_{i=1}^{n} y_i \right)}{\sum_{i=1}^{n} x_i^2 - \frac{1}{n} \left(\sum_{i=1}^{n} x_i \right)^2} \tag{3.16}$$

$$a_0 = \frac{\sum_{i=1}^{n} y_i}{n} - a_1 \frac{\sum_{i=1}^{n} x_i}{n} \tag{3.17}$$

Or:

$$a_0 = \bar{y} - a_1 \bar{x} \tag{3.18}$$

The required computations are done to obtain the parameters in the above linear system (see Table 3.3).

TABLE 3.3

Computations Needed for the Determination of the Linear Regression Coefficients a_0 and a_1 for Problem 3.11

	Density, k (vpm)	Speed, u_s (mph)	k^2	$(u_s)(k)$
	37.5	30.0	1406.3	1125.0
	22.5	35.0	506.3	787.5
	18.0	40.0	324.0	720.0
	15.0	44.0	225.0	660.0
	12.0	47.0	144.0	564.0
	9.0	50.0	81.0	450.0
	7.5	52.0	56.3	390.0
	6.0	54.0	36.0	324.0
	4.5	55.0	20.3	247.5
	3.0	58.0	9.0	174.0
SUM	135	465	2808.0	5442.0

Therefore:

$$\begin{bmatrix} 10 & 135 \\ 135 & 2808 \end{bmatrix} \begin{bmatrix} a_0 \\ a_1 \end{bmatrix} = \begin{bmatrix} 465 \\ 5442 \end{bmatrix} \tag{3.19}$$

The Gauss elimination method is used to solve the above linear system. The solution provides the following values of a_0 and a_1:

$$a_0 = 57.95 \cong 58.0$$

$$a_1 = -0.85$$

The MS Excel worksheet used to apply the Gauss elimination method in order to solve the linear system is shown in Figure 3.13.

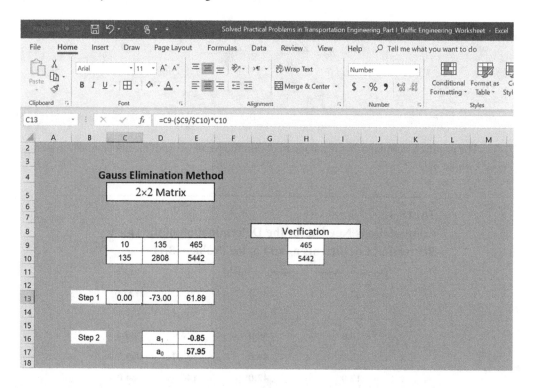

FIGURE 3.13 A screen image of the MS Excel worksheet used to apply the Gauss elimination method in the linear regression analysis for Problem 3.11.

Or:

$$a_1 = \frac{\sum_{i=1}^{n} x_i y_i - \frac{1}{n}\left(\sum_{i=1}^{n} x_i\right)\left(\sum_{i=1}^{n} y_i\right)}{\sum_{i=1}^{n} x_i^2 - \frac{1}{n}\left(\sum_{i=1}^{n} x_i\right)^2}$$

$$a_1 = \frac{5442 - \dfrac{1}{10}(135)(465)}{2808 - \dfrac{1}{10}(135)^2} = -0.85$$

$$a_0 = \frac{\sum_{i=1}^{n} y_i}{n} - a_1 \frac{\sum_{i=1}^{n} x_i}{n}$$

$$a_0 = \frac{465}{10} - (-0.85)\frac{135}{10} = 58$$

Therefore,

The Greenshields model that describes this data takes the following form:

$$u_s = 58.0 - 0.85\,k \tag{3.20}$$

The screen image of the MS Excel worksheet used to conduct the computations and the linear regression analysis of this problem is shown in Figure 3.14.

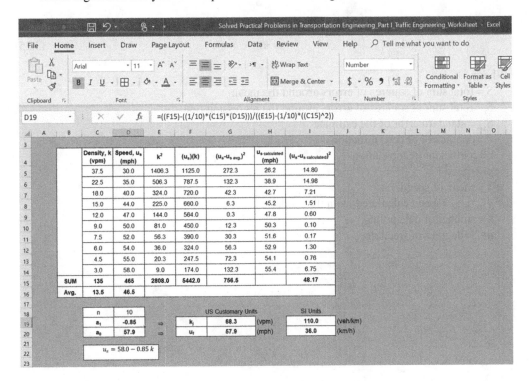

FIGURE 3.14   A screen image of the MS Excel worksheet used for the computations and the linear regression analysis for Problem 3.11.

3.12 For the Greenshields model determined in Problem 3.11, compute the following statistical parameters that quantify the "goodness" of fit of the model to the data and plot the density–speed relationship:
  (a) Total sum of squares of errors around the mean (S_t)
  (b) Total sum of squares of errors around the regression line (S_r)

(c) Standard error of the estimate ($S_{y/x}$)
(d) Coefficient of determination (r^2)
(e) Coefficient of correlation (r)

Solution:

$$S_t = \sum_{i=1}^{n} \left(y_i - \overline{y} \right)^2 \tag{3.21}$$

$$S_r = \sum_{i=1}^{n} \left(y_i - y_{i\text{-calculated}} \right)^2 \tag{3.22}$$

$$S_{y/x} = \sqrt{\frac{S_r}{n-2}} \tag{3.23}$$

$$r^2 = \frac{S_t - S_r}{S_t} \tag{3.24}$$

$$r = \sqrt{r^2} \tag{3.25}$$

Where:
S_t = total sum of squares of errors around the mean
S_r = total sum of squares of errors around the regression line
$S_{y/x}$ = standard error of the estimate
r^2 = coefficient of determination
r = coefficient of correlation
y_i = speed i
$\overline{y}$ = mean of speeds
$y_{i\text{-predicted}}$ = speed calculated from the model
n = number of data points
n−2 = degrees of freedom

The following computations are performed using the MS Excel worksheet shown in Table 3.4.

TABLE 3.4
Linear Regression Analysis and Computations of Linear Regression Parameters for Problem 3.11

Density, k (vpm)	Speed, u_s (mph)	k^2	$(u_s)(k)$	$(u_s - u_{s\,avg.})^2$	$u_{s\,calculated}$ (mph)	$(u_s - u_{s\,calculated})^2$
37.5	30.0	1406.3	1125.0	272.3	26.2	14.80
22.5	35.0	506.3	787.5	132.3	38.9	14.98
18.0	40.0	324.0	720.0	42.3	42.7	7.21
15.0	44.0	225.0	660.0	6.3	45.2	1.51
12.0	47.0	144.0	564.0	0.3	47.8	0.60
9.0	50.0	81.0	450.0	12.3	50.3	0.10
7.5	52.0	56.3	390.0	30.3	51.6	0.17
6.0	54.0	36.0	324.0	56.3	52.9	1.30
4.5	55.0	20.3	247.5	72.3	54.1	0.76
3.0	58.0	9.0	174.0	132.3	55.4	6.75
SUM 135	465	2808.0	5442.0	756.5		48.17

(a) $S_t = \sum_{i=1}^{n}(y_i - \bar{y})^2$

$\Rightarrow S_t = 756.5$

(b) $S_r = \sum_{i=1}^{n}(y_i - y_{i\text{-calculated}})^2$

$\Rightarrow S_r = 48.17$

(c) $S_{y/x} = \sqrt{\dfrac{S_r}{n-2}}$

$\Rightarrow$

$S_{y/x} = \sqrt{\dfrac{48.17}{10-2}} = 2.454$

(d) $r^2 = \dfrac{S_t - S_r}{S_t}$

$\Rightarrow$

$r^2 = \dfrac{756.5 - 48.17}{756.5} = 0.94$

(e) $r = \sqrt{r^2}$

$$\Rightarrow = \sqrt{0.94} = 0.97$$

The MS Excel worksheet used to perform the computations and to determine the linear regression parameters in this problem is shown in Figure 3.15.

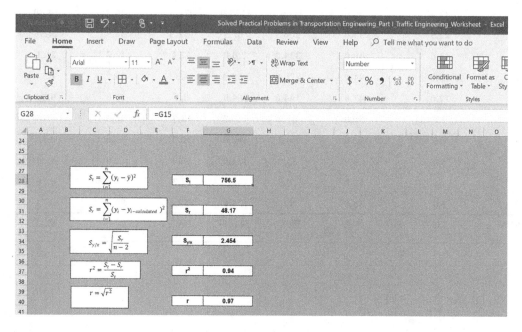

FIGURE 3.15 A screen image of the MS Excel worksheet used for determining the linear regression parameters for Problem 3.12.

Using the MS Excel worksheet, the density–speed relationship is also plotted as shown in Figure 3.16.

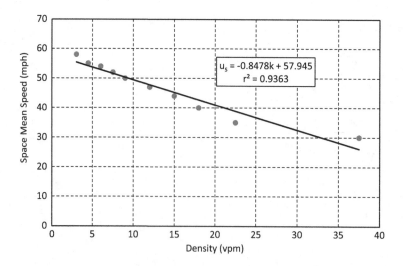

FIGURE 3.16 Density–speed relationship described by the Greenshields model for Problem 3.12.

3.13 In Problem 3.11, determine the mean free speed and the jam density for the traffic on the highway.

Solution:

By comparing the linear form of Greenshields model in Equation (3.12) with the linear form $y = a_0 + a_1 x$, the following are obtained:

$$a_0 = u_f$$

$$\Rightarrow u_f = a_0 = 58 \text{ mph}$$

$$a_1 = -\frac{u_f}{k_j}$$

$\Rightarrow$

$$k_j = -\frac{u_f}{a_1} = -\frac{58}{-0.85} = 68.3 \text{ vpm} \cong 69 \text{ vpm}$$

The MS Excel worksheet used to perform the computations in this problem is shown in Figure 3.14, in Problem 3.11.

3.14 If the model that describes the relationship between speed (mph) and density (vpm) on a highway segment is given as $u_s = 62 - 1.22\, k$, determine the following:
(a) Mean free speed
(b) Jam density
(c) Density at maximum flow
(d) Speed at maximum flow
(e) Maximum flow rate (capacity) of the highway segment

Solution:

(a) Since the given model that describes the speed-density relationship on the highway is linear, this is the Greenshields model, which takes the form:

$$u_s = u_f - \frac{u_f}{k_j} k.$$

Therefore, the mean free speed is equal to the intercept (62) in the linear model.

$$u_f = 62 \text{ mph} \left(\cong 100 \text{ kph} \right)$$

(b) The jam density is computed using the slope (−1.22) in the linear model, therefore:

$$-\frac{u_f}{k_j} = -1.22$$

$\Rightarrow$

$$k_j = \frac{62}{-1.22} = 50.8 \text{ vpm} \cong 51 \text{ vpm} \left(32 \text{ veh/km} \right)$$

(c)   The density at maximum flow is determined by taking the derivative of the flow with respect to density and equating the result to zero.

$$q = u_s k \qquad (3.26)$$

But:

$$u_s = u_f - \frac{u_f}{k_j} k$$

Hence:

$$q = \left(u_f - \frac{u_f}{k_j} k \right) k \qquad (3.27)$$

Or:

$$q = u_f k - \frac{u_f}{k_j} k^2 \qquad (3.28)$$

By deriving the above equation and equating it to zero, the following formula is obtained:

$$\frac{dq}{dk} = u_f - 2 \frac{u_f}{k_j} k = 0 \qquad (3.29)$$

$\Rightarrow$

$$k_m = \frac{k_j}{2} \qquad (3.30)$$

This is called the density at maximum flow.

$$\text{Density at maximum flow} = \frac{51}{2} = 25.5 \cong 26 \text{ vpm} \left(16 \text{ veh/km} \right)$$

(d)   The speed at maximum flow is determined following the same procedure: by taking the derivative of the flow with respect to speed and equating the result to zero.

$$q = u_s k$$

But:

$$u_s = u_f - \frac{u_f}{k_j} k$$

Reformulating the above formula to obtain k, the following formula for the density (k) is obtained:

$$k = k_j - \frac{k_j}{u_f} u_s$$

Hence:

$$q = \left(k_j - \frac{k_j}{u_f} u_s \right) u_s \qquad (3.31)$$

Or:

$$q = k_j u_s - \frac{k_j}{u_f} u_s^2 \tag{3.32}$$

By deriving the above equation and equating it to zero, the following formula is obtained:

$$\frac{dq}{du_s} = k_j - 2\frac{k_j}{u_f}u_s = 0 \tag{3.33}$$

$\Rightarrow$

$$u_m = \frac{u_f}{2} \tag{3.34}$$

This is called the speed at maximum flow.

$$\text{Speed at maximum flow} = \frac{62}{2} = 31 \text{ mph} (49.9 \text{ kph})$$

(e)   The maximum flow rate (capacity) of the highway segment is simply equal to the density at maximum flow multiplied by the speed at maximum flow. Therefore,

$$q_{max} = 26 \times 31 \cong 806 \text{ vph}$$

A screen image of the MS Excel worksheet used to perform the computations of this problem is shown in Figure 3.17.

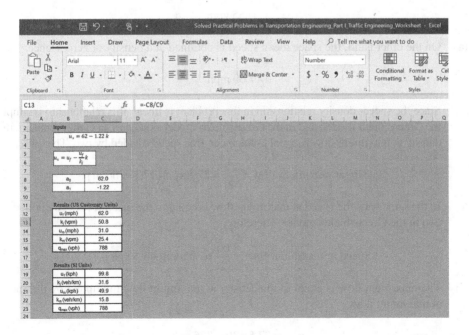

FIGURE 3.17   A screen image of the MS Excel worksheet used for the computations of Problem 3.14.

3.15 If the relationship between the density and the space mean speed for a traffic stream is given as shown in Figure 3.18, determine the following:
   (a) Jam density
   (b) Mean free speed
   (c) Density at maximum flow
   (d) Speed at maximum flow
   (e) Maximum flow

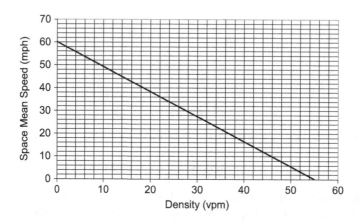

FIGURE 3.18   Density and space mean speed relationship for a traffic stream for Problem 3.15.

Solution:

From the speed–density relationship in the figure above, the jam density is obtained at a space mean speed of zero. Therefore:

$$k_j = 55 \text{ vpm}$$

(a) On the other hand, the space mean speed is obtained at a density of zero, Hence:

$$u_f = 60 \text{ mph.}$$

(b) Since the speed–density relationship in this problem is linear, this relationship is simply the Greenshields model; and in the Greenshields model as seen in an earlier problem, the density at maximum flow is equal to the mean free speed divided by two. Therefore,

$$\text{Density at maximum flow} = \frac{55}{2} = 27.5 \text{ vpm} \left(17.1 \text{ veh/km}\right)$$

(c) In the same way, the speed at maximum flow is equal to the mean fee speed divided by two in the Greenshields model. Therefore,

$$\text{Speed at maximum flow} = \frac{60}{2} = 30 \text{ mph} \left(48.3 \text{ kph}\right)$$

(d) The maximum flow is equal to the density at maximum flow multiplied by the speed at maximum flow.

$$q_{max} = 27.5 \times 30 = 825 \text{ vph}$$

The MS Excel worksheet used to obtain the results of this problem is shown in the screen image in Figure 3.19.

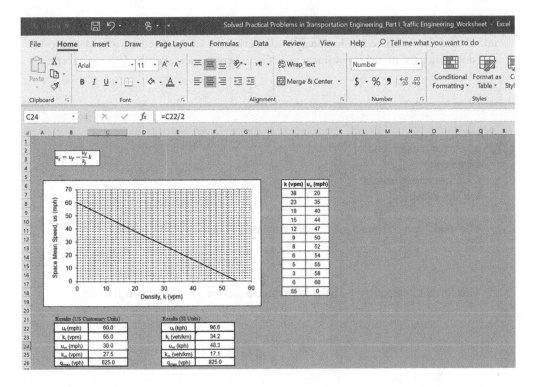

FIGURE 3.19 A screen image of the MS Excel worksheet used to obtain the results of Problem 3.15.

3.16 The traffic density–speed data shown in Table 3.5 is obtained on a highway segment. If the data can be described by the Greenberg model, use linear regression analysis to determine the following:

(a) The regression constants in the Greenberg model
(b) The coefficient of determination (r^2) for the model
(c) Plot the relationship between density and speed

TABLE 3.5
Traffic Density–Speed Data for a Highway Segment for Problem 3.16

Density, k (vpm)	Speed, u_s (mph)
100	20
92	21
84	25
75	27
70	29
64	33
60	35
54	38
42	41
35	46
28	53
22	61

Solution:

(a) To use the linear regression analysis, linearization of the Greenberg model is to be done since the model is not linear. The Greenberg model takes the following form:

$$u_s = C \ln \frac{k_j}{k} \tag{3.35}$$

Where:
u_s = space mean speed
k = density
C = regression constant
k_j = jam density

The model can be linearized and reformulated as in the following form:

$$u_s = C \ln k_j - C \ln k \tag{3.36}$$

This form is compatible with the linear line equation $y = a_0 + a_1 x$, such that:

$$y = u_s$$

$$a_0 = C \ln k_j$$

$$a_1 = -C$$

$$x = \ln k$$

Therefore, the normal equations (the linear system with matrix notation) that was used earlier for linear models can be used for the Greenberg non-linear model by replacing x and y with lnk and u_s, respectively as shown below:

$$\begin{bmatrix} n & \sum x \\ \sum x & \sum x^2 \end{bmatrix} \begin{bmatrix} a_0 \\ a_1 \end{bmatrix} = \begin{bmatrix} \sum y \\ \sum xy \end{bmatrix}$$

$$\begin{bmatrix} n & \sum \ln k \\ \sum \ln k & \sum (\ln k)^2 \end{bmatrix} \begin{bmatrix} a_0 \\ a_1 \end{bmatrix} = \begin{bmatrix} \sum u_s \\ \sum (\ln k) u_s \end{bmatrix} \tag{3.37}$$

The solution of the above linear system is:

$$a_1 = \frac{\sum_{i=1}^{n} x_i y_i - \frac{1}{n} \left(\sum_{i=1}^{n} x_i \right) \left(\sum_{i=1}^{n} y_i \right)}{\sum_{i=1}^{n} x_i^2 - \frac{1}{n} \left(\sum_{i=1}^{n} x_i \right)^2}$$

$$a_0 = \frac{\sum_{i=1}^{n} y_i}{n} - a_1 \frac{\sum_{i=1}^{n} x_i}{n}$$

Therefore,

$$a_1 = \frac{\sum_{i=1}^{n} (\ln k_i) u_{si} - \frac{1}{n} \sum_{i=1}^{n} (\ln k_i) \sum_{i=1}^{n} u_{si}}{\sum_{i=1}^{n} (\ln k_i)^2 - \frac{1}{n} \left(\sum_{i=1}^{n} (\ln k_i) \right)^2} \tag{3.38}$$

$$a_0 = \frac{\sum_{i=1}^{n} u_{si}}{n} - a_1 \frac{\sum_{i=1}^{n} (\ln k_i)}{n} \tag{3.39}$$

The computations of the results needed to determine the regression coefficients a_0 and a_1 are performed using the MS Excel worksheet and are shown in Table 3.6.

TABLE 3.6

Computations needed for the Determination of the Regression Coefficients a_0 and a_1 for Problem 3.16

Density, k (vpm)	Speed, u_s (mph)	lnk	(lnk)²	(u_s)(lnk)
100	20	4.61	21.21	92.1
92	21	4.52	20.45	95.0
84	25	4.43	19.63	110.8
75	27	4.32	18.64	116.6
70	29	4.25	18.05	123.2
64	33	4.16	17.30	137.2
60	35	4.09	16.76	143.3
54	38	3.99	15.91	151.6
42	41	3.74	13.97	153.2
35	46	3.56	12.64	163.5
28	53	3.33	11.10	176.6
22	61	3.09	9.55	188.6
SUM 726.00	429.00	48.08	195.22	1651.69

$$a_1 = \frac{1651.69 - \frac{1}{12}(48.08)(429.00)}{195.22 - \frac{1}{12}(48.08)^2} = -26.3$$

$$a_0 = \frac{429.00}{12} - (-)\frac{48.08}{12} = 141.1$$

$$-C = a_1$$

$\Rightarrow$

$$C = -a_1 = 26.3$$

$$C \ln k_j = a_0$$

$\Rightarrow$

$$k_j = e^{\left(\frac{a_0}{C}\right)} = e^{\left(\frac{141.1}{26.3}\right)} = 214 \text{ vpm} \left(133 \text{ veh/km}\right)$$

Therefore, the Greenberg model that describes this data takes the form:

$$u_s = 26.3 \ln \frac{214}{k} \tag{3.40}$$

(b) To compute the coefficient of determination (r^2) for the model, the speed calculated from the model should be determined. The computations are shown in Table 3.7.

TABLE 3.7

Regression Analysis and Computations of Linear Regression Parameters for Problem 3.16

Density, k (vpm)	Speed, u_s (mph)	lnk	$(\text{lnk})^2$	$(u_s)(\text{lnk})$	$(u_s - u_{s\,avg.})^2$	$u_{s\,calculated}$ (mph)	$(u_s - u_{s\,calculated})^2$
100	20	4.61	21.21	92.1	132.3	20.0	0.0007
92	21	4.52	20.45	95.0	110.3	22.2	1.4818
84	25	4.43	19.63	110.8	42.3	24.6	0.1536
75	27	4.32	18.64	116.6	20.3	27.6	0.3438
70	29	4.25	18.05	123.2	6.3	29.4	0.1596
64	33	4.16	17.30	137.2	2.3	31.8	1.5511
60	35	4.09	16.76	143.3	12.3	33.5	2.4004
54	38	3.99	15.91	151.6	42.3	36.2	3.1698
42	41	3.74	13.97	153.2	90.3	42.8	3.3279
35	46	3.56	12.64	163.5	210.3	47.6	2.6106
28	53	3.33	11.10	176.6	462.3	53.5	0.2304
22	61	3.09	9.55	188.6	870.3	59.8	1.3974
SUM 726.00	429.00	48.08	195.22	1651.69	2001.00	429.00	16.83

$$S_t = 2001.00$$

$$S_r = 16.83$$

$$r^2 = \frac{S_t - S_r}{S_t}$$

$\Rightarrow$

$$r^2 = \frac{2001.00 - 16.83}{2001.00} = 0.99$$

(c) The density-speed relationship described by the Greenberg model is plotted in Figure 3.20.

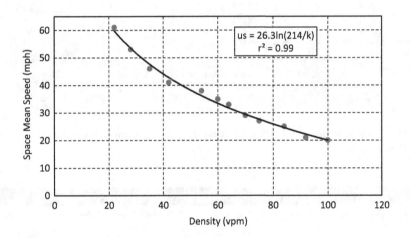

FIGURE 3.20 Density–speed relationship described by the Greenberg model for Problem 3.16.

An image of the MS Excel worksheet used to conduct the computations and the linear regression analysis of this problem is shown in Figures 3.21 and 3.22. Note that linearization is done for the non-linear model (Greenberg model) so that linear regression techniques can be used.

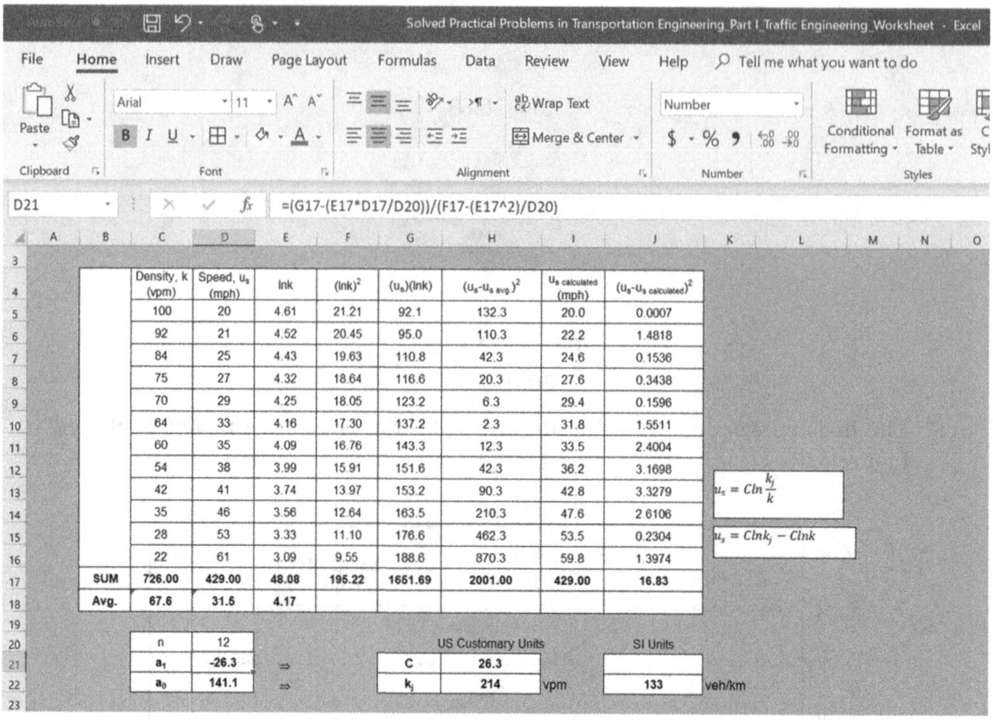

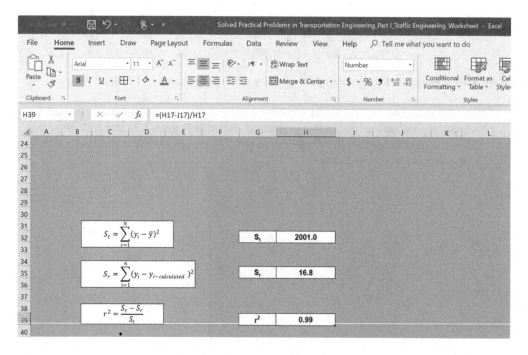

FIGURE 3.21 A screen image of the MS Excel worksheet for the computations and the linear regression analysis for Problem 3.16.

FIGURE 3.22 A screen image of the MS Excel worksheet used for determining the linear regression parameters for Problem 3.16.

3.17 In Problem 3.16, determine the density at maximum flow, the speed at maximum flow, and the maximum flow (capacity) of the highway.

Solution:

$$u_s = C \ln \frac{k_j}{k}$$

$$q = u_s k$$

But:

$$u_s = C \ln \frac{k_j}{k}$$

Hence:

$$q = \left(C \ln \frac{k_j}{k} \right) k$$

By deriving the above equation and equating it to zero, the following formula is obtained:

$$\frac{dq}{dk} = \left(C \ln \frac{k_j}{k} \right) + k \left(\frac{-c}{k} \right) = 0 \tag{3.41}$$

⇒

$$\ln \frac{k_j}{k} = 1 \tag{3.42}$$

$$\frac{k_j}{k} = e \tag{3.43}$$

⇒

$$k_m = \frac{k_j}{e} \tag{3.44}$$

This is called the density at maximum flow.

$$\text{Density at maximum flow} = \frac{214}{2} = 107 \text{ vpm} \left(\cong 67 \text{ veh/km} \right)$$

The speed at maximum flow is determined following the same procedure: by taking the derivative of the flow with respect to speed and equating the result to zero.

$$q = u_s k$$

But:

$$u_s = C \ln \frac{k_j}{k}$$

Reformulating the above formula to obtain k, the following formula for the density (k) is obtained:

$$k = \frac{k_j}{e^{\left(\frac{u_s}{C}\right)}} \tag{3.45}$$

Hence:

$$q = \left(\frac{k_j}{e^{\left(\frac{u_s}{C}\right)}}\right) u_s \tag{3.46}$$

By deriving the above equation and equating it to zero, the following formula is obtained:

$$\frac{dq}{du_s} = k_j e^{\left(\frac{u_s}{C}\right)} - \frac{1}{C} k_j u_s e^{\left(\frac{u_s}{C}\right)} = 0 \tag{3.47}$$

$\Rightarrow$

$$k_j e^{\left(\frac{u_s}{C}\right)} \left(1 - \frac{u_s}{C}\right) = 0 \tag{3.48}$$

$\Rightarrow$

$$u_m = C \tag{3.49}$$

This is called the speed at maximum flow.

$$\text{Speed at maximum flow} = 26.3 \text{ mph} \left(42.3 \text{ kph}\right)$$

And therefore, the maximum flow is equal to the density at maximum flow multiplied by the speed at maximum flow.

$$q = 107 \times 26.3 = 2814 \text{ vph}$$

3.18 If the model that describes the relationship between speed (mph) and density (vpm) on a highway is given as $u_s = 65e^{\left(\frac{-k}{52}\right)}$, determine the following:
(a) Density at maximum flow
(b) Speed at maximum flow
(c) Capacity of the highway

Solution:

(a) To determine the density at maximum flow, the flow as a function of the density is determined and derived with respect to the density.

$$q = u_s k$$

But according to the model given in this problem between speed and density, it is expressed in the following generalized form:

$$u_s = Ae^{\left(\frac{-k}{B}\right)}$$ (3.50)

Hence:

$$q = Ake^{\left(\frac{-k}{B}\right)}$$ (3.51)

By deriving the above equation and equating it to zero, the following formula is obtained:

$$\frac{dq}{dk} = \frac{-1}{B}Ake^{\left(\frac{-k}{B}\right)} + Ae^{\left(\frac{-k}{B}\right)} = 0$$ (3.52)

$\Rightarrow$

$$Ae^{\left(\frac{-k}{B}\right)}\left(1 - \frac{k}{B}\right) = 0$$ (3.53)

$\Rightarrow$

$$k_m = B$$ (3.54)

This is called the density at maximum flow.

$$\text{Density at maximum flow} = 52 \text{ vpm}\left(\cong 33 \text{ veh/km}\right)$$

(b) To determine the speed at maximum flow, the flow as a function of the speed is determined and derived with respect to the density.

$$q = u_s k$$

But:

$$u_s = Ae^{\left(\frac{-k}{B}\right)}$$

Reformulating the above equation to obtain k as a function of u_s:

$$k = -B\ln\left(\frac{u_s}{A}\right)$$ (3.55)

And therefore,

$$q = -Bu_s \ln\left(\frac{u_s}{A}\right)$$ (3.56)

By deriving the above equation and equating it to zero, the following formula is obtained:

$$\frac{dq}{du_s} = -B - B\ln\left(\frac{u_s}{A}\right) = 0 \tag{3.57}$$

$\Rightarrow$

$$-B\left(1 + \ln\left(\frac{u_s}{A}\right)\right) = 0 \tag{3.58}$$

$\Rightarrow$

$$1 + \ln\left(\frac{u_s}{A}\right) = 0 \tag{3.59}$$

$\Rightarrow$

$$u_m = \frac{A}{e} \tag{3.60}$$

This is called the speed at maximum flow.

$$\text{Speed at maximum flow} = \frac{65}{e} = 23.9\,\text{mph}\,(38.5\,\text{kph})$$

(c)   The maximum flow is determined by multiplying the density at maximum flow by the speed at maximum flow.

$$q = u_s k$$

$$q = 23.9 \times 52 = 1243\,\text{vph}$$

3.19 The relationship between the density and the space mean speed for a traffic stream is described by the Greenberg model. If the density at maximum flow is 32 vpm (20 veh/km), determine the jam density.

Solution:

Based on the Greenberg model, the density at maximum flow is given by:

$$k_m = \frac{k_j}{e}$$

Therefore, the jam density can be computed as:

$$k_j = e k_m$$

$$k_j = e \times 32 = 87\,\text{vpm}\,(54\,\text{veh/km})$$

3.20 If the model $\ln u_s = \ln u_f - \dfrac{k}{k_j}$ can be used to describe the relationship between speed (mph)

and density (vpm) on a highway segment; and using regression analysis, the constants of the model $Y = A + BX$ (where $Y = ln u_s$ and $X = k$) are $A = 3.9$ and $B = -0.018$, then estimate the mean free speed (u_f) and the jam density (k_j).

Solution:

Since the model is given in the following mathematical expression:

$$\ln u_s = \ln u_f - \frac{k}{k_j} \tag{3.61}$$

$\Rightarrow$

$$A = \ln u_f \tag{3.62}$$

$\Rightarrow$

$$u_f = e^A = e^{3.9} = 49.4\,\text{mph}\,(79.5\,\text{kph})$$

And:

$$B = -\frac{1}{k_j} \tag{3.63}$$

$\Rightarrow$

$$k_j = -\frac{1}{B} = -\frac{1}{-0.018} \cong 56\,\text{vpm}\,(35\,\text{veh/km})$$

3.21 If the model that describes the relationship between speed (mph) and density (vpm) on a

highway is given as $u_s = 45\ln\dfrac{80}{k}$, estimate the speed and the density at maximum flow.

Solution:

This relationship is simply the Greenberg model. In Problem 3.17, the derivations showed that the speed at maximum flow is equal to the constant, and the density at maximum flow is equal to the jam density divided by e.

Since:

$$u_s = 45\ln\frac{80}{k} \tag{3.64}$$

$$C = 45$$

And

$$k_j = 80\,\text{vpm}$$

Therefore:

$$\text{Speed at maximum flow} = C = 45\,\text{mph}\,(72.4\,\text{kph})$$

$$\text{Density at maximum flow} = \frac{k_j}{e} = \frac{80}{e} = 29.4 \text{ vpm} \left(18.3 \text{ veh/km}\right)$$

A screen image of the MS Excel worksheet used to perform the computations of this problem is shown in Figure 3.23.

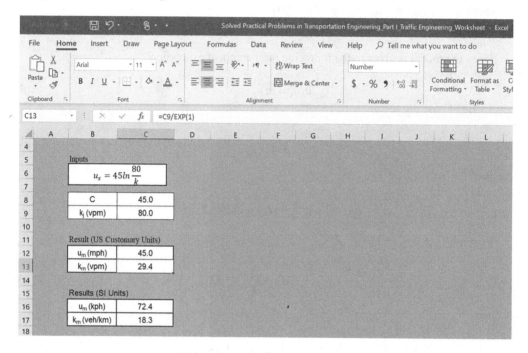

FIGURE 3.23   A screen image of the MS Excel worksheet used to compute the speed and density at maximum flow for Problem 3.21.

3.22 The density and speed of traffic on a two-lane two-way highway segment are 35 vpm (21.7 veh/km) and 50 mph (80.5 kph), respectively. A loaded truck traveling at a speed of 25 mph joins the traffic stream on the highway from an aggregate quarry in the vicinity of the area. The truck remains on the highway for a time period of 15 minutes before it exits the highway. The highway segment is located in a no-passing zone; and hence, vehicles are not permitted to make passing. If this situation creates a platoon of vehicles behind the truck traveling at the same speed of the truck and having a density of 120 vpm (74.6 veh/km), determine the following (see Figure 3.24):

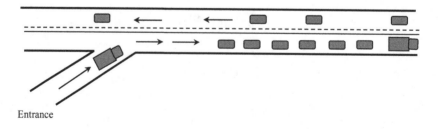

Entrance

FIGURE 3.24   A shockwave on a two-lane two-way highway due to a truck for Problem 3.22.

(a) The speed of the shockwave
(b) The length of the platoon that will be created during the existence of the truck on the highway
(c) The number of vehicles that will be affected by the incident

Solution:

(a) The speed of the shockwave is calculated using the formula shown below:

$$u_w = \frac{q_2 - q_1}{k_2 - k_1} \tag{3.65}$$

$$u_w = \frac{(25)(120) - (50)(35)}{120 - 35} = 14.7 \text{ mph} (23.7 \text{ kph})$$

(b) The length of the platoon of vehicles that will be created is equal to the relative speed multiplied by the time of the incident; it is computed using the following formula:

$$L = (u_{r2}) t \tag{3.66}$$

Where:
L = length of platoon
u_{r2} = relative speed after the shockwave condition = $(u_{s2}\text{-}u_w)$
t = time period of incident

Therefore:

$$L = (u_{s2} - u_w) t \tag{3.67}$$

$$L = (25 - 14.7)\frac{15}{60} = 2.57 \text{ mi} (4.14 \text{ km})$$

(c) The number of vehicles that will be affected by this incident is simply the number of vehicles in the platoon created behind the truck. Hence, it is calculated as:

$$\text{Number of vehicles in platoon} = Lk_2 \tag{3.68}$$

Or:

$$\text{Number of vehicles in platoon} = (u_{s2} - u_w) tk_2 \tag{3.69}$$

Therefore,

$$\text{Number of vehicles in platoon} = 2.57 \times 120 \cong 309 \text{ vehicles}$$

The screen images of the MS Excel worksheet used to conduct the computations of this problem are shown in Figures 3.25 and 3.26.

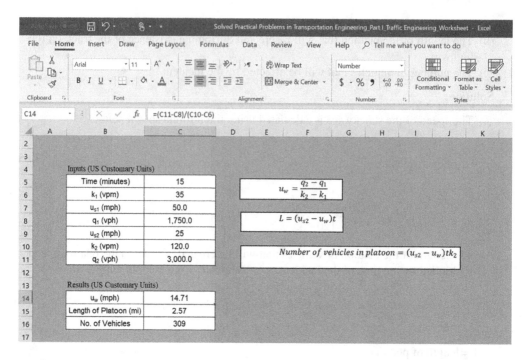

FIGURE 3.25 A screen image of the MS Excel worksheet used for the computations of Problem 3.22 (US Customary Units).

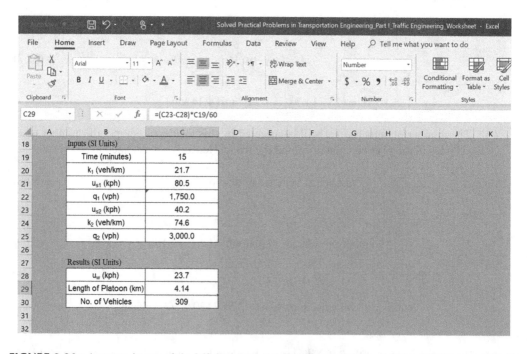

FIGURE 3.26 A screen image of the MS Excel worksheet used for the computations of Problem 3.22 (SI units).

3.23 The density on a two-lane two-way highway is 40 vpm (24.9 veh/km). An accident occurs on the highway blocking one lane completely. The incident remains on the highway for 20 minutes before it is cleared. This incident creates a queue of vehicles stopping behind the accident due to the heavy traffic volume in the other direction. If the traffic flow on this highway can be described by the Greenshields model with a mean free speed of 60 mph (96.6 kph) and a jam density of 100 vpm (62.1 veh/km), determine how many vehicles are affected by the occurrence of the accident (see Figure 3.27).

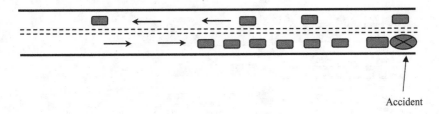

Accident

FIGURE 3.27   A shockwave on a two-lane two-way highway due to an accident for Problem 3.23.

Solution:

The speed of the shockwave is calculated using the formula shown below:
Using the Greenshields model, the speed of traffic can be computed as follows:

$$u_s = u_f - \frac{u_f}{k_j} k$$

⇒

$$u_s = 60 - \frac{60}{100} k$$

At $k_1 = 40$ vpm, the speed is equal to:

$$u_{s1} = 60 - \frac{60}{100}(40) = 36 \text{ mph} (57.9 \text{ kph})$$

Therefore,

$$u_w = \frac{q_2 - q_1}{k_2 - k_1}$$

$$u_w = \frac{0 - (36)(40)}{100 - 40}$$

$$u_w = -24 \text{ mph} (38.6 \text{ kph})$$

The number of vehicles that will be affected by this incident is equal to the length of the queue multiplied by the density of the traffic in the queue:

$$\text{Number of vehicles in platoon} = (u_{s2} - u_w) t k_2$$

$$\text{Number of vehicles in platoon} = (0 - (-24)) \frac{20}{60} \times 100 = 800 \text{ vehicles}$$

The screen images of the MS Excel worksheet used to perform the computations of this problem are illustrated in Figures 3.28 and 3.29.

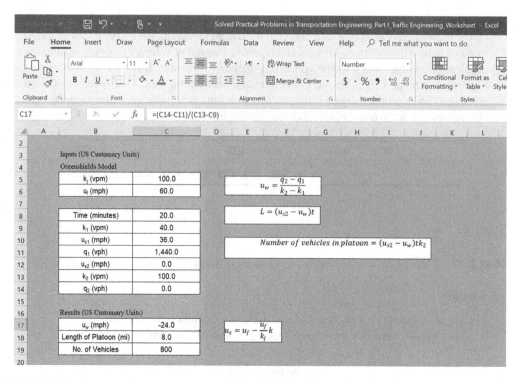

FIGURE 3.28   A screen image of the MS Excel worksheet used for the computations of Problem 3.23 (US Customary Units).

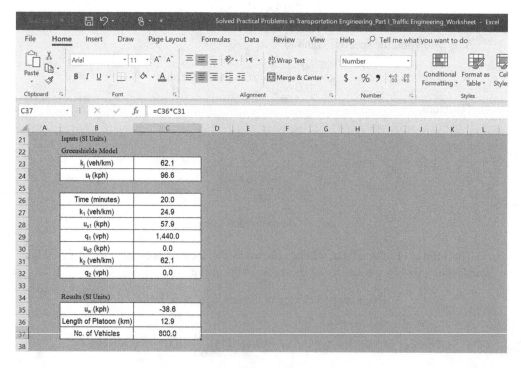

FIGURE 3.29   A screen image of the MS Excel worksheet used for the computations of Problem 3.23 (SI Units).

3.24 The traffic flow on a two-lane two-way highway segment is described by the Greenberg model with a speed at maximum flow of 50 mph (80.5 kph) and a jam density of 100 vpm (62.1 veh/km). The traffic speed on this highway segment is 45 mph (72.4 kph) before a loaded truck traveling at a speed of 20 mph (32.2 kph) joins the traffic stream on the highway for a time period of 20 minutes until it exits the highway. Vehicles are prohibited to make passing due to the fact that the highway segment is located in a no-passing zone. If this situation creates a platoon of vehicles behind the truck traveling at the same speed of the truck, determine the following (see Figure 3.30):

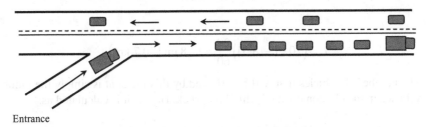

Entrance

FIGURE 3.30   A shockwave on a two-lane two-way highway due to a truck for Problem 3.24.

(a) The speed of the shockwave
(b) The length of the platoon that will be created during the existence of the truck on the highway
(c) The number of vehicles that will be affected by the incident

Solution:

(a)   The speed of the shockwave is calculated using the formula shown below:

$$u_w = \frac{q_2 - q_1}{k_2 - k_1}$$

Using the Greenberg model, the density of traffic can be computed as follows:

$$u_s = C \ln \frac{k_j}{k}$$

In Greenberg model, the speed at maximum flow is equal to the constant C in the model as shown in Problem 3.17; in other words:

$$C = 50$$

And therefore:

$$u_s = 50 \ln \frac{100}{k}$$

At $u_{s1} = 45$ mph, the density is equal to:

$$k_1 = \frac{100}{e^{\left(\frac{45}{50}\right)}} = 40.7 \text{ vpm} \left(25.3 \text{ kph}\right)$$

And at $u_{s2} = 20$ mph, the density is equal to:

$$k_2 = \frac{100}{e^{\left(\frac{20}{50}\right)}} = 67 \text{ vpm} \left(41.7 \text{ veh/km}\right)$$

Hence:

$$u_w = \frac{(20)(67) - (45)(40.7)}{67 - 40.7} = -18.5 \text{ mph} \left(-29.8 \text{ kph}\right)$$

(b) The length of the platoon of vehicles that will be created is equal to the relative speed multiplied by the time of the incident; it is computed using the following formula:

$$L = \left(u_{s2} - u_w\right)t$$

$$L = \left(20 - \left(-18.5\right)\right)\frac{20}{60} = 12.8 \text{ mi} \left(20.7 \text{ km}\right)$$

(c) The number of vehicles that will be affected by this incident is simply the number of vehicles in the platoon created behind the truck. Hence, it is calculated as:

$$\text{Number of vehicles in platoon} = Lk_2$$

Therefore,

$$\text{Number of vehicles in platoon} = 12.8 \times 67 \cong 861 \text{ vehicles}$$

The results in US customary units and SI units are shown in the following screen images of the MS Excel worksheet that is utilized to obtain the results (see Figures 3.31 and 3.32).

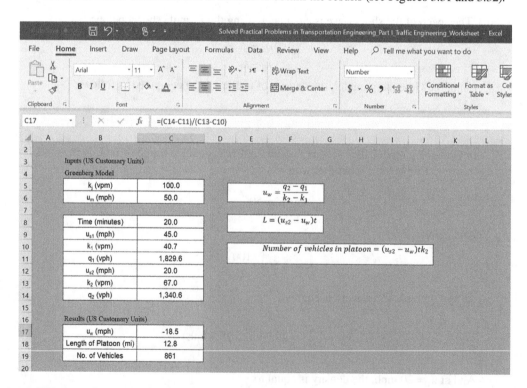

FIGURE 3.31 A screen image of the MS Excel worksheet used for the computations of Problem 3.24 (US customary units).

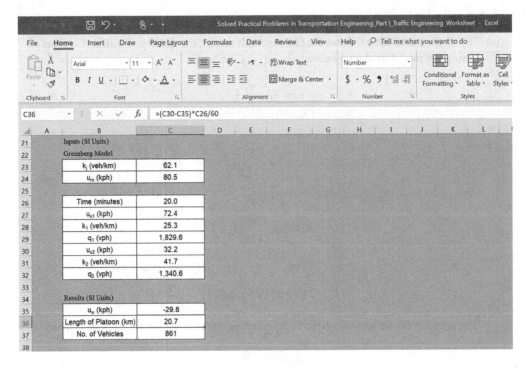

FIGURE 3.32   A screen image of the MS Excel worksheet used for the computations of Problem 3.24 (SI units).

3.25 Traffic flow on one of the single-lane approaches of a signalized intersection traveling at 40 mph (eastbound shown in Figure 3.33) can be described by the Greenshields model. The duration of the red phase for this approach is 25 seconds. If the jam density is 125 vpm and the mean free speed is 50 mph, determine the following:
   (a) Speed of shockwave
   (b) Length of the queue at the end of the red phase
   (c) Number of vehicles in the queue

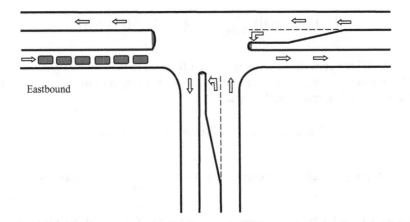

Eastbound

FIGURE 3.33   A shockwave on a single-lane approach at a signalized intersection for Problem 3.25

Solution:

(a)   Since the traffic flow follows the Greenshields model, the density of traffic is computed using the model as shown below:

$$u_s = u_f - \frac{u_f}{k_j} k$$

$\Rightarrow$

$$u_s = 50 - \frac{50}{125} k$$

At $u_{s1} = 40$ mph, the density is estimated as:

$$40 = 50 - \frac{50}{125} (k_1)$$

$\Rightarrow$

$$k_1 = 25 \text{ vpm} (15.5 \text{ veh/km})$$

The speed of the shockwave is then computed using the common formula shown below:

$$u_w = \frac{q_2 - q_1}{k_2 - k_1}$$

$$u_w = \frac{0 - (40)(25)}{125 - 25}$$

$$u_w = -10 \text{ mph} (-16.1 \text{ kph})$$

(b)   The length of the queue is calculated using the formula below:

$$L = (u_{s2} - u_w) t$$

$\Rightarrow$

$$L = (0 - (-10)) \times \frac{25}{3600} \times 5280 = 366.7 \text{ ft} (111.8 \text{ m})$$

The "5280" is a conversion factor to convert from mi to ft.

(c)   The number of vehicles in the queue is equal to the length of the queue multiplied by the density of the traffic in the queue:

$$\text{Number of vehicles in queue} = (u_{s2} - u_w) t k_2$$

$$\text{Number of vehicles in queue} = (0 - (-10)) \frac{25}{3600} \times 125 \cong 9 \text{ vehicles.}$$

Figures 3.34 and 3.35 show screen images of the MS Excel worksheet used to perform the computations of this problem.

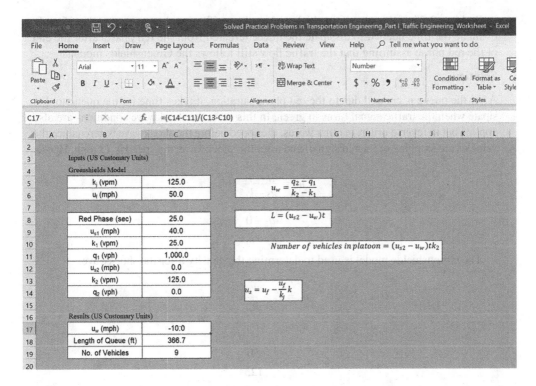

FIGURE 3.34 A screen image of the MS Excel worksheet used for the computations of Problem 3.25 (US customary units).

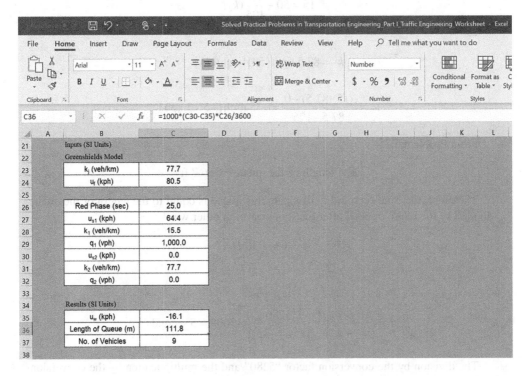

FIGURE 3.35 A screen image of the MS Excel worksheet used for the computations of Problem 3.25 (SI units).

3.26 In Problem 3.25, if the speed of vehicles when the traffic signal turns to green is 15 mph, estimate the time needed to dissipate the queue (created from the red phase) after the end of the red phase assuming that the traffic flow still follows the Greenshields model.

Solution:

After the queue is created due to the red phase duration, it takes time for the queue to dissipate when the traffic signal turns to green. In this case, the shockwave conditions are also applied but in reverse order; in other words, the speed before the shockwave is zero (since the vehicles are stopping at the red signal) and after the shockwave, the speed is 15 mph. Hence:

$$u_w = \frac{q_2 - q_1}{k_2 - k_1}$$

$$u_w = \frac{q_2 - 0}{k_2 - k_1}$$

The Greenshields model is used to determine the density after the shockwave:

$$u_s = u_f - \frac{u_f}{k_j} k$$

$\Rightarrow$

$$u_s = 50 - \frac{50}{125} k$$

At $u_{s2} = 15$ mph, the density is computed as:

$$15 = 50 - \frac{50}{125}(k_2)$$

$\Rightarrow$

$$k_2 = 87.5 \text{ vpm} \left(54.4 \text{ veh/km}\right)$$

$$u_w = \frac{(15)(87.5) - 0}{87.5 - 125} = -35 \text{ mph} \left(-56.3 \text{ kph}\right)$$

From Problem 3.25:

The length of the queue = 366.7 ft

Therefore, the time required to dissipate this queue is equal to the length of the queue divided by the relative speed after the shockwave; in other words:

$$L = \left(u_{s2} - u_w\right)t$$

$\Rightarrow$

$$t = \frac{L}{\left(u_{s2} - u_w\right)} = \frac{\left(\dfrac{366.7}{5280}\right)}{\left(15 - (-35)\right)} \times 3600 = 5 \text{ seconds}$$

The division by the conversion factor "5280" and the multiplication by the conversion factor "3600" are performed to convert from ft to mi and from hr to seconds, respectively.

The results (in US customary units and SI units) are computed using the MS Excel worksheet shown in Figures 3.36 and 3.37.

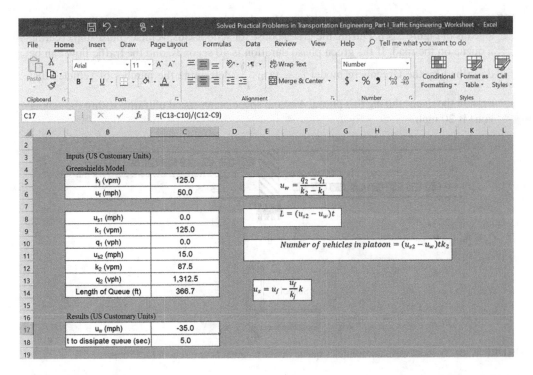

FIGURE 3.36 A screen image of the MS Excel worksheet used for the computations of Problem 3.26 (US customary units).

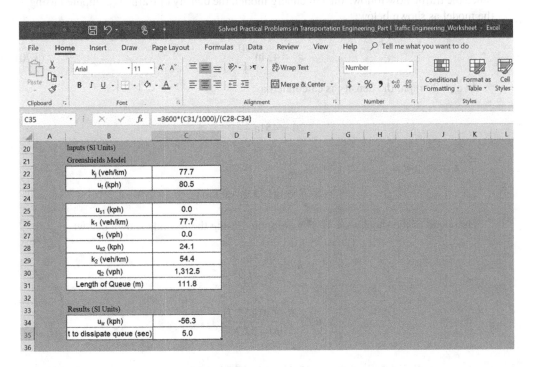

FIGURE 3.37 A screen image of the MS Excel worksheet used for the computations of Problem 3.26 (SI Units).

3.27 A traffic signal near a school is used to allow the school kids to cross the road to the other side (as in Figure 3.38). During the green phase for school kids, vehicles must stop creating a queue on the road. If the green phase duration is 40 seconds, and the traffic flow can be described by the Greenberg model $u_s = 45 \ln \dfrac{90}{k}$, determine the number of vehicles that exist in the queue during this phase. Assume space mean speed of traffic within the school zone is 20 mph.

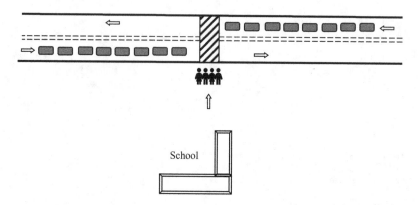

FIGURE 3.38   A shockwave and vehicles queue due to a signal near a school for Problem 3.27.

Solution:

Since the traffic flow follows the Greenberg model, the density of traffic is computed using the model as shown below:

$$u_s = 45 \ln \frac{90}{k} \tag{3.70}$$

At $u_{s1} = 20$ mph, the density is calculated as:

$$20 = 45 \ln \frac{90}{k_1}$$

$\Rightarrow$

$$k_1 = 57.7 \text{ vpm} \left(35.9 \text{ veh/km}\right)$$

The speed of the shockwave is then computed using the formula shown below:

$$u_w = \frac{q_2 - q_1}{k_2 - k_1}$$

$$u_w = \frac{0 - (20)(57.7)}{90 - 57.7}$$

$$u_w = -35.7 \text{ mph} \left(-57.5 \text{ kph}\right)$$

The length of the queue is calculated using the formula below:

$$L = (u_{s2} - u_w)t$$

$\Rightarrow$

$$L = (0 - (-35.7)) \times \frac{40}{3600} = 0.397 \text{ mi} (0.639 \text{ km})$$

The number of vehicles in the queue is equal to the length of the queue multiplied by the density of the traffic in the queue (jam density):

$$\text{Number of vehicles in queue} = Lk_2$$

$$\text{Number of vehicles in queue} = 0.397 \times 90 \cong 36 \text{ vehicles}$$

The screen images of the MS Excel worksheet that is used to perform the detailed computations of the results using the US customary units as well as the SI units are shown in Figures 3.39 and 3.40.

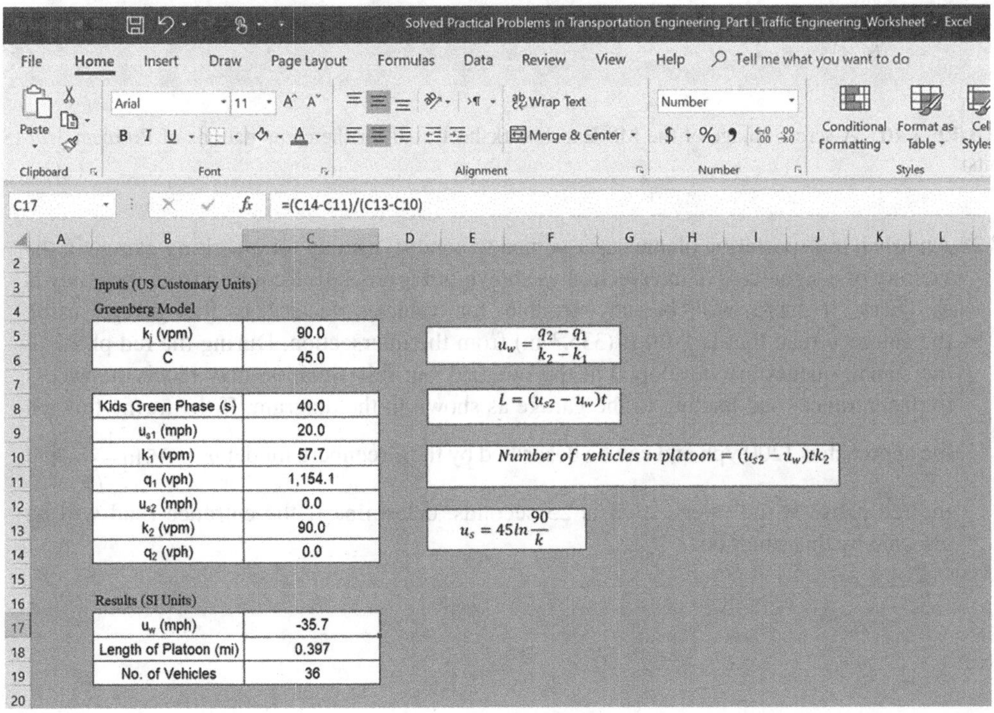

FIGURE 3.39 A screen image of the MS Excel worksheet used for the computations of Problem 3.27 (US customary units).

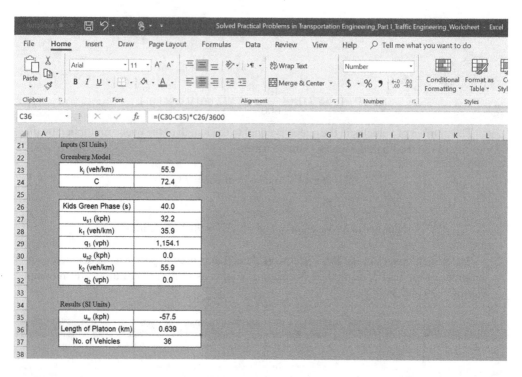

Inputs (SI Units)		
Greenberg Model		
k_j (veh/km)	55.9	
C	72.4	
Kids Green Phase (s)	40.0	
u_{s1} (kph)	32.2	
k_1 (veh/km)	35.9	
q_1 (vph)	1,154.1	
u_{s2} (kph)	0.0	
k_2 (veh/km)	55.9	
q_2 (vph)	0.0	
Results (SI Units)		
u_w (kph)	-57.5	
Length of Platoon (km)	0.639	
No. of Vehicles	36	

Cell C36 formula bar: =(C30-C35)*C26/3600

FIGURE 3.40 A screen image of the MS Excel worksheet used for the computations of Problem 3.27 (SI units).

3.28 An urban transportation planner is assigned to conduct a study for a parking garage in the vicinity of a signalized T-intersection as shown in Figure 3.41. Required from the study is to check if there will be any trouble for vehicles to access the garage using the entrance road that is 1100 ft (335.3 m) from the intersection. During the red phase of the signal, queues are developed at the intersection; this situation may block the access to the entrance road leading to the garage as shown in the diagram. If the traffic flow on the approach is 1200 vph and can be described by the Greenberg model $u_s = 35 \ln \dfrac{100}{k}$, and the red phase at the intersection is 25 seconds, determine if the entrance road will be affected by this situation.

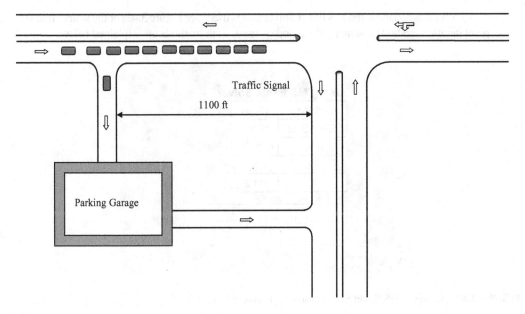

FIGURE 3.41 A parking garage in the vicinity of a signalized intersection with shockwave conditions for Problem 3.28.

Solution:

$$q = u_s k$$

But:

$$u_s = 35 \ln \frac{100}{k} \qquad (3.71)$$

Therefore:

$$q_1 = \left(35 \ln \frac{100}{k_1} \right) k_1 \qquad (3.72)$$

$\Rightarrow$

$$1200 = \left(35 \ln \frac{100}{k_1} \right) k_1 \qquad (3.73)$$

To solve this equation, the MS Excel Solver tool is used (see Figure 3.42) or the simple fixed-point iteration method (a numerical method) is used. Both methods are explained below.

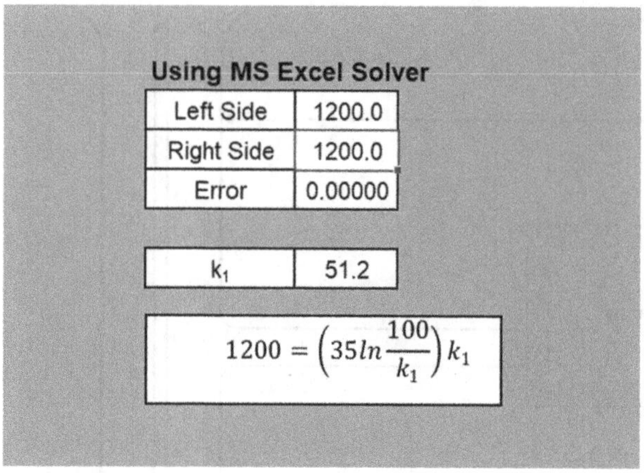

Using MS Excel Solver

Left Side	1200.0
Right Side	1200.0
Error	0.00000

k_1	51.2

$$1200 = \left(35\ln\frac{100}{k_1}\right)k_1$$

FIGURE 3.42 Image of MS Excel Solver result for Problem 3.28.

The left side of the equation is 1200, and the right side is written in terms of k_1. The error is defined as [log (left side)−log (right side)]². The MS Excel solver does an iterative approach by changing the value of k_1 until a minimum (or zero) error is obtained; i.e., the left side and the right side are approximately equal.

There are two solutions for this equation. Based on the initial value used in Excel Solver, one of the two solutions is obtained.

$$k_1 = 51.2 \text{ vpm } \left(\text{or } 24.1 \text{ vpm}\right)$$

Using the numerical method (simple fixed-point iteration), k_1 should be separated from the other terms in the above equation as below:

According to the simple fixed-point iteration method, the (i + 1)th k_1 value is equal to the value of the function on the right side at ith k_1 value. In other words:

$$k_{1(i+1)} = \frac{1200}{35\ln\left(\dfrac{100}{k_{1(i)}}\right)} \tag{3.74}$$

Using the MS Excel worksheet, this can be done for as many iterations as possible by copying and pasting the formula down in the cells to obtain a reasonable acceptable error (approximate percent relative error, ε_a).

The simple fixed-point iteration method converges to a value of k_1 equal to 24.1 vpm (see Table 3.8).

TABLE 3.8

The Solution for k_1 Using the Simple Fixed-Point Iteration Method after 16 Iterations for Problem 3.28

i	k_1	ε_a (%)
0	20	
1	21.3	6.116
2	22.2	3.921
3	22.8	2.587
4	23.2	1.740
5	23.4	1.186
6	23.6	0.816
7	23.8	0.565
8	23.9	0.393
9	23.9	0.274
10	24.0	0.191
11	24.0	0.134
12	24.0	0.094
13	24.0	0.066
14	24.1	0.046
15	24.1	0.032
16	24.1	0.023

Since the Greenberg model satisfies the traffic conditions of this intersection, and the Greenberg model can be applied only under congested conditions; therefore, the highest value of k_1 is selected (51.2 vpm).

When the traffic signal is red, the traffic flow, $q_2=0$ and the density, $k_2=k_j=100$ vpm. Hence:

$$u_w = \frac{q_2 - q_1}{k_2 - k_1}$$

$$u_w = \frac{0 - 1200}{100 - 51.2}$$

$$u_w = -24.6 \text{ mph} \left(-39.5 \text{ kph}\right)$$

The length of the queue is calculated using the formula below:

$$L = \left(u_{s2} - u_w\right)t$$

$\Rightarrow$

$$L = \left(0 - \left(-24.6\right)\right) \times \frac{25}{3600} \times 5280 = 901 \text{ ft} \left(274.6 \text{ m}\right)$$

Since the length of the queue stopping at the signalized intersection during the red phase (901 ft) < 1000 ft (the distance from the entrance road to the intersection), the entrance road will not be affected.

An image of the MS Excel worksheet used for the computations of the results (represented by the two systems of units: US Customary Units and SI Units) of this problem is shown in Figures 3.43 and 3.44.

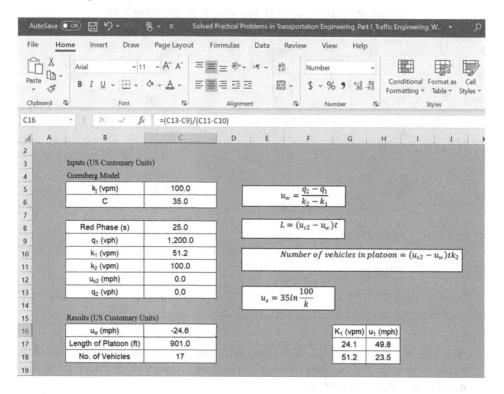

FIGURE 3.43 A screen image of the MS Excel worksheet used for the computations of Problem 3.28 (US customary units).

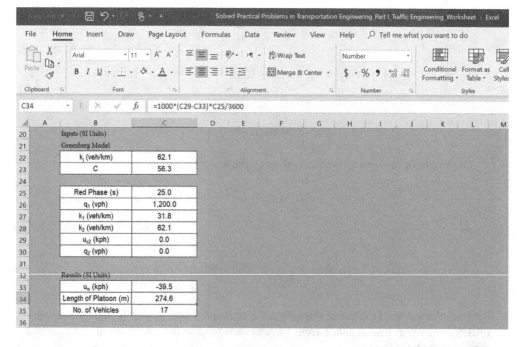

FIGURE 3.44 A screen image of the MS Excel worksheet used for the computations of Problem 3.28 (SI units).

3.29 The accepted and rejected gaps of vehicles on the minor approach of an unsignalized intersection are given as shown in Table 3.9. If the arrival of vehicles on the major approach (see Figure 3.45) can be described by the Poisson distribution, and the peak-hour volume is 1200 vph, determine the following:

TABLE 3.9
The Number of Accepted and Rejected Gaps on the Minor
Approach of an Unsignalized T-Intersection for Problem 3.29

Gap, t (sec)	Number of Accepted Gaps < t	Number of Rejected Gaps > t
1.0	4	100
2.0	22	58
3.0	32	36
4.0	66	15
5.0	96	3

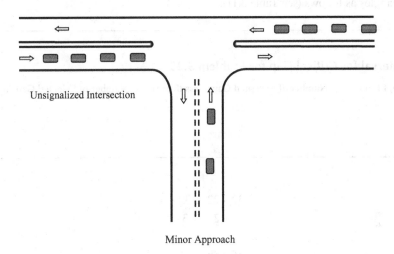

Unsignalized Intersection

Minor Approach

FIGURE 3.45 Gap acceptance at an unsignalized T-intersection for Problem 3.29.

(a) The critical gap using the algebraic method.
(b) The critical gap using the graphical method
(c) The number of accepted gaps that will be available for the minor approach vehicles during the peak hour
(d) The number of rejected gaps during the peak hour

Solution:

(a) To determine the critical gap using the algebraic method, the change in accepted gaps between two successive gaps and the change in rejected gaps between two successive gaps are computed. Then, the difference between the two changes is determined. The critical gap is located within the interval that has the smallest difference between the two changes as shown in Table 3.10.

TABLE 3.10

The Change in Accepted Gaps and Rejected Gaps between Two Successive Gaps for Problem 3.29

Successive Gaps (sec)	Change in Number of Accepted Gaps < t	Change in Number of Rejected Gaps > t	Difference Between the Two Changes
1.0–2.0	18	42	24
2.0–3.0	10	22	12
3.0–4.0	34	21	13
4.0–5.0	30	12	18

The lowest difference is 10, which refers to the gap interval of 2.0–3.0 seconds. Using the original data for the number of accepted gaps and the number of rejected gaps at 2.0 and 3.0 seconds, respectively; the critical gap (t_c) and the number of gaps (n = accepted gaps = rejected gaps) can be estimated using interpolation or similarity in triangles as follows (see Table 3.11):

TABLE 3.11

The Gap Interval for Critical Gap for Problem 3.29

Length of Gap, t (sec)	Number of Accepted Gaps (less than t)	Number of Rejected Gaps (more than t)
2.0	22	58
t_c	n	n
3.0	32	36

$$\frac{32-22}{3-2} = \frac{32-n}{3-t_c} \tag{3.75}$$

$$\frac{36-58}{3-2} = \frac{36-n}{3-t_c} \tag{3.76}$$

Solving these two equations simultaneously provides the following solution:

$$\frac{32-22}{32-n} = \frac{36-58}{36-n} \tag{3.77}$$

⇒

$$n = 33.25$$

$$t_c = 3.125 \text{ seconds}$$

(b) To determine the critical gap using the graphical method, the cumulative number of accepted gaps smaller the gap (t) and the cumulative number of rejected gaps greater than the gap (t) are calculated as shown in Table 3.12.

TABLE 3.12

The Cumulative Number of Accepted Gaps and Rejected Gaps for Problem 3.29

Gap, t (sec)	Cumulative Number of Accepted Gaps < t	Cumulative Number of Rejected Gaps > t
1.0	4	212
2.0	26	112
3.0	58	54
4.0	124	18
5.0	220	3

After that, the two sets of data are plotted against the gap (t) to obtain two curves in one figure. The intersection point of the two curves refers to the critical gap on the x-axis as shown in Figure 3.46.

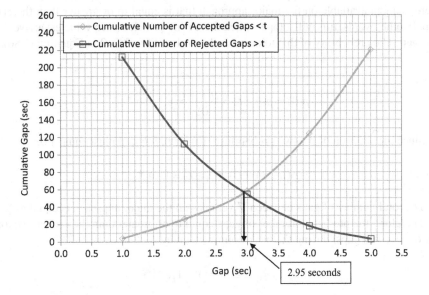

FIGURE 3.46 The graphical method for critical gap for Problem 3.29.

From the graph, $t_c = 2.95$ seconds.

(c) The number of accepted gaps that will be available for the vehicles in the minor approach during the peak hour is estimated using the following formula:

$$\text{Freq.}(h \geq t) = (V-1)e^{-\lambda t} \qquad (3.78)$$

Where :

V = peak hour volume (total number of vehicles arriving in a time period T)

λ = average number of vehicles arriving per time period in seconds = V/T; in other words, it is given by the following formula:

$$\lambda = \frac{V}{T} \qquad (3.79)$$

h = any gap
$e^{-\lambda t}$ = probability of $h \geq t$

Assuming that the arrival of the major approach traffic follows Poisson distribution, then:

$$p(x) = \frac{\mu^x e^{-\mu}}{x!}$$
(3.80)

Where:
p(x) = probability of x vehicles arriving in time t
μ = average number of vehicles arriving in time t

$$\mu = \lambda t$$
(3.81)

A vehicle in the minor approach will be able to merge into the major approach when there is a gap available on the major approach that is equal to or greater than the critical gap. In other words, this will happen when there is no vehicles (zero vehicles) arrive during this gap. The probability of this to occur is the probability of $x = 0$ in Equation (3.80).
⇒

$$p(0) = p(h \geq t) = \frac{\mu^0 e^{-\mu}}{0!} = e^{-\mu} = e^{-\lambda t}$$
(3.82)

Since the addition of the two probabilities ($p(h \geq t)$, $p(h < t)$) must equal 1.0:

$$p(h < t) = 1 - p(h \geq t) = 1 - e^{-\lambda t}$$
(3.83)

The total number of gaps between V vehicles on the major approach is equal to V−1.
Therefore, the predicted number of gaps equal to or greater than the critical gap (number of accepted gaps) during the peak hour is given by:

$$\text{Freq.}(h \geq t) = (V - 1)e^{-\lambda t}$$

V = 1200 vph
λ = 1200/3600 = 0.333 veh/sec
t = critical gap = 3.125 sec
Therefore:

$$\text{Freq.}(h \geq t) = (V - 1)e^{-\lambda t}$$

$$\text{Freq.}(h \geq t) = (1200 - 1)e^{-(0.333)(3.125)} = 423$$

(d) In a similar manner, the predicted number of gaps smaller than the critical gap (the number of rejected gaps) during the peak hour is calculated by multiplying the probability of its occurrence by the total number of gaps (V−1).

$$\text{Freq.}(h < t) = (V - 1)(1 - e^{-\lambda t})$$
(3.84)

Therefore:

$$\text{Freq.}\,(h < t) = (1200 - 1)\left(1 - e^{-(0.333)(3.125)}\right) = 776$$

Or simply:
The number of rejected gaps = the total number of gaps–the number of accepted gaps

$$\text{Freq.}\,(h < t) = 1199 - 423 = 776$$

Figures 3.47 and 3.48 show the screen images of the MS Excel worksheet used to conduct the computations and analyze the gap acceptance data for this problem.

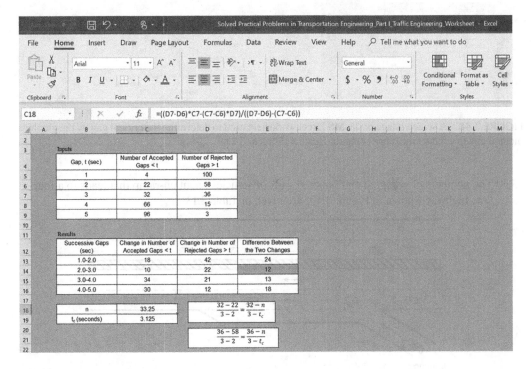

FIGURE 3.47 A screen image of the MS Excel worksheet used to compute the critical gap by the algebraic method for Problem 3.29.

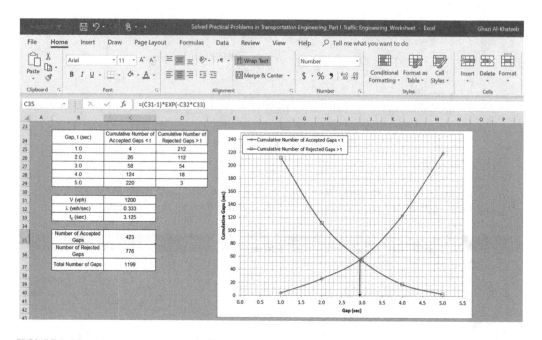

FIGURE 3.48 A screen image of the MS Excel worksheet used to determine the critical gap by the graphical method and to conduct the other computations for Problem 3.29.

3.30 The numbers of accepted and rejected gaps of vehicles on an on-ramp that leads to a four-lane highway (shown in Figure 3.49) are given in Table 3.13.

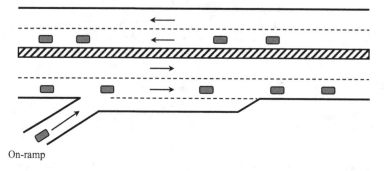

On-ramp

FIGURE 3.49 Gap acceptance on an on-ramp on a four-lane highway for Problem 3.30.

TABLE 3.13

The Number of Accepted and Rejected Gaps on the Minor Approach of an Unsignalized T-Intersection for Problem 3.30

Length of Gap, t (sec)	Number of Accepted Gaps (less than t)	Number of Rejected Gaps (more than t)
1.0	2	125
2.0	20	88
3.0	32	51
4.0	65	28
5.0	105	11
6.0	145	3

If the traffic flow rate during the peak hour on the highway is 1,800 vph, and the arrival of vehicles on the highway can be described by the Poisson distribution, determine the following:

(a) The critical gap using the algebraic method
(b) The critical gap using the graphical method
(c) The number of accepted gaps that will be available for the on-ramp vehicles during the peak hour
(d) The number of rejected gaps during the peak hour

Solution:

(a) Using the algebraic method (see Table 3.14):

TABLE 3.14

The Change in Accepted Gaps and Rejected Gaps between Two Successive Gaps for Problem 3.30

Successive Gaps (sec)	Change in Number of Accepted Gaps < t	Change in Number of Rejected Gaps > t	Difference Between the Two Changes
1.0–2.0	18	37	19
2.0–3.0	12	37	25
3.0–4.0	33	23	10
4.0–5.0	40	17	23
5.0–6.0	40	8	32

Using interpolation (similarity in triangles) between the two gaps 3 and 4 seconds, the critical gap along with the number of accepted/rejected gaps can be determined (see Table 3.15):

TABLE 3.15

The Gap Interval for Critical Gap for Problem 3.30

Length of Gap, t (sec)	Number of Accepted Gaps (less than t)	Number of Rejected Gaps (more than t)
3.0	32	51
t_c	n	n
4.0	65	28

$$\frac{65-32}{4-3} = \frac{65-n}{4-t_c} \tag{3.85}$$

$$\frac{28-51}{4-3} = \frac{28-n}{4-t_c} \tag{3.86}$$

Solving the above two equations simultaneously provides the solution shown below:

$$\frac{65-32}{65-n} = \frac{28-51}{28-n} \tag{3.87}$$

$\Rightarrow$

$$n = 43.20$$

$$t_c = 3.339 \text{ seconds}$$

(b)   Using the graphical method (see Table 3.16 and Figure 3.50)

TABLE 3.16

The Cumulative Number of Accepted Gaps and Rejected Gaps for Problem 3.30

Gap, t (sec)	Cumulative Number of Accepted Gaps < t	Cumulative Number of Rejected Gaps > t
1	2	306
2	22	181
3	54	93
4	119	42
5	224	14
6	369	3

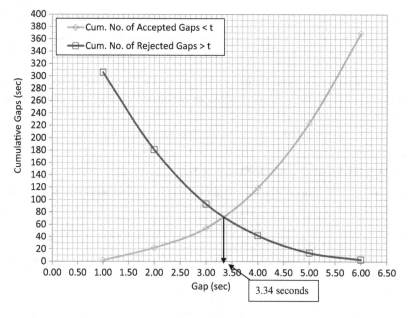

FIGURE 3.50   The graphical method for critical gap for Problem 3.30.

(c)   The number of accepted gaps is computed as follows:

$$\text{Freq.}\,(h \geq t) = (V-1)e^{-\lambda t}$$

$$\lambda = \frac{V}{T}$$

$$\lambda = \frac{1800}{3600} = 0.5 \text{ veh/sec}$$

$V = 1200$ vph
$t =$ critical gap $= 3.34$ sec
Therefore:

$$\text{Freq.}\left(h \geq t\right) = \left(V - 1\right)e^{-\lambda t}$$

$\Rightarrow$

$$\text{Freq.}\left(h \geq t\right) = \left(1800 - 1\right)e^{-(0.5)(3.34)} = 339$$

The number of rejected gaps is calculated as shown below:

$$\text{Freq.}\left(h < t\right) = \left(V - 1\right)\left(1 - e^{-\lambda t}\right)$$

$\Rightarrow$

$$\text{Freq.}\left(h < t\right) = \left(1800 - 1\right)\left(1 - e^{-(0.5)(3.34)}\right) = 1460$$

Note that the summation of the number of accepted gaps and the number of rejected gaps is equal to the total number of gaps within 1800 vehicles in the peak hour = 1800−1 = 1799.

Figures 3.51 and 3.52 show screen images of the MS Excel worksheet used to conduct the computations for this problem.

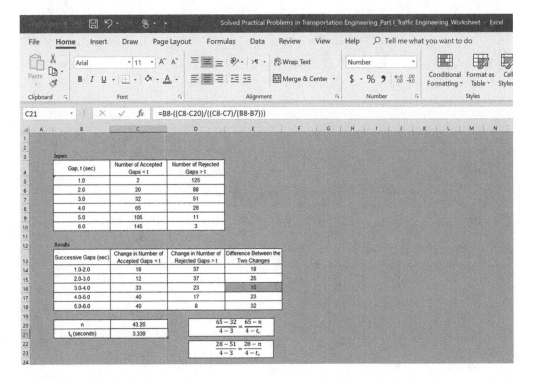

FIGURE 3.51 A screen image of the MS Excel worksheet used to compute the critical gap by the algebraic method for Problem 3.30.

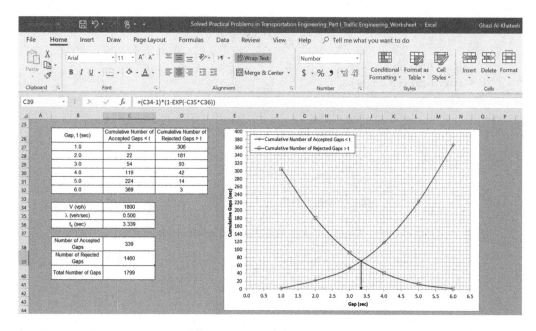

FIGURE 3.52 A screen image of the MS Excel worksheet used to determine the critical gap by the graphical method and to conduct the other computations for Problem 3.30.

4 Highway Capacity and Level of Service

Chapter 4 of this book discusses the capacity and the level of service (LOS) expected on a highway based on the existing conditions of the highway. Capacity refers typically to the maximum hourly flow rate expected on a roadway or a lane of roadway under existing traffic and road conditions. In this section, the traffic-related factors as well as road conditions are studied. Computation of related factors including the peak-hour factor and heavy vehicles adjustment factor are presented. The level of service, defined as a qualitative measure for the traffic service based on some performance indicators such as speed and density, is also presented in this part. The effect of the different existing traffic and road conditions on the determination of level of service for a highway is discussed. The level of service is determined for freeway segments including general segments and segments with specific grades (downgrades or upgrades). Additionally, the level of service for two-lane two-way highways and multi-lane highways is discussed. The performance measures required to determine the level of service are computed using the given traffic and roadway conditions and the standard tables and figures in the Highway Capacity Manual (HCM).

4.1 The traffic count on a roadway segment during the peak hour is shown in Table 4.1.

TABLE 4.1
Traffic Count During the Peak Hour for Problem 4.1

Time Period	Traffic Count
7:00–7:15am	800
7:15–7:30am	750
7:30–7:45am	710
7:45–8:00am	940

Determine the peak-hour factor (PHF) for this roadway segment.

Solution:

$$PHF = \frac{Hourly\ volume}{Peak\ 15 - minute\ hourly\ flow\ rate} \tag{4.1}$$

$$Peak\ 15 - minute\ hourly\ flow\ rate = 4 \times Traffic\ volume\ in\ peak\ 15\ minutes \tag{4.2}$$

Therefore:

$$PHF = \frac{800 + 750 + 710 + 940}{4 \times 940} = 0.85$$

The screen image of the MS Excel worksheet and the function (f_x) used to compute the PHF for this problem is shown in Figure 4.1.

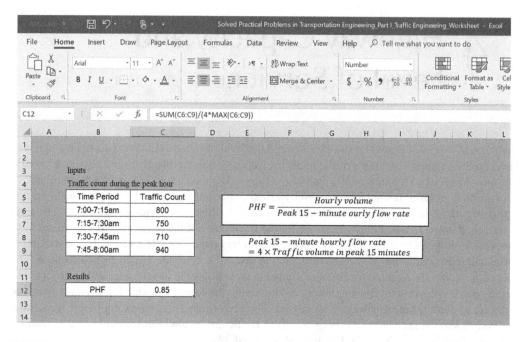

FIGURE 4.1 A screen image of the MS Excel worksheet used for the computations of Problem 4.1.

4.2 If the peak-hourly volume on a highway segment is 2,700 vph and the PHF is 0.90, compute the traffic count during the peak 15-minute period.

Solution:

The two formulas shown below for PHF and peak 15-minute hourly flow rate, respectively, are used:

$$PHF = \frac{Hourly\ volume}{Peak\ 15-minute\ hourly\ flow\ rate}$$

$$Peak\ 15-minute\ hourly\ flow\ rate = 4 \times Traffic\ volume\ in\ peak\ 15\ minutes$$

Therefore:

$$0.90 = \frac{2700}{4 \times Traffic\ volume\ in\ peak\ 15\ minutes} = 0.85$$

$$\Rightarrow$$

$$Traffic\ volume\ in\ peak\ 15\ minutes = \frac{2700}{4 \times 0.90} = 750\ vehicles$$

The screen image of the MS Excel worksheet with the function (f_x) used to obtain the traffic volume in the peak 15 minutes in this problem is shown in Figure 4.2.

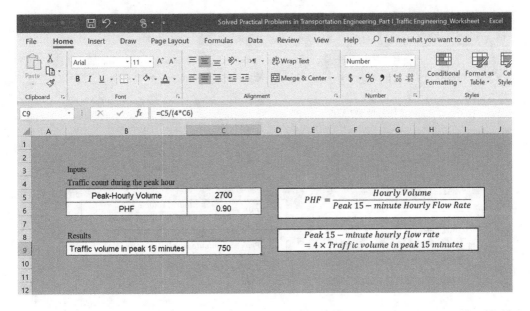

FIGURE 4.2 A screen image of the MS Excel worksheet used for the computations of Problem 4.2.

4.3 The traffic volume for a highway segment was collected in the peak hour for 15-minute time periods. If the difference between the peak 15-minute traffic count and the average 15-minute traffic count is 25%, compute the peak-hour factor (PHF).

Solution:

The peak-hour factor is calculated using the formula below:

$$PHF = \frac{\text{Hourly volume}}{\text{Peak 15-minute hourly flow rate}}$$

$$\text{Hourly volume} = 4 \times \text{Average 15-minute traffic count}$$

$$\text{Peak 15-minute hourly flow rate} = 4 \times \text{Peak 15-minute traffic count}$$

Since the difference between the peak 15-minute traffic count and the average 15-minute traffic count is 25%, then:

$$\text{Peak 15-min traffic count} = 1.25 \times \text{Average 15-min traffic count}$$

$$\Rightarrow$$

$$\text{Average 15-min traffic count} = \frac{\text{Peak 15-min traffic count}}{1.25}$$

Therefore:

$$PHF = \frac{4 \times \dfrac{\text{Peak 15-minute traffic count}}{1.25}}{4 \times \text{Peak 15-minute traffic count}} = \frac{1}{1.25} = 0.85$$

The screen image of the MS Excel worksheet with the function (f_x) used to obtain the PHF based on the given data in this problem is shown in Figure 4.3.

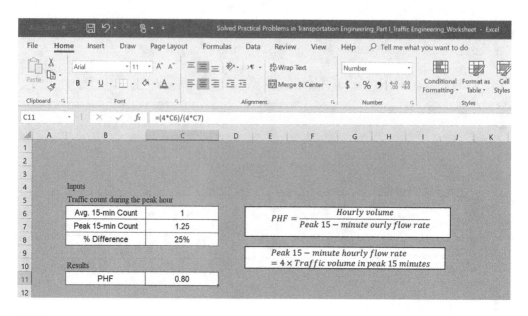

FIGURE 4.3   A screen image of the MS Excel worksheet used for the computations of Problem 4.3.

4.4   On a 10-lane freeway segment, the PHF and the peak-hourly volume in one direction are 0.90 and 4500 vph, respectively. If the percentage of heavy vehicles on this segment is 0% and drivers are all commuters, determine the 15-minute passenger car equivalent flow rate (pcphpl).

Solution:

The formula used to compute the 15-minute passenger car equivalent flow rate is shown in the equation below:

$$v_p = \frac{V}{(\text{PHF})(N)(f_p)(f_{\text{HV}})} \tag{4.3}$$

Where:
v_p = 15-minute passenger car equivalent flow rate (pcphpl)
V = peak-hour volume in one direction (vph)
PHF = peak-hour factor
N = number of lanes in one direction
f_p = driver population factor (0.85–1.00)
f_{HV} = adjustment factor for heavy vehicles

The peak-hour volume, V is given as 3600 vph, PHF = 0.90, N = 5 lanes (in one direction), f_p = 1.00 since drivers are all commuters, and f_{HV} = 1.00 since there is no heavy vehicles using the freeway.
   Therefore:

$$v_p = \frac{4500}{(0.90)(5)(1.00)(1.00)} = 1000 \text{ pcphpl}$$

A screen image of the MS Excel worksheet with the function (f_x) used to obtain the 15-minute passenger car equivalent flow rate (pcphpl) of this problem is shown in Figure 4.4.

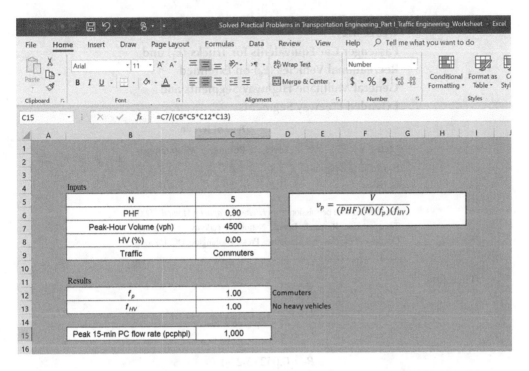

FIGURE 4.4 A screen image of the MS Excel worksheet used for the computations of Problem 4.4.

4.5 A 6-lane general extended freeway segment is designed in a level terrain to have a maximum service flow rate of 900 pcphpl. The designer intends to determine the percentage of trucks and buses that will be allowed to utilize this segment to keep the flow rate at this limit. If the PHF is 0.95, the peak-hourly volume is 2500 vph in one direction, and the drivers are all commuters, determine the percentage of trucks and buses.

Solution:

The formula for the 15-minute passenger car equivalent flow rate is used to back-calculate the adjustment factor for heavy vehicles (f_{HV}) as shown below:

$$v_p = \frac{V}{(\text{PHF})(N)(f_p)(f_{HV})}$$

$$900 = \frac{2500}{(0.95)(3)(1.00)(f_{HV})}$$

$\Rightarrow$

$$f_{HV} = \frac{2500}{(900)(0.95)(3)(1.00)} = 0.975$$

But the adjustment factor for heavy vehicles (f_{HV}) is given by the following formula:

$$f_{HV} = \frac{1}{1 + P_T\left(E_T - 1\right) + P_R\left(E_R - 1\right)} \tag{4.4}$$

Where:
    P_T and P_R = percentages of trucks/buses and recreational vehicles (RVs) in traffic, respectively (represented as decimal numbers)
    E_T and E_R = passenger car equivalent for trucks/buses and RVs, respectively
    $E_T = 1.5$ for level terrain from Table 4.1
    $P_R = 0\%$ (no recreational vehicles using the freeway) (see Table 4.2)

TABLE 4.2

Passenger Car Equivalents for Trucks (E_T) and Recreational Vehicles (E_R) on Extended General Multilane Highway Segments and Extended Freeway Segments

	Type of Terrain		
Factor	Level	Rolling	Mountainous
E_T (trucks and buses)	1.5	2.5	4.5
E_R (RVs)	1.2	2.0	4.0

Reproduced with permission from *Highway Capacity Manual*, 2000, Transportation Research Board, National Academy of Sciences, Courtesy of the National Academies Press, Washington, DC, USA.

Therefore:

$$0.975 = \frac{1}{1 + P_T\left(1.5 - 1\right)}$$

$\Rightarrow$

$$P_T = 0.052 \,(5.2\%)$$

A screen image of the MS Excel worksheet with the function (f_x) used to obtain the percentage of trucks/buses (P_T) of this problem is shown in Figure 4.5.

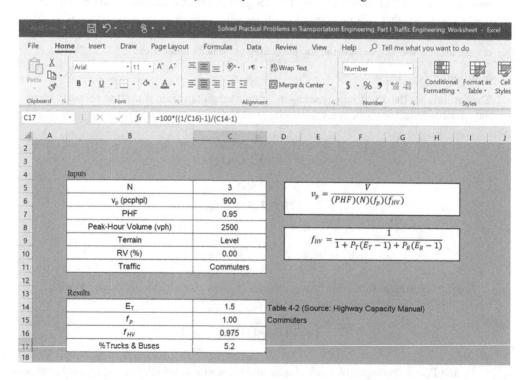

FIGURE 4.5 A screen image of the MS Excel worksheet used for the computation of the percentage of trucks/buses for Problem 4.5.

4.6 If the adjustment factor for heavy vehicles on a general extended freeway segment with level terrain is 0.956, determine the percentage of trucks and buses assuming that no recreational vehicles is using the freeway.

Solution:

The adjustment factor for heavy vehicles (f_{HV}) is given by the formula:

$$f_{HV} = \frac{1}{1 + P_T(E_T - 1) + P_R(E_R - 1)}$$

$E_T = 1.5$ (for level terrain from Table 4.2), and $P_R = 0$, therefore:

$$0.956 = \frac{1}{1 + P_T(1.5 - 1)}$$

⇒

$$P_T = \frac{1 - \dfrac{1}{0.956}}{(1.5 - 1)} = 0.092 = 9.2\%$$

A screen image of the MS Excel worksheet used to compute the percentage of trucks/buses (P_T) for this problem is shown in Figure 4.6.

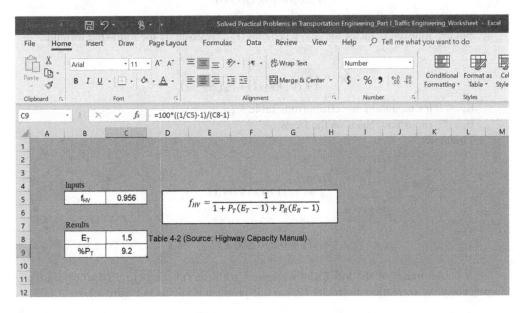

FIGURE 4.6 A screen image of the MS Excel worksheet used for the computation of the percentage of trucks/buses for Problem 4.6.

4.7 A highway traffic designer would like to maintain the service flow rate for a 4-lane freeway segment designed in a mountainous terrain to have a maximum service flow rate of 1200 pcphpl. The designer intends to determine the percentage of recreational vehicles that will be allowed to use this freeway to keep the flow rate at this limit as the freeway leads to national park and camping areas. If the PHF is 0.90, the peak-hourly volume is 1700 vph in one direction, and the drivers are assumed to be commuters, determine the percentage of recreational vehicles. No truck/buses are allowed to use the highway.

Solution:

To back-calculate the adjustment factor for heavy vehicles (f_{HV}), the formula for the 15-minute passenger car equivalent flow rate is used as shown below:

$$v_p = \frac{V}{(\text{PHF})(N)(f_p)(f_{HV})}$$

$$1200 = \frac{1700}{(0.90)(2)(1.00)(f_{HV})}$$

⇒

$$f_{HV} = \frac{1700}{(1200)(0.90)(2)(1.00)} = 0.787$$

But the adjustment factor for heavy vehicles (f_{HV}) is given by the following formula:

$$f_{HV} = \frac{1}{1 + P_T(E_T - 1) + P_R(E_R - 1)}$$

$E_T = 4.0$ for mountainous terrain from Table 4.2.
$P_T = 0\%$ (no trucks/buses on the freeway).
Therefore:

$$0.787 = \frac{1}{1 + P_R(4.0 - 1)}$$

⇒

$$P_R = 0.090 \ (9.0\%)$$

A screen image of the MS Excel worksheet with the function (f_x) used to obtain the percentage of recreational vehicles (P_R) in this problem is shown in Figure 4.7.

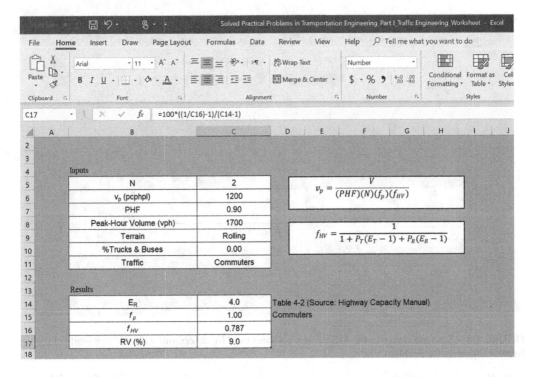

FIGURE 4.7 A screen image of the MS Excel worksheet used for the computation of the percentage of recreational vehicles for Problem 4.7.

4.8 If a freeway segment consists of two consecutive upgrades with grade and length of 3%–1200 ft and 2%–1800 ft, respectively, estimate the average grade of this segment.

Solution:

$$\%G_{\text{Average}} = 100 \frac{\sum_{i=1}^{n} G_i L_i}{\sum_{i=1}^{n} L_i} \qquad (4.5)$$

Where:
  G_{Average} = average grade (percent)
  G_i = grade for freeway segment i
  L_i = length of freeway segment i

Therefore:

$$\%G_{\text{Average}} = 100 \frac{G_1 L_1 + G_2 L_2}{L_1 + L_2}$$

$$\%G_{\text{Average}} = 100 \frac{(0.03)(1200) + (0.02)(1800)}{1200 + 1800} = 2.4\%$$

A screen image of the MS Excel worksheet with the function (f_x) used to obtain the average grade of the two freeway segments as a percentage is shown in Figure 4.8.

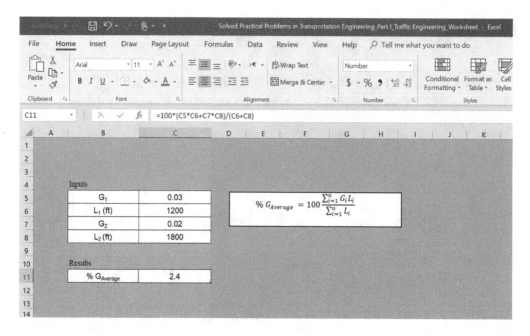

FIGURE 4.8 A screen image of the MS Excel worksheet used for determining the average grade of the two freeway segments for Problem 4.8.

4.9 If a freeway upgrade segment consists of two successive subsections of 2%, 3,000 ft long and 3%, 3,000 ft long, determine the equivalent grade for this segment (assume an entry truck speed is 55 mph).

Solution:

The equivalent grade for successive grades is determined using the performance curves for standard trucks in Figure 4.9. The procedure is described below:

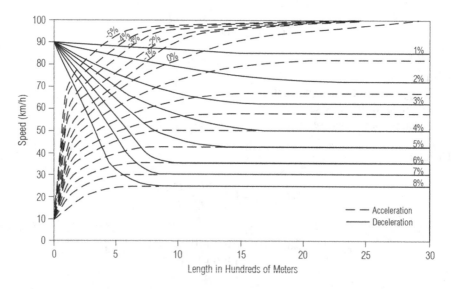

FIGURE 4.9 Performance curves for trucks (120 kg/kW = 200 lb/hp). Used with permission from *Highway Capacity Manual*, 2000, Transportation Research Board, National Academy of Sciences, Courtesy of the National Academies Press, Washington, DC, USA.

1. The truck speed of 55 mph (88.5 km/h) is used to enter the curve of the first grade (2% curve). Move on the curve for a distance equal to the length of the first grade (3000 ft = 914.4 m).io8
2. After that, the reduced speed is determined at this point by moving horizontally on the figure to the left to read the speed value (78 km/h = 48.5 mph).
3. From this speed value, the curve of second grade (3% curve) is entered by moving horizontally to the right until reaching the curve. From the intersection point with the 3%-curve, move on the curve for a distance equal to the length of the second grade (3000 ft = 914.4 m).
4. Read the new reduced speed by moving horizontally on the figure to the left. Read the new value of the speed (64 km/h = 39.8 mph).
5. From this speed value, make a horizontal line.
6. On the x-axis of the figure where the distance measurement (length of grade) is read, from the total length of the segment (3000 + 3000 = 6000 ft = 1828.8 m), make a vertical line.
7. The intersection point of the horizontal line and the vertical line represents the location of the equivalent grade of the freeway segment (2.8%).

The procedure is illustrated in Figure 4.10. From this figure, the equivalent grade for this segment is equal to 2.8%.

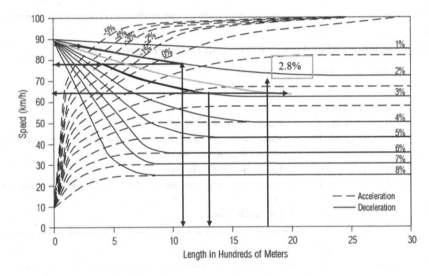

FIGURE 4.10   Determining the equivalent grade from performance curves for trucks (120 kg/kW = 200 lb/ hp). Used with permission from *Highway Capacity Manual*, 2000, Transportation Research Board, National Academy of Sciences, Courtesy of the National Academies Press, Washington, DC, USA.

4.10 A rural 10-lane freeway in rolling terrain was designed to operate at a level of service (LOS) B with the following characteristics:
Traffic data:
*   PHF = 0.95
*   Trucks = 8%
*   Commuter traffic
Geometric data:
*   Lane width = 12 ft
*   Shoulder width = 10 ft
*   Interchange spacing = 2/3 mile

Determine the maximum peak-hourly volume that this freeway can carry to maintain this LOS.

Solution:

The given data (inputs) are summarized in Table 4.3.

TABLE 4.3
Level of Service (LOS) Data for Problem 4.10

LOS	B
No. of Lanes, N	5
PHF	0.95
Trucks	8%
Drivers	Commuters
Lane Width (ft)	12
Shoulder Width (10 ft)	0.0
Interchange Spacing (mi)	2/3

The following data (outputs) shown in Table 4.4 are determined from standard Tables 4.2 and 4.5 through 4.8, respectively.

TABLE 4.4
Determining the Values of E_T, E_R, f_{LW}, f_{LC}, f_N, and f_{ID} for Problem 4.10

Item	Value	Table	Comment
E_T	2.5	Table 4.2	Using "rolling terrain"
E_R	---	No RVs	
f_{LW}	0.0	Table 4.5	Using "lane width of 12 ft"
f_{LC}	0.0	Table 4.6	Using "lanes in one direction ≥ 5 and right-shoulder lateral clearance of ≥ 6 ft (shoulder width 10 ft)"
f_N	0.0	Table 4.7	Using "lanes in one direction ≥ 5"
f_{ID}	5.0	Table 4.8	Using "interchanges per mile of 1.50" since the interchange spacing is 2/3 mile
BFFS (mph)	75.0		For rural freeways

TABLE 4.5
Adjustment for Lane Width (f_{LW})

Lane Width (ft)	Lane Width (m)	Reduction in FFS, f_{LW} (mi/h)	Reduction in FFS, f_{LW} (km/h)
12.0	3.6	0.0	0.0
11.5	3.5	0.6	1.0
11.2	3.4	1.3	2.1
11.0	3.3	1.9	3.1
10.5	3.2	3.5	5.6
10.2	3.1	5.0	8.1
10.0	3.0	6.6	10.6

Reproduced with permission from *Highway Capacity Manual*, 2000, Transportation Research Board, National Academy of Sciences, Courtesy of the National Academies Press, Washington, DC, USA.

TABLE 4.6
Adjustment for Right-Shoulder Lateral Clearance (f$_{LC}$)

Right-Shoulder Lateral Clearance (ft)	Right-Shoulder Lateral Clearance (m)	Reduction in FFS, f$_{LC}$ (mi/h)				Reduction in FFS, f$_{LC}$ (km/h)			
		Lanes in One Direction				Lanes in One Direction			
		2	3	4	≥ 5	2	3	4	≥ 5
≥ 6	≥ 1.8	0.0	0.0	0.0	0.0	0.0	0.0	0.0	0.0
5	1.5	0.6	0.4	0.2	0.1	1.0	0.7	0.3	0.2
4	1.2	1.2	0.8	0.4	0.2	1.9	1.3	0.7	0.4
3	0.9	1.8	1.2	0.6	0.3	2.9	1.9	1.0	0.6
2	0.6	2.4	1.6	0.8	0.4	3.9	2.6	1.3	0.8
1	0.3	3.0	2.0	1.0	0.5	4.8	3.2	1.6	1.1
0	0.0	3.6	2.4	1.2	0.6	5.8	3.9	1.9	1.3

Reproduced with permission from *Highway Capacity Manual*, 2000, Transportation Research Board, National Academy of Sciences, Courtesy of the National Academies Press, Washington, DC, USA.

TABLE 4.7
Adjustment for Number of Lanes (f$_N$)

Number of Lanes	Reduction in FFS, f$_N$ (mi/h)	Reduction in FFS, f$_N$ (km/h)
≥ 5	0.0	0.0
4	1.5	2.4
3	3.0	4.8
2	4.5	7.3

Reproduced with permission from Highway Capacity Manual, 2000, Transportation Research Board, National Academy of Sciences, Courtesy of the National Academies Press, Washington, DC, USA.

TABLE 4.8
Adjustment for Interchange Density (f$_{ID}$)

Interchanges per Mile	Interchanges per Kilometer	Reduction in FFS, f$_N$ (mi/h)	Reduction in FFS, f$_N$ (km/h)
≤ 0.50	≤ 0.3	0.0	0.0
0.65	0.4	0.7	1.1
0.75	0.5	1.3	2.1
1.00	0.6	2.5	3.9
1.15	0.7	3.1	5.0
1.25	0.8	3.7	6.0
1.50	0.9	5.0	8.1
1.60	1.0	5.7	9.2
1.75	1.1	6.3	10.2
2.00	1.2	7.5	12.1

Reproduced with permission from *Highway Capacity Manual*, 2000, Transportation Research Board, National Academy of Sciences, Courtesy of the National Academies Press, Washington, DC, USA.

The following parameters and factors (outputs) shown in Table 4.9 are computed/determined from standard tables.

TABLE 4.9
Results for Problem 4.10

Item	Value	Table
f_{HV}	0.893	Computed
f_p	1.00	Commuters
FFS (mph)	75.0	Computed
v_p (pcphpl)	1,210	Table 4.10
V (vph)	5,344	Computed

TABLE 4.10
Level-of-Service (LOS) Criteria for Basic Freeway Segments

Criteria	LOS				
	A	B	C	D	E
FFS = 75 mi/h (120 km/h)*					
Maximum density (pc/mi/ln, pc/km/ln)	11 (7)	18 (11)	26 (16)	35 (22)	45 (28)
Maximum speed (mi/h, km/h)	75 (120.0)	75 (120.0)	68.2 (114.6)	61.5 (99.6)	53.3 (85.7)
Maximum v/c	0.35	0.55	0.77	0.92	1
Maximum service flow rate (pc/h/ln)	840	1320	1840	2200	2400
FFS = 70 mi/h (110 km/h)					
Maximum density (pc/mi/ln, pc/km/ln)	11 (7)	18 (11)	26 (16)	35 (22)	45 (28)
Maximum speed (mi/h, km/h)	70.0 (110.0)	70.0 (110.0)	68.2 (108.5)	61.5 (97.2)	53.3 (83.9)
Maximum v/c	0.33	0.51	0.74	0.91	1
Maximum service flow rate (pc/h/ln)	770	1210	1740	2135	2350
FFS = 65 mi/h (105 km/h)					
Maximum density (pc/mi/ln, pc/km/ln)	11 (7)	18 (11)	26 (16)	35 (22)	45 (28)
Maximum speed (mi/h, km/h)	65.0 (105.0)	65.0 (105.0)	64.6 (104.0)	59.7 (96.1)	52.2 (84.0)
Maximum v/c	0.30	0.50	0.71	0.89	1.00
Maximum service flow rate (pc/h/ln)	710	1170	1680	2090	2350
FFS = 60 mi/h (100 km/h)					
Maximum density (pc/mi/ln, pc/km/ln)	11 (7)	18 (11)	26 (16)	35 (22)	45 (28)
Maximum speed (mi/h, km/h)	60.0 (100.0)	60.0 (100.0)	60.0 (100.0)	57.6 (93.8)	51.1 (82.1)
Maximum v/c	0.30	0.48	0.70	0.90	1.00
Maximum service flow rate (pc/h/ln)	700	1100	1600	2065	2300
FFS = 55 mi/h (90 km/h)					
Maximum density (pc/mi/ln, pc/km/ln)	11 (7)	18 (11)	26 (16)	35 (22)	45 (28)
Maximum speed (mi/h, km/h)	55.0 (90.0)	55.0 (90.0)	55.0 (90.0)	54.7 (89.1)	50.0 (80.4)
Maximum v/c	0.28	0.44	0.64	0.87	1.00
Maximum service flow rate (pc/h/ln)	630	990	1440	1955	2250

* Number between brackets are in SI units.

Reproduced with permission from *Highway Capacity Manual*, 2000, Transportation Research Board, National Academy of Sciences, Courtesy of the National Academies Press, Washington, DC, USA.

The adjustment factor for heavy vehicles (f_{HV}) is computed using the following formula:

$$f_{HV} = \frac{1}{1 + P_T\left(E_T - 1\right) + P_R\left(E_R - 1\right)}$$

$$f_{HV} = \frac{1}{1 + 0.08\left(2.5 - 1\right)} = 0.893$$

The free-flow speed is computed from the base free-flow speed (BFFS) using the following formula:

$$FFS = BFFS - f_{LW} - f_{LC} - f_N - f_{ID} \tag{4.6}$$

Where:
FFS = free-flow speed
BFFS = base free-flow speed
f_{LW} = adjustment factor for lane width
f_{LC} = adjustment factor for lateral clearance
f_N = adjustment factor for number of lanes
f_{ID} = adjustment factor for interchange spacing

$\Rightarrow$

$$FFS = 75.0 - 0.0 - 0.0 - 0.0 - 5.0 = 70 \text{ mph}$$

From Table 4.10, at LOS = B and FFS = 70 mph, the maximum service flow rate (v_p) = 1210 pcphpl. The peak-hourly volume is then calculated using the formula below:

$$v_p = \frac{V}{(PHF)(N)(f_p)(f_{HV})}$$

$\Rightarrow$

$$V = (1210)\,(0.95)\,(5)\,(1.00)\,(0.893) = 5132 \text{ vph}$$

The MS Excel worksheet used to determine the maximum peak-hourly volume based on the given LOS data in this problem is shown in Figure 4.11.

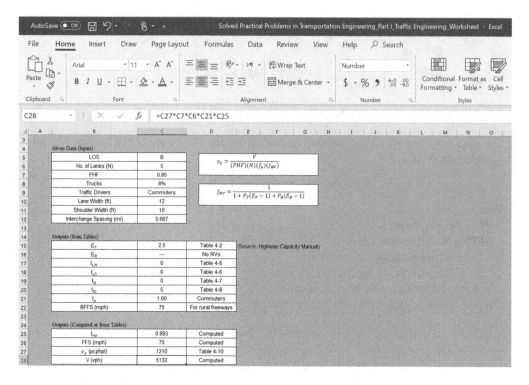

FIGURE 4.11 A screen image of the MS Excel worksheet used to compute the maximum peak-hourly volume in Problem 4.10.

4.11 For a 6-lane highway segment with the following characteristics (see Table 4.11):

TABLE 4.11

Traffic Count During the Peak Hour for Problem 4.11

Time Period	Traffic Count
7:00–7:15am	600
7:15–7:30am	580
7:30–7:45am	630
7:45–8:00am	700

- Traffic composition: Trucks = 8%, RVs = 4%
- Commuter traffic
- Upgrade = 4%
- Length = 0.30 mile

Determine the following:
- (a) Peak-hour factor (PHF)
- (b) Passenger car equivalent values for trucks and RVs
- (c) Adjustment factor for heavy vehicles (f_{HV})
- (d) 15-minute passenger car equivalent flow rate (v_p)

Solution:

(a) The peak-hour factor (PHF) is estimated based on the following formula:

$$PHF = \frac{\text{Hourly volume}}{\text{Peak 15} - \text{minute hourly flow rate}}$$

$\Rightarrow$

$$PHF = \frac{600 + 580 + 630 + 700}{4 \times 700} = 0.90$$

(b) The passenger car equivalent values for trucks and RVs in Table 4.12 are determined from Tables 4.13 and 4.14. First, the freeway segment should be classified whether it is a specific upgrade or general segment. This is based on the following condition:

A freeway/highway segment having a grade $\geq 3\%$ and length $> \frac{1}{4}$ mile or a grade $< 3\%$ and a length $> \frac{1}{2}$ mile is considered a specific upgrade.

Since the given freeway segment has an upgrade of $4\% \geq 3\%$ and length of 0.30 mile $> \frac{1}{4}$ mile $\Rightarrow$ the segment is considered a specific upgrade.

Therefore, the tables for specific upgrades are used. E_T and E_R are determined as shown in Table 4.12.

TABLE 4.12

Determining the E_T and E_R Values for a Specific Upgrade for Problem 4.11

Item	Value	Table	Comment
E_T	2.0	Table 4.13	Using 8% Trucks and G of 4% > 3–4%, and L = 0.30 mile > 0.25–0.50 mile
E_R	2.5	Table 4.14	Using 4% RVs and G of 4% > 3–4%, and L = 0.30 mile > 0.25–0.50 mile

TABLE 4.13

Passenger-Car Equivalents for Trucks and Buses (E_T) on Upgrades, Multilane Highways, and Basic Freeway Segments

Upgrade (%)	Length (mi)	Length (km)	E_T Percentage of Trucks and Buses								
			2	4	5	6	8	10	15	20	25
< 2	All	All	1.5	1.5	1.5	1.5	1.5	1.5	1.5	1.5	1.5
≥ 2–3	> 0.00–0.25	> 0.0–0.4	1.5	1.5	1.5	1.5	1.5	1.5	1.5	1.5	1.5
	> 0.25–0.50	> 0.4–0.8	1.5	1.5	1.5	1.5	1.5	1.5	1.5	1.5	1.5
	> 0.50–0.75	> 0.8–1.2	1.5	1.5	1.5	1.5	1.5	1.5	1.5	1.5	1.5
	> 0.75–1.00	> 1.2–1.6	2.0	2.0	2.0	2.0	1.5	1.5	1.5	1.5	1.5
	> 1.00–1.50	> 1.6–2.4	2.5	2.5	2.5	2.5	2.0	2.0	2.0	2.0	2.0
	> 1.50	> 2.4	3.0	3.0	2.5	2.5	2.0	2.0	2.0	2.0	2.0
> 3–4	> 0.00–0.25	> 0.0–0.4	1.5	1.5	1.5	1.5	1.5	1.5	1.5	1.5	1.5
	> 0.25–0.50	> 0.4–0.8	2.0	2.0	2.0	2.0	2.0	2.0	1.5	1.5	1.5
	> 0.50–0.75	> 0.8–1.2	2.5	2.5	2.0	2.0	2.0	2.0	2.0	2.0	2.0
	> 0.75–1.00	> 1.2–1.6	3.0	3.0	2.5	2.5	2.5	2.5	2.0	2.0	2.0
	> 1.00–1.50	> 1.6–2.4	3.5	3.5	3.0	3.0	3.0	3.0	2.5	2.5	2.5
	> 1.50	> 2.4	4.0	3.5	3.0	3.0	3.0	3.0	2.5	2.5	2.5
> 4–5	> 0.00–0.25	> 0.0–0.4	1.5	1.5	1.5	1.5	1.5	1.5	1.5	1.5	1.5
	> 0.25–0.50	> 0.4–0.8	3.0	2.5	2.5	2.5	2.0	2.0	2.0	2.0	2.0
	> 0.50–0.75	> 0.8–1.2	3.5	3.0	3.0	3.0	2.5	2.5	2.5	2.5	2.5
	> 0.75–1.00	> 1.2–1.6	4.0	3.5	3.5	3.5	3.0	3.0	3.0	3.0	3.0
	> 1.00	> 1.6	5.0	4.0	4.0	4.0	3.5	3.5	3.0	3.0	3.0
> 5–6	> 0.00–0.25	> 0.0–0.4	2.0	2.0	1.5	1.5	1.5	1.5	1.5	1.5	1.5
	> 0.25–0.30	> 0.4–0.5	4.0	3.0	2.5	2.5	2.0	2.0	2.0	2.0	2.0
	> 0.30–0.50	> 0.5–0.8	4.5	4.0	3.5	3.0	2.5	2.5	2.5	2.5	2.5
	> 0.50–0.75	> 0.8–1.2	5.0	4.5	4.0	3.5	3.0	3.0	3.0	3.0	3.0
	> 0.75–1.00	> 1.2–1.6	5.5	5.0	4.5	4.0	3.0	3.0	3.0	3.0	3.0
	> 1.00	> 1.6	6.0	5.0	5.0	4.5	3.5	3.5	3.5	3.5	3.5
> 6	> 0.00–0.25	> 0.0–0.4	4.0	3.0	2.5	2.5	2.5	2.5	2.0	2.0	2.0
	> 0.25–0.30	> 0.4–0.5	4.5	4.0	3.5	3.5	3.5	3.0	2.5	2.5	2.5
	> 0.30–0.50	> 0.5–0.8	5.0	4.5	4.0	4.0	3.5	3.0	2.5	2.5	2.5
	> 0.50–0.75	> 0.8–1.2	5.5	5.0	4.5	4.5	4.0	3.5	3.0	3.0	3.0
	> 0.75–1.00	> 1.2–1.6	6.0	5.5	5.0	5.0	4.5	4.0	3.5	3.5	3.5
	> 1.00	> 1.6	7.0	6.0	5.5	5.5	5.0	4.5	4.0	4.0	4.0

Reproduced with permission from *Highway Capacity Manual*, 2000, Transportation Research Board, National Academy of Sciences, Courtesy of the National Academies Press, Washington, DC, USA.

TABLE 4.14

Passenger-Car Equivalents for Recreational Vehicles (E_R) on Uniform Upgrades, Multilane Highways, and Basic Freeway Segments

Upgrade (%)	Length (mi)	Length (km)	E_R Percentage of RVs								
			2	4	5	6	8	10	15	20	25
≤2	All	All	1.2	1.2	1.2	1.2	1.2	1.2	1.2	1.2	1.2
> 2–3	> 0.00–0.50	0.0–0.8	1.2	1.2	1.2	1.2	1.2	1.2	1.2	1.2	1.2
	> 0.50	> 0.8	3.0	1.5	1.5	1.5	1.5	1.5	1.2	1.2	1.2
> 3–4	> 0.00–0.25	0.0–0.4	1.2	1.2	1.2	1.2	1.2	1.2	1.2	1.2	1.2
	> 0.25–0.50	> 0.4–0.8	2.5	2.5	2.0	2.0	2.0	2.0	1.5	1.5	1.5
	> 0.50	> 0.8	3.0	2.5	2.5	2.5	2.0	2.0	2.0	1.5	1.5
> 4–5	> 0.00–0.25	> 0.0–0.4	2.5	2.0	2.0	2.0	1.5	1.5	1.5	1.5	1.5
	> 0.25–0.50	> 0.4–0.8	4.0	3.0	3.0	3.0	2.5	2.5	2.0	2.0	2.0
	> 0.50	> 0.8	4.5	3.5	3.0	3.0	3.0	2.5	2.5	2.0	2.0
> 5	> 0.00–0.25	> 0.0–0.4	4.0	3.0	2.5	2.5	2.5	2.0	2.0	2.0	1.5
	> 0.25–0.50	> 0.4–0.8	6.0	4.0	4.0	3.5	3.0	3.0	2.5	2.5	2.0
	> 0.50	> 0.8	6.0	4.5	4.0	4.5	3.5	3.0	3.0	2.5	2.0

Reproduced with permission from *Highway Capacity Manual*, 2000, Transportation Research Board, National Academy of Sciences, Courtesy of the National Academies Press, Washington, DC, USA.

(c) The adjustment factor for heavy vehicles (f_{HV}) is computed using the E_T and E_R values determined in part (b) above.

$$f_{HV} = \frac{1}{1 + P_T(E_T - 1) + P_R(E_R - 1)}$$

$\Rightarrow$

$$f_{HV} = \frac{1}{1 + 0.08(2.0 - 1) + 0.04(2.5 - 1)} = 0.877$$

(d) The 15-minute passenger car equivalent flow rate (v_p) is calculated using the formula shown below:

$$v_p = \frac{V}{(PHF)(N)(f_p)(f_{HV})}$$

$\Rightarrow$

$$v_p = \frac{(600 + 580 + 630 + 700)}{(0.90)(3)(1.00)(0.877)} = 1064 \text{ pcphpl}$$

A screen image of the MS Excel worksheet with the function (f_x) used to compute the 15-minute passenger car equivalent flow rate (v_p) for this problem is shown in Figure 4.12.

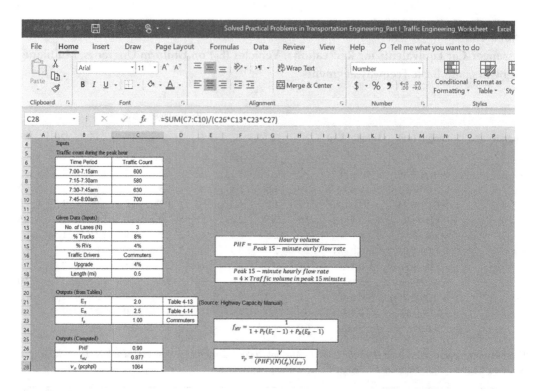

FIGURE 4.12 A screen image of the MS Excel worksheet used to compute the 15-minute passenger car equivalent flow rate for Problem 4.11.

4.12 A freeway segment was expanded from 4 to 5 lanes in each direction. What is the effect of this lane expansion on the 15-minute hourly flow rate assuming all other factors are the same?

Solution:

The 15-minute passenger car equivalent flow rate (v_p) is determined using the following formula:

$$v_p = \frac{V}{(\text{PHF})(N)(f_p)(f_{\text{HV}})}$$

At number of lanes, N=4:

$$v_p = \frac{V}{4(\text{PHF})(f_p)(f_{\text{HV}})}$$

And at number of lanes, N=5:

$$v_p = \frac{V}{5(\text{PHF})(f_p)(f_{\text{HV}})}$$

$$\text{Reduction in } v_p = 100\frac{\left(\dfrac{V}{(\text{PHF})(f_p)(f_{\text{HV}})}\right)\left(\dfrac{1}{4}-\dfrac{1}{5}\right)}{\left(\dfrac{V}{(\text{PHF})(f_p)(f_{\text{HV}})}\right)\left(\dfrac{1}{4}\right)} = 20.0\%$$

The screen image of the MS Excel worksheet used to compute the reduction in the 15-minute hourly flow rate for this problem is shown in Figure 4.13.

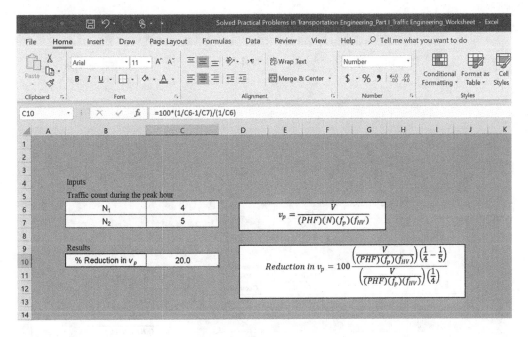

FIGURE 4.13 A screen image of the MS Excel worksheet used to compute the reduction in the 15-minute hourly flow rate for Problem 4.12.

4.13 A freeway segment is designed to have a free-flow speed (FFS) of 65 mph and operate at level of service (LOS) B. If the freeway segment with a peak-hourly volume of 3600 vph and PHF of 0.90 is intended to operate at the maximum service flow rate, determine the minimum number of lanes needed for each direction (assume commuter traffic and no heavy vehicles exist).

Solution:

Since the traffic drivers are commuters and no heavy vehicles will use the freeway, the driver population factor (fp) is 1.00 and the adjustment factor for heavy vehicles (f_{HV}) is also 1.00.

For a level of service (LOS) B, and a free-flow speed (FFS) of 65 mph, the maximum service flow rate (v_p) is determined from Table 4.10:

$v_p = 1170$ pcphpl.

The required number of lanes to maintain a LOS B and this flow rate is computed using the formula:

$$v_p = \frac{V}{(PHF)(N)(f_p)(f_{HV})}$$

$\Rightarrow$

$$N = \frac{V}{(v_p)(PHF)(f_p)(f_{HV})} = \frac{3600}{(1170)(0.90)(1.00)(1.00)} = 3.4 \cong 4 \text{ lanes}$$

Therefore, for this freeway segment to have a LOS B and operate at the maximum service flow rate, at least 4 lanes are required in each direction under the existing conditions.

The MS Excel worksheet with a function (f_x) is used to compute the required number of lanes in this problem. A screen image of the MS Excel worksheet along with the function used is shown in Figure 4.14.

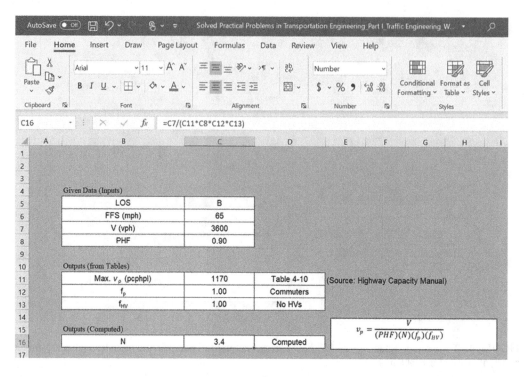

FIGURE 4.14 A screen image of the MS Excel worksheet used to compute the required number of lanes for Problem 4.13.

4.14 Determine the free flow speed (FFS) of a rural 8-lane freeway segment having a lane width of 10 ft, no obstruction, and an interchange density of 1.0 interchange/mile.

Solution:

The free-flow speed is computed from the base free-flow speed (BFFS) using the following formula:

$$FFS = BFFS - f_{LW} - f_{LC} - f_N - f_{ID}$$

From Tables 4.5 through 4.8:
BFFS = 75 mph (for rural highways).
f_{LW} = 6.6 mph (using Table 4.5 at lane width = 10 ft)
f_{LC} = 0.0 mph (using Table 4.6 at right-shoulder lateral clearance ≥ 6 ft and number of lanes in one direction = 4 lanes)
f_N = 1.5 mph (using Table 4.7 at number of lanes in one direction = 4 lanes)
f_{ID} = 2.5 mph (using Table 4.8 at interchange density = 1 interchange/mile)
⇒

$$FFS = 75.0 - 6.6 - 0.0 - 1.5 - 2.5 = 64.4 \text{ mph (103.6 kph)}$$

The MS Excel worksheet is used to compute the free flow speed (FFS) of the freeway in this problem. A screen image of the MS Excel worksheet along with the function used is shown in Figure 4.15.

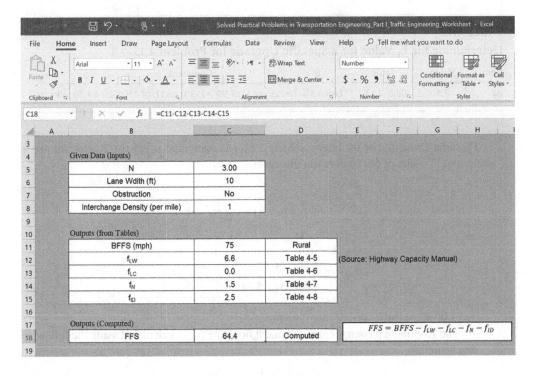

FIGURE 4.15 A screen image of the MS Excel worksheet used to compute the free flow speed on the freeway for Problem 4.14.

4.15 A highway traffic designer is conducting a study to check if the level of service (LOS) on a rural 6-lane freeway segment with the following characteristics located in a mountainous terrain is maintained at C after an increase in the number of trucks on the freeway segment from 5% to 10% due to the opening of a new nearby factory. Determine whether the LOS is maintained at C or not. If not, what measures or solutions can the designer take to keep the LOS at C?

Traffic data:
- PHF = 0.92
- Peak-hourly volume = 4200 vph (in one direction)
- Trucks = 5% (increased to 10%)
- No recreational vehicles
- Commuter traffic

Geometric data:
- Length = 2 miles
- Upgrade = 3%
- Lane width = 10 ft
- Shoulder width = 5 ft
- Interchange spacing = 2.0 miles

Solution:

Since the freeway segment has a grade of 3% ≥ 3% and le ngth of 2.0 miles > ¼ mile, the segment is considered a specific upgrade.

Therefore, the tables for specific upgrades are used. E_T is determined as shown in Table 4.15.

TABLE 4.15

Determining the E_T Value for a Specific Upgrade for Problem 4.15

Item	Value	Table	Comment
E_T	2.5	Table 4.13	Using 5% Trucks and G of 3% ≥ 2–3%, and L of 2.0 miles > 1.50 mile

The adjustment factor for heavy vehicles (f_{HV}) is computed using the E_T determined above and P_T value of 5% (before the increase).

$$f_{HV} = \frac{1}{1 + P_T(E_T - 1) + P_R(E_R - 1)}$$

⇒

$$f_{HV} = \frac{1}{1 + 0.05(2.5 - 1)} = 0.930$$

$f_p = 1.00$ since traffic drivers are commuters.

The 15-minute passenger car equivalent flow rate (v_p) is calculated using the following formula:

$$v_p = \frac{V}{(PHF)(N)(f_p)(f_{HV})}$$

⇒

$$v_p = \frac{4200}{(0.92)(3)(1.00)(0.930)} \cong 1636 \text{ pcphpl}$$

$$FFS = BFFS - f_{LW} - f_{LC} - f_N - f_{ID}$$

From standard tables, the four adjustment factors in the formula are determined as shown below:

BFFS = 75 mph (for rural highways).

$f_{LW} = 6.6$ mph (using Table 4.5 at lane width = 10 ft)

$f_{LC} = 0.4$ mph (using Table 4.6 at right-shoulder lateral clearance = 5 ft and number of lanes in one direction = 3 lanes)

$f_N = 3.0$ mph (using Table 4.7 at number of lanes in one direction = 3 lanes)

$f_{ID} = 0.0$ mph (using Table 4.8 at interchange density = 0.50 ≤ 0.50 interchange/mile)

⇒

$$FFS = 75.0 - 6.6 - 0.4 - 3.0 - 0.0 = 65.0 \text{ mph (104.6 kph)}$$

The average passenger car speed (S) can be computed using one of the formulas shone below:

For FFS > 70 mph:

$$S = FFS - \left[\left(FFS - \frac{160}{3}\right)\left(\frac{v_p + 30FFS - 3400}{30FFS - 1000}\right)^{2.6}\right] \tag{4.7}$$

For FFS ≤ 70 mph:

$$S = \text{FFS} - \left[\frac{1}{9}\left(7\text{FFS} - 340\right)\left(\frac{v_p + 30\text{FFS} - 3400}{40\text{FFS} - 1700}\right)^{2.6}\right] \tag{4.8}$$

Since the FFS $= 65$ mph ≤ 70 mph, the following formula is used to determine the average passenger car speed (S):

$$S = \text{FFS} - \left[\frac{1}{9}\left(7\text{FFS} - 340\right)\left(\frac{v_p + 30\text{FFS} - 3400}{40\text{FFS} - 1700}\right)^{2.6}\right]$$

$$S = 65 - \left[\frac{1}{9}\left(7(65) - 340\right)\left(\frac{1636 + 30(65) - 3400}{40(65) - 1700}\right)^{2.6}\right] = 64.2 \text{ mph} \left(103.3 \text{ kph}\right)$$

The density on the freeway segment is computed using the formula below:

$$D = \frac{v_p}{S} \tag{4.9}$$

Where:
D = density on the freeway segment (pc/mi/lane).
Hence:

$$D = \frac{1636}{64.2} \cong 26 \text{ pc/mi/lane}$$

Since the density (D) is in the range of $18 \leq D = 26 \leq 26$, where: 18 is the maximum density for LOS B and 26 is the maximum density for LOS C, the LOS is C.

After the increase when $P_T = 10\%$, the adjustment factor for heavy vehicles (f_{HV}) is computed as:

$$f_{HV} = \frac{1}{1 + 0.10\,(2.5 - 1)} = 0.870$$

Therefore,

$$v_p = \frac{4200}{(0.92)(3)(1.00)(0.870)} \cong 1750 \text{ pcphpl}$$

Since the FFS $= 65$ mph ≤ 70 mph, the same formula is used:

$$S = 65 - \left[\frac{1}{9}\left(7(65) - 340\right)\left(\frac{1750 + 30(65) - 3400}{40(65) - 1700}\right)^{2.6}\right] = 62.3 \text{ mph} \left(100.3 \text{ kph}\right)$$

The density is computed using the same formula:

$$D = \frac{v_p}{S}$$

$$D = \frac{1750}{62.3} \cong 29 \text{ pc/mi/lane}$$

Since the density (D) is in the range of $26 \leq D = 29 \leq 35$, where: 26 is the maximum density for LOS C and 35 is the maximum density for LOS D, the LOS is D.

After the increase in the percentage of trucks from 5% to 10%, the computed flow rate has increased from 1636 to 1750 pcphpl, the density has also increased from 26 to 29 pc/mi/lane and the LOS is reduced from C to D.

The scenarios that can improve the LOS under the existing traffic conditions should be related to the existing geometric conditions. In other words, a change in one of the existing geometric conditions would possibly improve the LOS. Two possible scenarios can be done to maintain the LOS at C:

Scenario #1: An additional lane on this freeway segment can be constructed. The additional lane reduces the maximum flow rate on the segment as shown below:

$$v_p = \frac{4200}{(0.92)(4)(1.00)(0.870)} \cong 1312 \text{ pcphpl}$$

$$FFS = BFFS - f_{LW} - f_{LC} - f_N - f_{ID}$$

The adjustment factors f_{LW}, f_{LC}, and f_{ID} stay the same, but the adjustment factor f_N changes since the number of lanes is changed from 3 to 4 lanes. Therefore:

BFFS = 75 mph (for rural highways).

f_{LW} = 6.6 mph (using Table 4.5 at lane width = 10 ft)

f_{LC} = 0.4 mph (using Table 4.6 at right-shoulder lateral clearance = 5 ft and number of lanes in one direction = 3 lanes)

f_N = 1.5 mph (using Table 4.7 at number of lanes in one direction = 3 lanes)

f_{ID} = 0.0 mph (using Table 4.8 at interchange density = $0.50 \leq 0.50$ interchange/mile)

$\Rightarrow$

$$FFS = 75.0 - 6.6 - 0.4 - 1.5 - 0.0 = 66.5 \text{ mph } (107.0 \text{ kph})$$

Since the FFS = 66.5 mph $\leq$ 70 mph, the following formula is used to determine the average passenger car speed (S):

$$S = FFS - \left[\frac{1}{9}(7FFS - 340)\left(\frac{v_p + 30FFS - 3400}{40FFS - 1700} \right)^{2.6} \right]$$

But since $v_p = 1312 < 3400 - 30FFS = 3400 - 30(66.5) = 1405$ pcphpl, then S = FFS. Therefore,

$$S = 66.5 \text{ mph} \left(107.0 \text{ kph} \right)$$

The density on the freeway segment is computed using the formula below:

$$D = \frac{v_p}{S}$$

Hence:

$$D = \frac{1312}{66.5} \cong 20 \text{ pc/mi/lane}$$

Since the density (D) is in the range of $18 \leq D = 20 \leq 26$, where: 18 is the maximum density for LOS B and 26 is the maximum density for LOS C, the LOS is C.

Scenario #2: Increase the lane width only 1 ft from 10 to 11 ft; that is a total of 3 ft for the 3 lanes (in one direction). The reduction in the service flow rate and the effect on the LOS based on the 1-ft lane increase is computed as shown below:

The increase in the lane width will affect the free-flow speed (FFS):

$f_{LW} = 1.9$ mph (using Table 4.5 at lane width $= 11$ ft)

All other factors are the same as before.

$\Rightarrow$

$$FFS = 75.0 - 1.9 - 0.4 - 3.0 - 0.0 = 69.7 \text{ mph} \cong 70 \text{ mph} (112.6 \text{ kph})$$

Since the FFS $= 69.7$ mph ≤ 70 mph, the following formula is used to determine the average passenger car speed (S):

$$S = FFS - \left[\frac{1}{9}(7FFS - 340)\left(\frac{v_p + 30FFS - 3400}{40FFS - 1700} \right)^{2.6} \right]$$

$$S = 69.7 - \left[\frac{1}{9}(7(69.7) - 340)\left(\frac{1312 + 30(69.7) - 3400}{40(69.7) - 1700} \right)^{2.6} \right] = 69.7 \text{ mph}$$

$$v_p = 1750 \text{ pcphpl}$$

The density is computed using the formula:

$$D = \frac{v_p}{S}$$

$$D = \frac{1750}{69.7} \cong 26 \text{ pc/mi/lane}$$

Since the density (D) is in the range of $18 \leq D = 26 \leq 26$, where: 18 is the maximum density for LOS B and 26 is the maximum density for LOS C, the LOS is C.

The scenario of increasing the interchange spacing on the freeway would not benefit improving the LOS. In other words, looking at Table 4.8, the zero reduction in the free-flow speed (FFS) corresponds to an interchange density of ≤ 0.50 mile. In this problem, the interchange spacing of 2 miles (i.e., interchange density of 0.50 mile) provides a zero reduction in the FFS. Hence, no further increase in the spacing (reduction in density) would improve the FFS and hence the LOS on this freeway segment.

4.16 A designer would like to assess which of the following three scenarios would mostly impact the LOS of a freeway segment with the following characteristics:
Classification of freeway:
- Urban
- Rolling terrain

Traffic data:
- PHF = 0.90
- Peak-hourly volume = 7020 vph (in one direction)
- Trucks = 8%
- RVs = 5%
- Commuter traffic

Geometric data:
- Number of lanes = 4 lanes (in one direction)
- Length = 2 miles
- Upgrade = 2%
- Lane width = 12 ft
- Shoulder width = 0 ft (no lateral clearance)
- Interchange spacing = 2.0 miles
- Rolling terrain

Scenario #1: increasing the number of lanes from 4 to 5 lanes.

Scenario #2: prohibiting large-size trucks with heavy loads from using the freeway during normal hours of the day and allow them to use the freeway after midnight when traffic is not heavy. This will reduce the percentage of trucks from 8% to 4%.

Scenario #3: adding a shoulder having a width of 6 ft to the freeway.

Solution:

Since the freeway segment has a grade of 2% < 3% and length of 2.0 miles > ½ mile, the segment is considered a specific upgrade.

Therefore, the tables for specific upgrades are used. E_T is determined as shown in Table 4.16.

TABLE 4.16

Determining the E_T and E_R Values for a Specific Upgrade for Problem 4.16

Item	Value	Table	Comment
E_T	2.0	Table 4.13	Using 8% Trucks and G of 2% ≥ 2–3%, and L of 2.0 miles > 1.50 mile
E_R	1.2	Table 4.14	Using 5% RVs and G of 2% ≤ 2%, and L of 2.0 miles ("All" in the table)

The adjustment factor for heavy vehicles (f_{HV}) is computed using the E_T determined above and P_T value of 8% (before the decrease).

$$f_{HV} = \frac{1}{1 + P_T\left(E_T - 1\right) + P_R\left(E_R - 1\right)}$$

⇒

$$f_{HV} = \frac{1}{1 + 0.08\left(2.0 - 1\right) + 0.05\left(1.2 - 1\right)} = 0.917$$

$f_p = 1.00$ since traffic drivers are commuters.

The 15-minute passenger car equivalent flow rate (v_p) is calculated using the following formula:

$$v_p = \frac{V}{(PHF)(N)(f_p)(f_{HV})}$$

$\Rightarrow$

$$v_p = \frac{7020}{(0.90)(4)(1.00)(0.917)} \cong 2126 \text{ pcphpl}$$

$$FFS = BFFS - f_{LW} - f_{LC} - f_N - f_{ID}$$

BFFS = 70 mph (for urban highways)

f_{LW} = 0.0 mph (using Table 4.5 at lane width = 12 ft)

f_{LC} = 1.2 mph (using Table 4.6 at right-shoulder lateral clearance = 0 ft and number of lanes in one direction = 4 lanes)

f_N = 1.5 mph (using Table 4.7 at number of lanes in one direction = 4 lanes)

f_{ID} = 0.0 mph (using Table 4.8 at interchange density = 0.50 ≤ 0.50 interchange/mile since spacing = 2.0 miles)

$\Rightarrow$

$$FFS = 70.0 - 0.0 - 1.2 - 1.5 - 0.0 = 67.3 \text{ mph (108.3 kph)}$$

Since the FFS = 67.3 mph ≤ 70 mph, the following formula is used to determine the average passenger car speed (S):

$$S = FFS - \left[\frac{1}{9}(7FFS - 340)\left(\frac{v_p + 30FFS - 3400}{40FFS - 1700} \right)^{2.6} \right]$$

$$S = 67.3 - \left[\frac{1}{9}(7(67.3) - 340)\left(\frac{2126 + 30(67.3) - 3400}{40(67.3) - 1700} \right)^{2.6} \right] = 60.4 \text{ mph}$$

The density is computed using the formula:

$$D = \frac{v_p}{S}$$

$$D = \frac{2126}{60.4} \cong 36 \text{ pc/mi/lane}$$

Since the density (D) is in the range of 35 ≤ D = 36 ≤ 45, where: 35 is the maximum density for LOS D and 45 is the maximum density for LOS E, the LOS is E.

Assessment of scenarios:

Scenario #1: increasing the number of lanes from 4 to 5 lanes.

The increase of the number of lanes from 4 to 5 lanes affects the following:

$$v_p = \frac{7020}{(0.90)(5)(1.00)(0.917)} \cong 1701 \text{ pcphpl}$$

f_{LC} = 0.6 mph (using Table 4.6 at right-shoulder lateral clearance = 0 ft and number of lanes in one direction = 5 lanes)

f_N = 0.0 mph (using Table 4.7 at number of lanes in one direction = 5 lanes)

$\Rightarrow$

$$FFS = 70.0 - 0.0 - 0.6 - 0.0 - 0.0 = 69.4 \text{ mph } (111.7 \text{ kph})$$

Since the FFS = 69.4 mph ≤ 70 mph, the following formula is used to determine the average passenger car speed (S):

$$S = FFS - \left[\frac{1}{9}(7FFS - 340)\left(\frac{v_p + 30FFS - 3400}{40FFS - 1700} \right)^{2.6} \right]$$

$$S = 69.4 - \left[\frac{1}{9}(7(69.4) - 340)\left(\frac{1700 + 30(69.4) - 3400}{40(69.4) - 1700} \right)^{2.6} \right] = 68.3 \text{ mph}$$

The density is computed using the formula:

$$D = \frac{v_p}{S}$$

$$D = \frac{1701}{68.3} \cong 25 \text{ pc/mi/lane}$$

Since the density (D) is in the range of $18 \leq D = 25 \leq 26$, where: 18 is the maximum density for LOS B and 26 is the maximum density for LOS C, the LOS is C.

Scenario #2: prohibiting large-size trucks with heavy loads from using the freeway during normal hours of the day and allow them to use the freeway after midnight when traffic is not heavy. This will reduce the percentage of trucks from 8% to 4%.

In this case, the E_T value is changed (see Table 4.17).

TABLE 4.17

Determining the E_T Value for a Specific Upgrade for Problem 4.16

Item	Value	Table	Comment
E_T	3.0	Table 4.13	Using 4% Trucks and G of 2% ≥ 2–3%, and L of 2.0 miles > 1.50 mile

---

The adjustment factor for heavy vehicles is computed accordingly as shown below:

$$f_{HV} = \frac{1}{1 + 0.04(3.0 - 1) + 0.05(1.2 - 1)} = 0.917$$

$$v_p = \frac{V}{(PHF)(N)(f_p)(f_{HV})}$$

$\Rightarrow$

$$v_p = \frac{7020}{(0.90)(4)(1.00)(0.917)} \cong 2126 \text{ pcphpl}$$

The free-flow speed is the same as before applying any scenario since the change in the percentage in trucks does not affect the FFS adjustment factors:

$$FFS = 70.0 - 0.0 - 1.2 - 1.5 - 0.0 = 67.3 \text{ mph} (108.3 \text{ kph})$$

Since the FFS = 67.3 mph ≤ 70 mph, the following formula is used to determine the average passenger car speed (S):

$$S = FFS - \left[\frac{1}{9}(7FFS - 340) \left(\frac{v_p + 30FFS - 3400}{40FFS - 1700} \right)^{2.6} \right]$$

$$S = 67.3 - \left[\frac{1}{9}(7(67.3) - 340) \left(\frac{2126 + 30(67.3) - 3400}{40(67.3) - 1700} \right)^{2.6} \right] = 60.4 \text{ mph}$$

The density is computed using the formula:

$$D = \frac{v_p}{S}$$

$$D = \frac{2126}{60.4} \cong 36 \text{ pc/mi/lane}$$

Since the density (D) is in the range of $35 \leq D = 25 \leq 45$, where: 35 is the maximum density for LOS D and 45 is the maximum density for LOS E, the LOS is E.

Scenario #3: adding a shoulder having a width of 6 ft to the freeway.

The service flow rate is the same as before applying any scenario since the change in the shoulder width does not affect the flow rate:

$$v_p = \frac{7020}{(0.90)(4)(1.00)(0.917)} \cong 2126 \text{ pcphpl}$$

$f_{LC} = 0.0$ mph (using Table 4.6 at right-shoulder lateral clearance = 6 ft and number of lanes in one direction = 4 lanes)

⇒

$$FFS = 70.0 - 0.0 - 0.0 - 1.5 - 0.0 = 68.5 \text{ mph} (110.2 \text{ kph})$$

Since the FFS = 68.5 mph ≤ 70 mph, the following formula is used to determine the average passenger car speed (S):

$$S = FFS - \left[\frac{1}{9}(7FFS - 340) \left(\frac{v_p + 30FFS - 3400}{40FFS - 1700} \right)^{2.6} \right]$$

$$S = 68.5 - \left[\frac{1}{9}(7(68.5) - 340) \left(\frac{2126 + 30(68.5) - 3400}{40(68.5) - 1700} \right)^{2.6} \right] = 61.1 \text{ mph}$$

The density is computed using the formula:

$$D = \frac{v_p}{S}$$

$$D = \frac{2126}{61.1} \cong 35 \text{ pc/mi/lane}$$

Since the density (D) is in the range of $35 \leq D = 35 \leq 45$, where: 35 is the maximum density for LOS D and 45 is the maximum density for LOS E, the LOS is D.

Investigating the three scenarios, Scenarios #1 (increasing the number of lanes from 4 to 5 lanes) improved the LOS from E to C and Scenario #3 (adding a shoulder having a width of 6 ft to the freeway) improved the LOS from E to D. On the other hand, Scenario #2 (reducing the percentage of trucks from 8% to 4%) did not improve the LOS of E. In conclusion, the improvement in Scenario #1 by having an additional lane is more significant (improvement by two levels) than the improvement by adding a 6-ft shoulder (improvement by one level). Yet, adding a shoulder of 6-ft width is less costly than adding another 12-ft lane to the freeway segment of 2.0 miles long.

4.17 Determine the LOS for the peak direction of a Class I two-lane highway with the following characteristics:

Traffic data:
- PHF=0.85
- Hourly volume=850 vph (in the peak direction)
- Opposing volume=650 vph
- Trucks=6%
- RVs=3%
- BFFS=55 mph

Geometric data:
- Lane width=12 ft
- Shoulder width=6 ft
- 20 access points per mile
- No-passing zones=20%
- Rolling terrain

Solution:

To determine the level of service of the analysis direction on Class I two-lane highways, the percent time-spent following (PTSF) and the average travel speed (ATS) in the analysis direction should be computed.

PTSF Computation:

A trial initial value for the flow rate in the peak direction can be computed using the following formula:

$$v = \frac{V}{\text{PHF}}$$

$$v = \frac{850}{0.85} = 1000 \text{ pcph}$$

The values of f_G (adjustment factor for grade), E_T, and E_R are determined from Table 4.18. The value of f_G is determined from Table 4.19 and the values of E_T and E_R are determined from Table 4.20.

TABLE 4.18

Determining f_G, E_T, and E_R for the Peak Direction (PTSF Method) for Problem 4.17

Item	Value	Table	Comment
f_G	1.00	Table 4.19	Using directional flow rate = 1000 > 600 pcph, and rolling terrain
E_T	1.0	Table 4.20	Using directional flow rate = 1000 > 600 pcph, and rolling terrain
E_R	1.0	Table 4.20	Using directional flow rate = 1000 > 600 pcph, and rolling terrain

TABLE 4.19

Grade Adjustment Factor (f_G) to Determine Percent Time-Spent-Following on Two-Way and Directional Segments

Range of Two-Way Flow Rates (pc/h)	Range of Directional Flow Rates (pc/h)	Type of Terrain Level	Rolling
0–600	0–300	1.00	0.77
> 600–1200	> 300–600	1.00	0.94
> 1200	> 600	1.00	1.00

Reproduced with permission from *Highway Capacity Manual*, 2000, Transportation Research Board, National Academy of Sciences, Courtesy of the National Academies Press, Washington, DC, USA.

TABLE 4.20

Passenger-Car Equivalents for Trucks (E_T) and Recreational Vehicles (E_R) to Determine Percent Time-Spent Following on Two-Way and Directional Segments

Vehicle Type	Range of Two-Way Flow Rates (pc/h)	Range of Directional Flow Rates (pc/h)	Type of Terrain Level	Rolling
Trucks, E_T	0–600	0–300	1.1	1.8
	> 600–1200	> 300–600	1.1	1.5
	> 1200	> 600	1.0	1.0
RVs, E_R	0–600	0–300	1.0	1.0
	> 600–1200	> 300–600	1.0	1.0
	> 1200	> 600	1.0	1.0

Reproduced with permission from *Highway Capacity Manual*, 2000, Transportation Research Board, National Academy of Sciences, Courtesy of the National Academies Press, Washington, DC, USA.

$$f_{HV} = \frac{1}{1 + P_T(E_T - 1) + P_R(E_R - 1)}$$

$\Rightarrow$

$$f_{HV} = \frac{1}{1 + 0.06(1.0 - 1) + 0.03(1.0 - 1)} = 1.00$$

$$v = \frac{V}{(PHF)(f_G)(f_{HV})}$$

$\Rightarrow$

$$v = \frac{850}{(0.85)(1.00)(1.00)} = 1000 \text{ pcph}$$

The opposing flow rate is calculated using the procedure shown below:
A trial initial value for the opposing flow rate can be computed using the following formula:

$$v_o = \frac{V}{\text{PHF}} \tag{4.10}$$

$$v_o = \frac{650}{0.85} \cong 765 \text{ pcph}$$

The values of f_G (adjustment factor for grade), E_T, and E_R are determined as shown in Table 4.21. The value of f_G is determined from Table 4.19 and the values of E_T and E_R are determined from Table 4.20.

TABLE 4.21

Determining f_G, E_T, and E_R for the Opposing Direction (PTSF Method) for Problem 4.17

Item	Value	Table	Comment
f_G	1.00	Table 4.18	Using directional opposing flow rate = 765 > 600 pcph, and rolling terrain
E_T	1.0	Table 4.19	Using directional opposing flow rate = 765 > 600 pcph, and rolling terrain
E_R	1.0	Table 4.19	Using directional opposing flow rate = 765 > 600 pcph, and rolling terrain

---

$$f_{HV} = \frac{1}{1 + P_T (E_T - 1) + P_R (E_R - 1)}$$

$\Rightarrow$

$$f_{HV} = \frac{1}{1 + 0.06(1.0 - 1) + 0.03(1.0 - 1)} = 1.00$$

$$v_o = \frac{V}{(\text{PHF})(f_G)(f_{HV})}$$

$\Rightarrow$

$$v_o = \frac{650}{(0.85)(1.00)(1.00)} = 765 \text{ pcph}$$

The percent time-spent following (PTSF) is calculated using the following formula:

$$\text{PTSF} = \text{BPTSF} + f_{np} \tag{4.11}$$

Where:
BPTSF = base percent time-spent following.
f_{np} = adjustment factor for percentage of no-passing zones in the analysis direction.

The base percent time-spent-following (BPTSF) is computed using the following equation:

$$\text{BPTSF} = 100\left[1 - e^{av^b}\right] \tag{4.12}$$

Where:
v = passenger car equivalent flow rate for the peak 15-minute period in the analysis direction.
a and b = coefficients based on peak 15-minute passenger car equivalent opposing flow rate.

The coefficients a and b for an opposing demand flow rate = 765 pcph are calculated using interpolation between the values of a and b at opposing flow rates of 600 and 800 pcph as shown in Table 4.22. The values of the coefficients a and b are determined from Table 4.23.

TABLE 4.22

Tabulating the Values of a and b for the Opposing Direction for Problem 4.17

Opposing Flow Rate (pcph)	a	b
600	−0.100	0.413
765	?	?
800	−0.173	0.349

TABLE 4.23

Values of Coefficients (a, b) Used in Estimating Percent Time-Spent-Following for Directional Segments

Opposing Demand Flow Rate (pc/h)	a	b
≤ 200	−0.013	0.668
400	−0.057	0.479
600	−0.100	0.413
800	−0.173	0.349
1000	−0.320	0.276
1200	−0.430	0.242
1400	−0.522	0.225
≥ 1600	−0.665	0.199

Reproduced with permission from *Highway Capacity Manual*, 2000, Transportation Research Board, National Academy of Sciences, Courtesy of the National Academies Press, Washington, DC, USA.

$\Rightarrow$ at 765 pcph,

$$a = -0.173 - \left\{ (800 - 765) \times \frac{(-0.173 - (-0.100))}{(800 - 600)} \right\} = -0.160$$

and

$$b = 0.349 - \left\{ (800 - 765) \times \frac{(0.349 - 0.413)}{(800 - 600)} \right\} = -0.160$$

Therefore,

$$\text{BPTSF} = 100 \left[1 - e^{-0.160(1000)^{0.360}} \right] = 85.4\%$$

$$\text{FFS} = \text{BFFS} - f_{LS} - f_A \qquad (4.13)$$

Where:
FFS = free-flow speed (mph)
BFFS = base free-flow speed (mph)
f_{LS} = adjustment factor for lane and shoulder widths
f_A = adjustment factor for number of access points
f_{LS} = 0.0 (from Table 4.24 for lane width = 12 ft ≥ 12 ft and shoulder width = 6 ft ≥ 6 ft)
f_A = 5.0 (from Table 4.25 for number of access point = 20 per mile)

TABLE 4.24
Adjustment for Lane Width and Shoulder Width (f_{LS})

		Reduction in FFS (mi/h, km/h)*			
		Shoulder Width			
		≥ 0 – <2 (ft)	≥ 2 – <4 (ft)	≥ 4 – <6 (ft)	≥ 6 (ft)
Lane Width (ft)	Lane Width (m)	≥ 0.0 – <0.6 (m)	≥ 0.6 – <1.2 (m)	≥ 1.2 – <1.8 (m)	≥ 1.8 (m)
9 – <10	2.7 – <3.0	6.4 (10.3)	4.8 (7.7)	3.5 (5.6)	2.2 (3.5)
≥ 10 – <11	≥ 3.0 – <3.3	5.3 (8.5)	3.7 (5.9)	2.4 (3.8)	1.1 (1.7)
≥ 11 – <12	≥ 3.3 – <3.6	4.7 (7.5)	3.0 (4.9)	1.7 (2.8)	0.4 (0.7)
≥ 12	≥ 3.6	4.2 (6.8)	2.6 (4.2)	1.3 (2.1)	0.0 (0.0)

* Number between brackets are in SI units.

Reproduced with permission from *Highway Capacity Manual*, 2000, Transportation Research Board, National Academy of Sciences, Courtesy of the National Academies Press, Washington, DC, USA.

TABLE 4.25
Adjustment for Access-Point Density (fA)

Access Points per Mile	Access Points per Kilometer	Reduction in FFS (mi/h)	Reduction in FFS (km/h)
0	0	0.0	0.0
10	6	2.5	4.0
20	12	5.0	8.0
30	18	7.5	12.0
≥ 40	≥ 24	10.0	16.0

Reproduced with permission from *Highway Capacity Manual*, 2000, Transportation Research Board, National Academy of Sciences, Courtesy of the National Academies Press, Washington, DC, USA.

Hence,

$$FFS = 55 - 0.0 - 5.0 = 50 \text{ mph}$$

From Table 4.26 by interpolation between the values at opposing flow rates of 600 and 800 pcph, and using the data: no-passing zones = 20% $\leq$ 20%, opposing demand flow rate = 765 pcph and FFS = 50 mph, the adjustment factor (f_{np}) is calculated as shown below (see Table 4.27):

TABLE 4.26
Adjustment for No-Passing Zones (f_{np}) to Percent Time-Spent Following in Directional Segments

Opposing Demand Flow Rate (pc/h)	No Passing Zones (%)				
	≤ 20	40	60	80	100
FFS = 65 mi/h (110 km/h)					
≤ 100	10.1	17.2	20.2	21.0	21.8
200	12.4	19.0	22.7	23.8	24.8
400	9.0	12.3	14.1	14.4	15.4
600	5.3	7.7	9.2	9.7	10.4
800	3.0	4.6	5.7	6.2	6.7
1000	1.8	2.9	3.7	4.1	4.4
1200	1.3	2.0	2.6	2.9	3.1
1400	0.9	1.4	1.7	1.9	2.1
≥ 1600	0.7	0.9	1.1	1.2	1.4
FFS = 60 mi/h (100 km/h)					
≤ 100	8.4	14.9	20.9	22.8	26.6
200	11.5	18.2	24.1	26.2	29.7
400	8.6	12.1	14.8	15.9	18.1
600	5.1	7.5	9.6	10.6	12.1
800	2.8	4.5	5.9	6.7	7.7
1000	1.6	2.8	3.7	4.3	4.9
1200	1.2	1.9	2.6	3.0	3.4
1400	0.8	1.3	1.7	2.0	2.3

(Continued)

TABLE 4.26   (CONTINUED)

Adjustment for No-Passing Zones (f_{np}) to Percent Time-Spent Following in Directional Segments

Opposing Demand Flow Rate (pc/h)	No Passing Zones (%)				
	≤ 20	40	60	80	100
≥ 1600	0.6	0.9	1.1	1.2	1.5
FFS = 55 mi/h (90 km/h)					
≤ 100	6.7	12.7	21.7	24.5	31.3
200	10.5	17.5	25.4	28.6	34.7
400	8.3	11.8	15.5	17.5	20.7
600	4.9	7.3	10.0	11.5	13.9
800	2.7	4.3	6.1	7.2	8.8
1000	1.5	2.7	3.8	4.5	5.4
1200	1.0	1.8	2.6	3.1	3.8
1400	0.7	1.2	1.7	2.0	2.4
≥ 1600	0.6	0.9	1.2	1.3	1.5
FFS = 50 mi/h (80 km/h)					
≤ 100	5.0	10.4	22.4	26.3	36.1
200	9.6	16.7	26.8	31.0	39.6
400	7.9	11.6	16.2	19.0	23.4
600	4.7	7.1	10.4	12.4	15.6
800	2.5	4.2	6.3	7.7	9.8
1000	1.3	2.6	3.8	4.7	5.9
1200	0.9	1.7	2.6	3.2	4.1
1400	0.6	1.1	1.7	2.1	2.6
≥ 1600	0.5	0.9	1.2	1.3	1.6
FFS = 45 mi/h (70 km/h)					
≤ 100	3.7	8.5	23.2	28.2	41.6
200	8.7	16.0	28.2	33.6	45.2
400	7.5	11.4	16.9	20.7	26.4
600	4.5	6.9	10.8	13.4	17.6
800	2.3	4.1	6.5	8.2	11.0
1000	1.2	2.5	3.8	4.9	6.4
1200	0.8	1.6	2.6	3.3	4.5
1400	0.5	1.0	1.7	2.2	2.8
≥ 1600	0.4	0.9	1.2	1.3	1.7

Reproduced with permission from *Highway Capacity Manual*, 2000, Transportation Research Board, National Academy of Sciences, Courtesy of the National Academies Press, Washington, DC, USA.

TABLE 4.27

Tabulating the Values of f_{np} for the Opposing Direction (PTSF Method) for Problem 4.17

Opposing Flow Rate (pcph)	f_{np}
600	4.7
765	?
800	2.5

$\Rightarrow$

$$f_{np} = 2.5 - \left\{ (800 - 765) \times \frac{(2.5 - 4.7)}{(800 - 600)} \right\} = 2.885$$

Therefore:

$$PTSF = 85.4 + 2.885 \cong 88.3\%$$

ATS Computation:

A trial initial value for the flow rate in the peak direction can be computed using the following formula:

$$v = \frac{V}{PHF}$$

$$v = \frac{850}{0.85} = 1000 \text{ pcph}$$

The values of f_G (adjustment factor for grade), E_T, and E_R are determined as shown in Table 4.28. The value of f_G is determined from Table 4.29 and the values of E_T and E_R are determined from Table 4.30.

TABLE 4.28
Determining the Values of f_G, E_T, and E_R for the Peak Direction (ATS Method) for Problem 4.17

Item	Value	Table	Comment
f_G	0.99	Table 4.29	Using directional flow rate = 1000 > 600 pcph, and rolling terrain
E_T	1.5	Table 4.30	Using directional flow rate = 1000 > 600 pcph, and rolling terrain
E_R	1.1	Table 4.30	Using directional flow rate = 1000 > 600 pcph, and rolling terrain

---

TABLE 4.29
Grade Adjustment Factor (f_G) to Determine Speeds on Two-Way and Directional Segments

Range of Two-Way Flow Rates (pc/h)	Range of Directional Flow Rates (pc/h)	Type of Terrain	
		Level	Rolling
0–600	0–300	1.00	0.71
> 600–1200	> 300–600	1.00	0.93
> 1200	> 600	1.00	0.99

Reproduced with permission from *Highway Capacity Manual*, 2000, Transportation Research Board, National Academy of Sciences, Courtesy of the National Academies Press, Washington, DC, USA.

TABLE 4.30

Passenger-Car Equivalents for Trucks (E_T) and Recreational Vehicles (E_R) to Determine Speeds on Two-Way and Directional Segments

Vehicle Type	Range of Two-Way Flow Rates (pc/h)	Range of Directional Flow Rates (pc/h)	Type of Terrain	
			Level	Rolling
Trucks, E_T	0–600	0–300	1.7	2.5
	> 600–1200	> 300–600	1.2	1.9
	> 1200	> 600	1.1	1.5
RVs, E_R	0–600	0–300	1.0	1.1
	> 600–1200	> 300–600	1.0	1.1
	> 1200	> 600	1.0	1.1

Reproduced with permission from *Highway Capacity Manual*, 2000, Transportation Research Board, National Academy of Sciences, Courtesy of the National Academies Press, Washington, DC, USA.

$$f_{HV} = \frac{1}{1 + P_T\left(E_T - 1\right) + P_R\left(E_R - 1\right)}$$

$\Rightarrow$

$$f_{HV} = \frac{1}{1 + 0.06\left(1.5 - 1\right) + 0.03\left(1.1 - 1\right)} = 0.968$$

$$v = \frac{V}{\left(\text{PHF}\right)\left(f_G\right)\left(f_{HV}\right)}$$

$\Rightarrow$

$$v = \frac{850}{\left(0.85\right)\left(0.99\right)\left(0.968\right)} \cong 1044 \text{ pcph}$$

The opposing flow rate is computed using the procedure shown below:

A trial initial value for the opposing flow rate can be computed using the following formula:

$$v_o = \frac{V}{\text{PHF}}$$

$$v_o = \frac{650}{0.85} \cong 765 \text{ pcph}$$

The values of f_G (adjustment factor for grade), E_T, and E_R are determined as shown in Table 4.31. The value of f_G is determined from Table 4.29 and the values of E_T and E_R are determined from Table 4.30.

TABLE 4.31

Determining the Values of f_G, E_T, and E_R for the Opposing Direction (ATS Method) for Problem 4.17

Item	Value	Table	Comment
f_G	0.99	Table 4.29	Using directional opposing flow rate = 765 > 600 pcph, and rolling terrain
E_T	1.5	Table 4.30	Using directional opposing flow rate = 765 > 600 pcph, and rolling terrain
E_R	1.1	Table 4.30	Using directional flow rate = 1000 > 600 pcph, and rolling terrain

$$f_{HV} = \frac{1}{1 + P_T\left(E_T - 1\right) + P_R\left(E_R - 1\right)}$$

$\Rightarrow$

$$f_{HV} = \frac{1}{1 + 0.06\left(1.5 - 1\right) + 0.03\left(1.1 - 1\right)} = 0.968$$

$$v_o = \frac{V}{\left(PHF\right)\left(f_G\right)\left(f_{HV}\right)}$$

$\Rightarrow$

$$v_o = \frac{650}{\left(0.85\right)\left(0.99\right)\left(0.968\right)} \cong 798 \text{ pcph}$$

$$ATS = FFS - 0.00776\left(v + v_o\right) - f_{np} \qquad (4.14)$$

Where:

ATS = average travel speed in the analysis direction

FFS = free-flow speed in the analysis direction

v = passenger car equivalent flow rate for the peak 15-minute period in the analysis direction

v_o = passenger car equivalent flow rate for the peak 15-minute period in the opposing direction

f_{np} = adjustment factor for the percentage of no-passing zones in the analysis direction

From Table 4.32 by interpolation between the values at opposing flow rates of 600 and 800 pcph, and using the data: no-passing zones = 20% ≤ 20%, opposing demand flow rate = 798 pcph and FFS = 50 mph, the adjustment factor (f_{np}) is calculated as shown below (see Table 4.33):

TABLE 4.32

Adjustment for No-Passing Zones (f_{np}) to Average Travel Speed in Directional Segments

Opposing Demand Flow Rate (pc/h)	No Passing Zones (%)				
	≤ 20	40	60	80	100
FFS = 65 mi/h (110 km/h)*					
≤ 100	1.1 (1.7)	2.2 (3.5)	2.8 (4.5)	3.0 (4.8)	3.1 (5.0)
200	2.2 (3.5)	3.3 (5.3)	3.9 (6.2)	4.0 (6.5)	4.2 (6.8)
400	1.6 (2.6)	2.3 (3.7)	2.7 (4.4)	2.8 (4.5)	2.9 (4.7)
600	1.4 (2.2)	1.5 (2.4)	1.7 (2.8)	1.9 (3.1)	2.0 (3.3)
800	0.7 (1.1)	1.0 (1.6)	1.2 (2.0)	1.4 (2.2)	1.5 (2.4)
1000	0.6 (1.0)	0.8 (1.3)	1.1 (1.7)	1.1 (1.8)	1.2 (1.9)
1200	0.6 (0.9)	0.8 (1.3)	0.9 (1.5)	1.0 (1.6)	1.1 (1.7)
1400	0.6 (0.9)	0.7 (1.2)	0.9 (1.4)	0.9 (1.4)	0.9 (1.5)
≥ 1600	0.6 (0.9)	0.7 (1.1)	0.7 (1.2)	0.7 (1.2)	0.8 (1.3)
FFS = 60 mi/h (100 km/h)					
≤ 100	0.7 (1.2)	1.7 (2.7)	2.5 (4.0)	2.8 (4.5)	2.9 (4.7)
200	1.9 (3.0)	2.9 (4.6)	3.7 (5.9)	4.0 (6.4)	4.2 (6.7)
400	1.4 (2.3)	2.0 (3.3)	2.5 (4.1)	2.7 (4.4)	2.9 (4.6)
600	1.1 (1.8)	1.3 (2.1)	1.6 (2.6)	1.9 (3.0)	2.0 (3.2)
800	0.6 (0.9)	0.9 (1.4)	1.1 (1.8)	1.3 (2.1)	1.4 (2.3)
1000	0.6 (0.9)	0.7 (1.1)	0.9 (1.5)	1.1 (1.7)	1.2 (1.9)
1200	0.5 (0.8)	0.7 (1.1)	0.9 (1.4)	0.9 (1.5)	1.1 (1.7)
1400	0.5 (0.8)	0.6 (1.0)	0.8 (1.3)	0.8 (1.3)	0.9 (1.4)
≥ 1600	0.5 (0.8)	0.6 (1.0)	0.7 (1.1)	0.7 (1.1)	0.7 (1.2)
FFS = 55 mi/h (90 km/h)					
≤ 100	0.5 (0.8)	1.2 (1.9)	2.2 (3.6)	2.6 (4.2)	2.7 (4.4)
200	1.5 (2.4)	2.4 (3.9)	3.5 (5.6)	3.9 (6.3)	4.1 (6.6)
400	1.3 (2.1)	1.9 (3.0)	2.4 (3.8)	2.7 (4.3)	2.8 (4.5)
600	0.9 (1.4)	1.1 (1.8)	1.6 (2.5)	1.8 (2.9)	1.9 (3.1)
800	0.5 (0.8)	0.7 (1.1)	1.1 (1.7)	1.2 (2.0)	1.4 (2.2)
1000	0.5 (0.8)	0.6 (0.9)	0.8 (1.3)	0.9 (1.5)	1.1 (1.8)
1200	0.5 (0.8)	0.6 (0.9)	0.7 (1.2)	0.9 (1.4)	1.0 (1.6)
1400	0.5 (0.8)	0.6 (0.9)	0.7 (1.1)	0.7 (1.2)	0.9 (1.4)
≥ 1600	0.5 (0.8)	0.5 (0.8)	0.6 (0.9)	0.6 (0.9)	0.7 (1.1)
FFS = 50 mi/h (80 km/h)					
≤ 100	0.2 (0.3)	0.7 (1.1)	1.9 (3.1)	2.4 (3.9)	2.5 (4.1)
200	1.2 (1.9)	2.0 (3.2)	3.3 (5.3)	3.9 (6.2)	4.0 (6.5)
400	1.1 (1.8)	1.6 (2.6)	2.2 (3.5)	2.6 (4.2)	2.7 (4.4)
600	0.6 (1.0)	0.9 (1.5)	1.4 (2.3)	1.7 (2.8)	1.9 (3.0)
800	0.4 (0.6)	0.6 (0.9)	0.9 (1.5)	1.2 (1.9)	1.3 (2.1)
1000	0.4 (0.6)	0.4 (0.7)	0.7 (1.1)	0.9 (1.4)	1.1 (1.8)
1200	0.4 (0.6)	0.4 (0.7)	0.7 (1.1)	0.8 (1.3)	1.0 (1.6)
1400	0.4 (0.6)	0.4 (0.7)	0.6 (1.0)	0.7 (1.1)	0.8 (1.3)
≥ 1600	0.4 (0.6)	0.4 (0.7)	0.5 (0.8)	0.5 (0.8)	0.6 (1.0)
FFS = 45 mi/h (70 km/h)					
≤ 100	0.1 (0.1)	0.4 (0.6)	1.7 (2.7)	2.2 (3.6)	2.4 (3.8)
200	0.9 (1.5)	1.6 (2.6)	3.1 (5.0)	3.8 (6.1)	4.0 (6.4)
400	0.9 (1.5)	0.5 (0.8)	2.0 (3.2)	2.5 (4.1)	2.7 (4.3)
600	0.4 (0.7)	0.3 (0.5)	1.3 (2.1)	1.7 (2.7)	1.8 (2.9)
800	0.3 (0.5)	0.3 (0.5)	0.8 (1.3)	1.1 (1.8)	1.2 (2.0)

(Continued)

TABLE 4.32 (CONTINUED)
Adjustment for No-Passing Zones (f_{np}) to Average Travel Speed in Directional Segments

Opposing Demand Flow Rate (pc/h)	No Passing Zones (%)				
	≤ 20	40	60	80	100
1000	0.3 (0.5)	0.3 (0.5)	0.6 (1.0)	0.8 (1.3)	1.1 (1.8)
1200	0.3 (0.5)	0.3 (0.5)	0.6 (1.0)	0.7 (1.2)	1.0 (1.6)
1400	0.3 (0.5)	0.3 (1.5)	0.6 (1.0)	0.6 (1.0)	0.7 (1.2)
≥ 1600	0.3 (0.5)	0.3 (0.5)	0.4 (0.7)	0.4 (0.7)	0.6 (0.9)

* Number between brackets are in SI units.
Reproduced with permission from *Highway Capacity Manual*, 2000, Transportation Research Board, National Academy of Sciences, Courtesy of the National Academies Press, Washington, DC, USA.

TABLE 4.33
Tabulating the Values of f_{np} for the Opposing Direction (ATS Method) for Problem 4.17

Opposing Flow Rate (pcph)	f_{np}
600	0.6
798	?
800	0.4

$\Rightarrow$

$$f_{np} = 0.4 - \left\{ (800 - 765) \times \frac{(0.4 - 0.6)}{(800 - 600)} \right\} = 0.402$$

$\Rightarrow$

$$ATS = 50 - 0.00776(1044 + 798) - 0.402 = 35.3 \text{ mph}$$

From Table 4.34 for Class I two-lane highways and using the PTSF value of 88.3% > 80%, and the ATS value of 35.3 mph ≤ 40 mph, the level of service (LOS) for the analysis direction of this highway is E.

TABLE 4.34

Level-of-Service (LOS) Criteria for Two-Lane Highways in Class I

LOS	Percent Time-Spent-Following	Average Travel Speed (mi/h)	Average Travel Speed (km/h)
A	≤ 35	> 55	> 90
B	> 35–50	> 50–55	> 80–90
C	> 50–65	> 45–50	> 70–80
D	> 65–80	> 40–45	> 60–70
E	> 80	≤ 40	≤ 60

Note: LOS F applies whenever the flow rate exceeds the segment capacity.
Reproduced with permission from *Highway Capacity Manual*, 2000, Transportation Research Board, National Academy of Sciences, Courtesy of the National Academies Press, Washington, DC, USA.

4.18 Solve Problem 4.17 but using a Class II two-lane highway with the same characteristics.

Solution:

For a Class II two-lane highway, only the PTSF is needed to determine the LOS.

Therefore, from Table 4.35 and using the PTSF value (that was computed in Problem 4.17) of 88.3% > 85%, the LOS for the peak direction of this highway is E.

TABLE 4.35

Level-of-Service (LOS) Criteria for Two-Lane Highways in Class II

LOS	Percent Time-Spent-Following
A	≤ 40
B	> 40–55
C	> 55–70
D	> 70–85
E	> 85

Note: LOS F applies whenever the flow rate exceeds the segment capacity.
Reproduced with permission from *Highway Capacity Manual*, 2000, Transportation Research Board, National Academy of Sciences, Courtesy of the National Academies Press, Washington, DC, USA.

4.19 Determine the LOS of service for a Class I two-way two-lane highway with the following characteristics:

Traffic data:
- PHF=0.90
- Volume=1500 vph (two ways)
- Directional split=50/50
- Trucks=12%
- RVs=2%
- BFFS=60 mph

Geometric data:
- Length of segment=5 miles
- Lane width=12 ft
- Shoulder width=8 ft
- 40 access points per mile
- No-passing zones=40%
- Level terrain

Solution:

To determine the level of service of the two-way two-lane highway segment classified as Class I, the percent time-spent following (PTSF) and the average travel speed (ATS) should be computed.

PTSF Computation:

A trial value for the 15-minute passenger car equivalent flow rate (v_p) can be computed using the following formula:

$$v_p = \frac{V}{\text{PHF}}$$

$$v_p = \frac{1500}{0.90} \cong 1667 \text{ pcph}$$

The values of f_G (adjustment factor for grade), E_T, and E_R are determined as shown in Table 4.36. The value of f_G is determined from Table 4.19 and the values of E_T and E_R are determined from Table 4.20.

TABLE 4.36

Determining f_G, E_T, and E_R for the Peak Direction (PTSF Method) for Problem 4.19

Item	Value	Table	Comment
f_G	1.00	Table 4.19	Using two-way flow rate=1667 > 1200 pcph, and rolling terrain
E_T	1.0	Table 4.20	Using two-way flow rate=1667 > 1200 pcph, and rolling terrain
E_R	1.0	Table 4.20	Using two-way flow rate=1667 > 1200 pcph, and rolling terrain

---

$$f_{HV} = \frac{1}{1 + P_T\left(E_T - 1\right) + P_R\left(E_R - 1\right)}$$

$$\Rightarrow$$

$$f_{HV} = \frac{1}{1 + 0.12\left(1.0 - 1\right) + 0.02\left(1.0 - 1\right)} = 1.00$$

$$v_p = \frac{V}{\left(PHF\right)\left(f_G\right)\left(f_{HV}\right)} \tag{4.15}$$

$$\Rightarrow$$

$$v_p = \frac{1500}{\left(0.90\right)\left(1.00\right)\left(1.00\right)} \cong 1667 \text{ pcph}$$

The percent time-spent following (PTSF) is calculated using the following formula:

$$PTSF = BPTSF + f_{d/np} \tag{4.16}$$

Where:

$f_{d/np}$ = adjustment factor for the combined effect of directional split and percentage of no-passing zones

The base percent time-spent-following (BPTSF) is computed using the following equation:

$$BPTSF = 100\left[1 - e^{-0.000879 v_p}\right] \tag{4.17}$$

Where:

v_p = the 15-minute passenger car equivalent flow rate

Therefore,

$$BPTSF = 100\left[1 - e^{-0.000879\left(1667\right)}\right] = 76.9\%$$

From Table 4.37 by interpolation between the values at two-way flow rates of 1400 and 2000 pcph, and using the data: no-passing zones = 40%, and the directional split of 50/50, the adjustment factor ($f_{d/np}$) is calculated as shown below (see Table 4.38):

TABLE 4.37

Adjustment for Combined Effect of Directional Distribution of Traffic and Percentage of No-Passing Zones ($f_{d/np}$) on Percent Time-Spent Following on Two-Way Segments

| Two-Way Flow Rate, v_p (pc/h) | Increase in Percent Time-Spent-Following (%) | | | | | |
| | No Passing Zones (%) | | | | | |
	0	20	40	60	80	100
Directional Split = 50/50						
≤ 200	0.0	10.1	17.2	20.2	21.0	21.8
400	0.0	12.4	19.0	22.7	23.8	24.8
600	0.0	11.2	16.0	18.7	19.7	20.5
800	0.0	9.0	12.3	14.1	14.5	15.4
1400	0.0	3.6	5.5	6.7	7.3	7.9
2000	0.0	1.8	2.9	3.7	4.1	4.4
2600	0.0	1.1	1.6	2.0	2.3	2.4
3200	0.0	0.7	0.9	1.1	1.2	1.4
Directional Split = 60/40						
≤ 200	1.6	11.8	17.2	22.5	23.1	23.7
400	0.5	11.7	16.2	20.7	21.5	22.2
600	0.0	11.5	15.2	18.9	19.8	20.7
800	0.0	7.6	10.3	13.0	13.7	14.4
1400	0.0	3.7	5.4	7.1	7.6	8.1
2000	0.0	2.3	3.4	3.6	4.0	4.3
≥ 2600	0.0	0.9	1.4	1.9	2.1	2.2
Directional Split = 70/30						
≤ 200	2.8	13.4	19.1	24.8	25.2	25.5
400	1.1	12.5	17.3	22.0	22.6	23.2
600	0.0	11.6	15.4	19.1	20.0	20.9
800	0.0	7.7	10.5	13.3	14.0	14.6
1400	0.0	3.8	5.6	7.4	7.9	8.3
≥ 2000	0.0	1.4	4.9	3.5	3.9	4.2
Directional Split = 80/20						
≤ 200	5.1	17.5	24.3	31.0	31.3	31.6
400	2.5	15.8	21.5	27.1	27.6	28.0
600	0.0	14.0	18.6	23.2	23.9	24.5
800	0.0	9.3	12.7	16.0	16.5	17.0
1400	0.0	4.6	6.7	8.7	9.1	9.5
≥ 2000	0.0	2.4	3.4	4.5	4.7	4.9
Directional Split = 90/10						
≤ 200	5.6	21.6	29.4	37.2	37.4	37.6
400	2.4	19.0	25.6	32.2	32.5	32.8
600	0.0	16.3	21.8	27.2	27.6	28.0
800	0.0	10.9	14.8	18.6	19.0	19.4
≥ 1400	0.0	5.5	7.8	10.0	10.4	10.7

Reproduced with permission from *Highway Capacity Manual*, 2000, Transportation Research Board, National Academy of Sciences, Courtesy of the National Academies Press, Washington, DC, USA.

TABLE 4.38

Tabulating the Values of $f_{d/np}$ for Two-Way Flow Rates (PTSF Method) for Problem 4.19

Opposing Flow Rate (pcph)	$f_{d/np}$
1400	5.5
1667	?
2000	2.9

$\Rightarrow$

$$f_{d/np} = 2.9 - \left\{ \left(2000 - 1667\right) \times \frac{\left(2.9 - 5.5\right)}{\left(2000 - 1400\right)} \right\} = 4.343$$

Therefore:

$$\text{PTSF} = 76.9 + 4.343 \cong 81.2\%$$

ATS Computation:

$$\text{FFS} = \text{BFFS} - f_{LS} - f_A \qquad (4.18)$$

Where:
FFS = free-flow speed (mph)
BFFS = base free-flow speed (mph)
f_{LS} = adjustment factor for lane and shoulder widths
f_A = adjustment factor for number of access points
$f_{LS} = 0.0$ (from Table 4.24 for lane width = 12 ft ≥ 12 ft and shoulder width = 8 ft ≥ 6 ft)
$f_A = 10.0$ (from Table 4.25 for number of access point = 40 per mile)

Hence,

$$\text{FFS} = 60 - 0.0 - 10.0 = 50 \text{ mph (80.5 kph)}$$

$$\text{ATS} = \text{FFS} - 0.00776 v_p - f_{np} \qquad (4.19)$$

Where:
ATS = average travel speed for both directions combined (mph)
FFS = free-flow speed (when volume < 200 pcph)
v_p = passenger car equivalent flow rate for the peak 15-minute period
f_{np} = adjustment factor for the percentage of no-passing zones

From Table 4.39 by interpolation between the values at two-way flow rates of 1400 and 2000 pcph, and using the data: no-passing zones = 40%, the adjustment factor (f_{np}) is calculated as shown below (see Table 4.40).

TABLE 4.39
Adjustment for Effect of No-Passing Zones (f_{np}) on Average Travel Speed on Two-Way Segments

Two-Way Demand Flow Rate, v_p (pc/h)	Reduction in Average Travel Speed (mi/h, km/h)					
	No Passing Zones (%)					
	0	20	40	60	80	100
0	0.0 (0.0)*	0.0 (0.0)	0.0 (0.0)	0.0 (0.0)	0.0 (0.0)	0.0 (0.0)
200	0.0 (0.0)	0.6 (1.0)	1.4 (2.3)	2.4 (3.8)	2.6 (4.2)	3.5 (5.6)
400	0.0 (0.0)	1.7 (2.7)	2.7 (4.3)	3.5 (5.7)	3.9 (6.3)	4.5 (7.3)
600	0.0 (0.0)	1.6 (2.5)	2.4 (3.8)	3.0 (4.9)	3.4 (5.5)	3.9 (6.2)
800	0.0 (0.0)	1.4 (2.2)	1.9 (3.1)	2.4 (3.9)	2.7 (4.3)	3.0 (4.9)
1000	0.0 (0.0)	1.1 (1.8)	1.6 (2.5)	2.0 (3.2)	2.2 (3.6)	2.6 (4.2)
1200	0.0 (0.0)	0.8 (1.3)	1.2 (2.0)	1.6 (2.6)	1.9 (3.0)	2.1 (3.4)
1400	0.0 (0.0)	0.6 (0.9)	0.9 (1.4)	1.2 (1.9)	1.4 (2.3)	1.7 (2.7)
1600	0.0 (0.0)	0.6 (0.9)	0.8 (1.3)	1.1 (1.7)	1.3 (2.1)	1.5 (2.4)
1800	0.0 (0.0)	0.5 (0.8)	0.7 (1.1)	1.0 (1.6)	1.1 (1.8)	1.3 (2.1)
2000	0.0 (0.0)	0.5 (0.8)	0.6 (1.0)	0.9 (1.4)	1.0 (1.6)	1.1 (1.8)
2200	0.0 (0.0)	0.5 (0.8)	0.6 (1.0)	0.9 (1.4)	0.9 (1.5)	1.1 (1.7)
2400	0.0 (0.0)	0.5 (0.8)	0.6 (1.0)	0.8 (1.3)	0.9 (1.5)	1.1 (1.7)
2600	0.0 (0.0)	0.5 (0.8)	0.6 (1.0)	0.8 (1.3)	0.9 (1.4)	1.0 (1.6)
2800	0.0 (0.0)	0.5 (0.8)	0.6 (1.0)	0.7 (1.2)	0.8 (1.3)	0.9 (1.4)
3000	0.0 (0.0)	0.5 (0.8)	0.6 (0.9)	0.7 (1.1)	0.7 (1.1)	0.8 (1.3)
3200	0.0 (0.0)	·0.5 (0.8)	0.6 (0.9)	0.6 (1.0)	0.6 (1.0)	0.7 (1.1)

* Numbers between brackets are in SI units.

Reproduced with permission from *Highway Capacity Manual*, 2000, Transportation Research Board, National Academy of Sciences, Courtesy of the National Academies Press, Washington, DC, USA.

TABLE 4.40
Tabulating the Values of $f_{d/np}$ for Two-Way Flow Rates (ATS Method) for Problem 4.19

Two-Way Flow Rate (pcph)	f_{np}
1600	0.8
1667	?
1800	0.7

$\Rightarrow$

$$f_{np} = 0.7 - \left\{ (1800 - 1600) \times \frac{(0.7 - 0.8)}{(1800 - 1667)} \right\} = 0.85$$

$\Rightarrow$

$$ATS = 50 - 0.00776 (1667) - 0.85 = 36.2 \text{ mph } (58.3 \text{ kph})$$

From Table 4.34 for Class I two-lane highways and using the PTSF value of 81.2% > 80%, and the ATS value of 36.2 mph ≤ 40 mph, the level of service (LOS) for the analysis direction of this highway is E.

4.20 Determine the LOS for the peak direction of a Class I two-lane highway with the following characteristics:

Traffic data:
- PHF=0.90
- Hourly volume=900 vph (in the peak direction)
- Opposing volume=500 vph
- Trucks=4% (of which 10% are semi-trailers)
- RVs=2%
- BFFS=60 mph
- Difference between FFS and truck crawl speed=40 mph

Geometric data:
- Length of segment=2 miles
- Upgrade=4%
- Lane width=11 ft
- Shoulder width=5 ft
- 40 access points per mile
- No-passing zones=10%

Solution:

To determine the level of service of the peak direction on Class I two-lane highways, the percent time-spent following (PTSF) and the average travel speed (ATS) in the analysis direction should be computed.

The following rules apply:
- Any segment having a grade $\geq 3\%$ and a length ≥ 0.6 mile is considered a specific upgrade.
- Segments in mountainous terrain are considered specific upgrades.
- When a segment has multiple grades, a composite grade as a percentage is computed as the total change in elevation divided by the total length.

In this case, since the segment has a length of 2 miles ≥ 0.6 mile and a grade of $4\% \geq 3\%$, the segment is analyzed as a specific upgrade.

PTSF Computation:

A trial initial value for the flow rate in the peak direction can be computed using the following formula:

$$v = \frac{V}{PHF}$$

$$v = \frac{900}{0.90} = 1000 \text{ pcph}$$

The values of f_G (adjustment factor for grade), E_T, and E_R are determined from Tables 4.41 and 4.42, respectively as shown in Table 4.43:

TABLE 4.41

Grade Adjustment Factor (f$_G$) for Estimating Percent Time-Spent-Following on Specific Upgrades

Grade (%)	Length of Grade (mi)	Length of Grade (km)	Grade Adjustment Factor (f$_G$) Range of Directional Flow Rates (pc/h)		
			0–300	> 300–600	> 600
≥ 3 < 3.5	0.25	0.4	1.00	0.92	0.92
	0.50	0.8	1.00	0.93	0.93
	0.75	1.2	1.00	0.93	0.93
	1.00	1.6	1.00	0.93	0.93
	1.50	2.4	1.00	0.94	0.94
	2.00	3.2	1.00	0.95	0.95
	3.00	4.8	1.00	0.97	0.96
	≥ 4.00	≥ 6.4	1.00	1.00	0.97
≥ 3.5 < 4.5	0.25	0.4	1.00	0.94	0.92
	0.50	0.8	1.00	0.97	0.96
	0.75	1.2	1.00	0.97	0.96
	1.00	1.6	1.00	0.97	0.97
	1.50	2.4	1.00	0.97	0.97
	2.00	3.2	1.00	0.98	0.98
	3.00	4.8	1.00	1.00	1.00
	≥ 4.00	≥ 6.4	1.00	1.00	1.00
≥ 4.5 < 5.5	0.25	0.4	1.00	1.00	0.97
	0.50	0.8	1.00	1.00	1.00
	0.75	1.2	1.00	1.00	1.00
	1.00	1.6	1.00	1.00	1.00
	1.50	2.4	1.00	1.00	1.00
	2.00	3.2	1.00	1.00	1.00
	3.00	4.8	1.00	1.00	1.00
	≥ 4.00	≥ 6.4	1.00	1.00	1.00
≥ 5.5 < 6.5	0.25	0.4	1.00	1.00	1.00
	0.50	0.8	1.00	1.00	1.00
	0.75	1.2	1.00	1.00	1.00
	1.00	1.6	1.00	1.00	1.00
	1.50	2.4	1.00	1.00	1.00
	2.00	3.2	1.00	1.00	1.00
	3.00	4.8	1.00	1.00	1.00
	≥ 4.00	≥ 6.4	1.00	1.00	1.00
≥ 6.5	0.25	0.4	1.00	1.00	1.00
	0.50	0.8	1.00	1.00	1.00
	0.75	1.2	1.00	1.00	1.00
	1.00	1.6	1.00	1.00	1.00
	1.50	2.4	1.00	1.00	1.00
	2.00	3.2	1.00	1.00	1.00
	3.00	4.8	1.00	1.00	1.00
	≥ 4.00	≥ 6.4	1.00	1.00	1.00

TABLE 4.42

Passenger-Car Equivalents for Trucks (E_T) and RVs (E_R) for Estimating Percent Time-Spent-Following on Specific Upgrades

			Passenger-Car Equivalent for Trucks (E_T)			
			Range of Directional Flow Rates (pc/h)			
Grade (%)	Length of Grade (mi)	Length of Grade (km)	0–300	> 300–600	> 600	RVs, E_R
$\geq 3 < 3.5$	0.25	0.4	1.0	1.0	1.0	1.0
	0.50	0.8	1.0	1.0	1.0	1.0
	0.75	1.2	1.0	1.0	1.0	1.0
	1.00	1.6	1.0	1.0	1.0	1.0
	1.50	2.4	1.0	1.0	1.0	1.0
	2.00	3.2	1.0	1.0	1.0	1.0
	3.00	4.8	1.4	1.0	1.0	1.0
	≥ 4.00	≥ 6.4	1.5	1.0	1.0	1.0
$\geq 3.5 < 4.5$	0.25	0.4	1.0	1.0	1.0	1.0
	0.50	0.8	1.0	1.0	1.0	1.0
	0.75	1.2	1.0	1.0	1.0	1.0
	1.00	1.6	1.0	1.0	1.0	1.0
	1.50	2.4	1.1	1.0	1.0	1.0
	2.00	3.2	1.4	1.0	1.0	1.0
	3.00	4.8	1.7	1.1	1.2	1.0
	≥ 4.00	≥ 6.4	2.0	1.5	1.4	1.0
$\geq 4.5 < 5.5$	0.25	0.4	1.0	1.0	1.0	1.0
	0.50	0.8	1.0	1.0	1.0	1.0
	0.75	1.2	1.0	1.0	1.0	1.0
	1.00	1.6	1.0	1.0	1.0	1.0
	1.50	2.4	1.1	1.2	1.2	1.0
	2.00	3.2	1.6	1.3	1.5	1.0
	3.00	4.8	2.3	1.9	1.7	1.0
	≥ 4.00	≥ 6.4	3.3	2.1	1.8	1.0
$\geq 5.5 < 6.5$	0.25	0.4	1.0	1.0	1.0	1.0
	0.50	0.8	1.0	1.0	1.0	1.0
	0.75	1.2	1.0	1.0	1.0	1.0
	1.00	1.6	1.0	1.2	1.2	1.0
	1.50	2.4	1.5	1.6	1.6	1.0
	2.00	3.2	1.9	1.9	1.8	1.0
	3.00	4.8	3.3	2.5	2.0	1.0
	≥ 4.00	≥ 6.4	4.3	3.1	2.0	1.0
≥ 6.5	0.25	0.4	1.0	1.0	1.0	1.0
	0.50	0.8	1.0	1.0	1.0	1.0
	0.75	1.2	1.0	1.0	1.3	1.0
	1.00	1.6	1.3	1.4	1.6	1.0
	1.50	2.4	2.1	2.0	2.0	1.0
	2.00	3.2	2.8	2.5	2.1	1.0
	3.00	4.8	4.0	3.1	2.2	1.0
	≥ 4.00	≥ 6.4	4.8	3.5	2.3	1.0

Reproduced with permission from *Highway Capacity Manual*, 2000, Transportation Research Board, National Academy of Sciences, Courtesy of the National Academies Press, Washington, DC, USA.

TABLE 4.43

Determining f_G, E_T, and E_R for a Specific Upgrade on Two-Lane Highway (PTSF Method) for Problem 4.20

Item	Value	Table	Comment
f_G	0.98	Table 4.41	Using directional flow rate = 1000 > 600 pcph, and grade = 4% ≥ 3.5 < 4.5, and length of grade = 2 miles
E_T	1.0	Table 4.42	Using directional flow rate = 1000 > 600 pcph, and grade = 4% ≥ 3.5 < 4.5, and length of grade = 2 miles
E_R	1.0	Table 4.42	For specific upgrades

$$f_{HV} = \frac{1}{1 + P_T\left(E_T - 1\right) + P_R\left(E_R - 1\right)}$$

$\Rightarrow$

$$f_{HV} = \frac{1}{1 + 0.04\left(1.0 - 1\right) + 0.02\left(1.0 - 1\right)} = 1.00$$

$$v = \frac{V}{\left(PHF\right)\left(f_G\right)\left(f_{HV}\right)}$$

$\Rightarrow$

$$v = \frac{900}{\left(0.90\right)\left(0.98\right)\left(1.00\right)} \cong 1021 \text{ pcph}$$

The opposing flow rate is calculated using the procedure shown below:

A trial initial value for the opposing flow rate can be computed using the following formula:

$$v_o = \frac{V}{PHF}$$

$$v_o = \frac{500}{0.90} \cong 556 \text{ pcph}$$

The values of f_G (adjustment factor for grade) for downgrades = 1.00

E_T and E_R for the downgrade direction is determined as if it is a directional segment in level terrain from Table 4.20, and E_{TC} is determined from Table 4.44 as shown in Table 4.45:

TABLE 4.44

Passenger-Car Equivalents (E_{TC}) for Estimating the Effect on Average Travel Speed of Trucks that Operate at Crawl Speeds on Long Steep Downgrades

Difference Between FFS and Truck Crawl Speeds (mi/h)	Difference Between FFS and Truck Crawl Speeds (km/h)	Passenger-Car Equivalents for Trucks at Crawl Speeds (E_{TC}) Range of Directional Flow Rates (pc/h)		
		0–300	**> 300–600**	**> 600**
≤ 15	≤ 20	4.4	2.8	1.4
25	40	14.3	9.6	5.7
≥ 40	≥ 60	34.1	23.1	13.0

Reproduced with permission from *Highway Capacity Manual*, 2000, Transportation Research Board, National Academy of Sciences, Courtesy of the National Academies Press, Washington, DC, USA.

TABLE 4.45

Determining E_T, E_R, and E_{TC} for Downgrade Direction on Two-Lane Highway (PTSF Method) for Problem 4.20

Item	Value	Table	Comment
E_T	1.1	Table 4.20	Using directional opposing flow rate = 556 > 300–600 pcph, and level terrain
E_R	1.0	Table 4.20	Using directional opposing flow rate = 556 > 300–600 pcph, and level terrain
E_{TC}	23.1	Table 4.44	Using directional opposing flow rate = 556 > 300–600 pcph, and difference between FFS and truck crawl speed = 40 mph ≥ 40 mph

$$f_{HV} = \frac{1}{1 + P_{TC}P_T\left(E_{TC}-1\right) + \left(1-P_{TC}\right)P_T\left(E_T-1\right) + P_R\left(E_R-1\right)} \qquad (4.20)$$

Where:

P_{TC} = percentage of trucks that travel at crawl speed from the total percentage of trucks
E_{TC} = passenger car equivalent for trucks traveling at crawl speed

$\Rightarrow$

$$f_{HV} = \frac{1}{1 + 0.10\left(0.04\right)\left(23.1-1\right) + \left(1-0.10\right)\left(0.04\right)\left(1.1-1\right) + 0.02\left(1.0-1\right)}$$

$$f_{HV} = 0.916$$

$$v_o = \frac{V}{\left(PHF\right)\left(f_G\right)\left(f_{HV}\right)} \qquad (4.21)$$

$\Rightarrow$

$$v_o = \frac{500}{(0.90)(1.00)(0.916)} \cong 607 \text{ pcph}$$

The percent time-spent following (PTSF) is calculated using the following formula:

$$\text{PTSF} = \text{BPTSF} + f_{np}$$

The base percent time-spent-following (BPTSF) is computed using the following equation:

$$\text{BPTSF} = 100\left[1 - e^{av^b}\right]$$

From Table 4.23, the coefficients a and b for an opposing demand flow rate = 607 pcph are calculated using interpolation between the values of a and b at opposing flow rates of 600 and 800 pcph as shown below (see Table 4.46).

TABLE 4.46
Tabulating the Values of a and b for the
Opposing Direction for Problem 4.20

Opposing Flow Rate (pcph)	a	b
600	−0.100	0.413
607	?	?
800	−0.173	0.349

$\Rightarrow$ at 607 pcph,

$$a = -0.173 - \left\{(800-607) \times \frac{(-0.173 - (-0.100))}{(800-600)}\right\} = -0.103$$

and

$$b = -0.349 - \left\{(800-607) \times \frac{(0.349 - 0.413)}{(800-600)}\right\} = 0.411$$

Therefore,

$$\text{BPTSF} = 100\left[1 - e^{-0.103(1021)^{0.411}}\right] = 83.1\%$$

$$\text{FFS} = \text{BFFS} - f_{LS} - f_A$$

$f_{LS} = 1.7$ (from Table 4.24 for lane width = 11 ft > $11 \leq 12$ ft and shoulder width = 5 ft $\geq 4 < 6$ ft)
$f_A = 10.0$ (from Table 4.25 for number of access point = 40 per mile)
Hence,

$$\text{FFS} = 60 - 1.7 - 10.0 = 48.3 \text{ mph } (77.7 \text{ kph})$$

The adjustment factor (f_{np}) is determined using interpolation between opposing flow rates of 600 and 800 pcph for FFS = 50 mph, and again another interpolation between opposing flow rates 600 and 800 pcph for FFS = 45 mph. After that, a third interpolation is performed between FFS of 45 and 50 mph to calculate the value of at FFS = 48.3 mph.

From Table 4.26 by interpolation between the values at opposing flow rates of 600 and 800 pcph, and using the data: no-passing zones = 10% ≤ 20%, opposing demand flow rate = 607 pcph and FFS = 50 mph, the adjustment factor (f_{np}) is calculated as shown below (see Table 4.47):

TABLE 4.47

Tabulating the Values of f_{np} for the Opposing Direction (PTSF Method) for Problem 4.20

Opposing Flow Rate (pcph)	f_{np}
600	4.7
607	?
800	2.5

⇒

$$f_{np} = 2.5 - \left\{ (800 - 607) \times \frac{(2.5 - 4.7)}{(800 - 600)} \right\} = 4.623$$

And again, from Table 4.26 by interpolation between the values at opposing flow rates of 600 and 800 pcph and using the data: no-passing zones = 10% ≤ 20%, opposing demand flow rate = 607 pcph and FFS = 45 mph, the adjustment factor (f_{np}) is calculated as shown below (see Table 4.48). Then, by interpolation between the values at FFS of 45 mph and 50 mph, the adjustment factor (f_{np}) is calculated (see Table 4.49):

TABLE 4.48

Tabulating the Values of f_{np} for the Opposing Direction (PTSF Method) for Problem 4.20

Opposing Flow Rate (pcph)	f_{np}
600	4.5
607	?
800	2.3

$\Rightarrow$

$$f_{np} = 2.3 - \left\{ (800 - 607) \times \frac{(2.3 - 4.5)}{(800 - 600)} \right\} = 4.423$$

TABLE 4.49
Tabulating the Values of f_{np} for the Opposing Direction (PTSF Method) for Problem 4.20

FFS (mph	f_{np}
45	4.423
48.3	?
50	4.623

$\Rightarrow$

$$f_{np} = 4.623 - \left\{ (50 - 48.3) \times \frac{(4.623 - 4.423)}{(50 - 45)} \right\} = 4.555$$

Therefore:

$$PTSF = 83.1 + 4.555 \cong 87.7\%$$

<u>ATS Computation:</u>
A trial initial value for the flow rate in the peak direction can be computed using the following formula:

$$v = \frac{V}{PHF}$$

$$v = \frac{900}{0.90} = 1000 \text{ pcph}$$

The values of f_G (adjustment factor for grade), E_T, and E_R are determined from Tables 4.50 through 4.52, respectively as shown in Table 4.53:

TABLE 4.50

Grade Adjustment Factor (f_G) for Estimating Average Travel Speed on Specific Upgrades

Grade (%)	Length of Grade (mi)	Length of Grade (km)	Grade Adjustment Factor (f_G) Range of Directional Flow Rates (pc/h)		
			0–300	> 300–600	> 600
≥ 3 < 3.5	0.25	0.4	0.81	1.00	1.00
	0.50	0.8	0.79	1.00	1.00
	0.75	1.2	0.77	1.00	1.00
	1.00	1.6	0.76	1.00	1.00
	1.50	2.4	0.75	0.99	1.00
	2.00	3.2	0.75	0.97	1.00
	3.00	4.8	0.75	0.95	0.97
	≥ 4.00	≥ 6.4	0.75	0.94	0.95
≥ 3.5 < 4.5	0.25	0.4	0.79	1.00	1.00
	0.50	0.8	0.76	1.00	1.00
	0.75	1.2	0.72	1.00	1.00
	1.00	1.6	0.69	0.93	1.00
	1.50	2.4	0.68	0.92	1.00
	2.00	3.2	0.66	0.91	1.00
	3.00	4.8	0.65	0.91	0.96
	≥ 4.00	≥ 6.4	0.65	0.90	0.96
≥ 4.5 < 5.5	0.25	0.4	0.75	1.00	1.00
	0.50	0.8	0.65	0.93	1.00
	0.75	1.2	0.60	0.89	1.00
	1.00	1.6	0.59	0.89	1.00
	1.50	2.4	0.57	0.86	0.99
	2.00	3.2	0.56	0.85	0.98
	3.00	4.8	0.56	0.84	0.97
	≥ 4.00	≥ 6.4	0.55	0.82	0.93
≥ 5.5 < 6.5	0.25	0.4	0.63	0.91	1.00
	0.50	0.8	0.57	0.85	0.99
	0.75	1.2	0.52	0.83	0.97
	1.00	1.6	0.51	0.79	0.97
	1.50	2.4	0.49	0.78	0.95
	2.00	3.2	0.48	0.78	0.94
	3.00	4.8	0.46	0.76	0.93
	≥ 4.00	≥ 6.4	0.45	0.76	0.93
≥ 6.5	0.25	0.4	0.59	0.86	0.98
	0.50	0.8	0.48	0.76	0.94
	0.75	1.2	0.44	0.74	0.91
	1.00	1.6	0.41	0.70	0.91
	1.50	2.4	0.40	0.67	0.91
	2.00	3.2	0.39	0.67	0.89
	3.00	4.8	0.39	0.66	0.88
	≥ 4.00	≥ 6.4	0.38	0.66	0.87

Reproduced with permission from *Highway Capacity Manual*, 2000, Transportation Research Board, National Academy of Sciences, Courtesy of the National Academies Press, Washington, DC, USA.

TABLE 4.51

Passenger-Car Equivalents for Trucks (E_T) for Estimating Average Travel Speed on Specific Upgrades

Grade (%)	Length of Grade (mi)	Length of Grade (km)	Passenger-Car Equivalent for Trucks (E_T) Range of Directional Flow Rates (pc/h)		
			0–300	> 300–600	> 600
≥ 3 < 3.5	0.25	0.4	2.5	1.9	1.5
	0.50	0.8	3.5	2.8	2.3
	0.75	1.2	4.5	3.9	2.9
	1.00	1.6	5.1	4.6	3.5
	1.50	2.4	6.1	5.5	4.1
	2.00	3.2	7.1	5.9	4.7
	3.00	4.8	8.2	6.7	5.3
	≥ 4.00	≥ 6.4	9.1	7.5	5.7
≥ 3.5 < 4.5	0.25	0.4	3.6	2.4	1.9
	0.50	0.8	5.4	4.6	3.4
	0.75	1.2	6.4	6.6	4.6
	1.00	1.6	7.7	6.9	5.9
	1.50	2.4	9.4	8.3	7.1
	2.00	3.2	10.2	9.6	8.1
	3.00	4.8	11.3	11.0	8.9
	≥ 4.00	≥ 6.4	12.3	11.9	9.7
≥ 4.5 < 5.5	0.25	0.4	4.2	3.7	2.6
	0.50	0.8	6.0	6.0	5.1
	0.75	1.2	7.5	7.5	7.5
	1.00	1.6	9.2	9.0	8.9
	1.50	2.4	10.6	10.5	10.3
	2.00	3.2	11.8	11.7	11.3
	3.00	4.8	13.7	13.5	12.4
	≥ 4.00	≥ 6.4	15.3	15.0	12.5
≥ 5.5 < 6.5	0.25	0.4	4.7	4.1	3.5
	0.50	0.8	7.2	7.2	7.2
	0.75	1.2	9.1	9.1	9.1
	1.00	1.6	10.3	10.3	10.2
	1.50	2.4	11.9	11.8	11.7
	2.00	3.2	12.8	12.7	12.6
	3.00	4.8	14.4	14.3	14.2
	≥ 4.00	≥ 6.4	15.4	15.2	15.0
≥ 6.5	0.25	0.4	5.1	4.8	4.6
	0.50	0.8	7.8	7.8	7.8
	0.75	1.2	9.8	9.8	9.8
	1.00	1.6	10.4	10.4	10.3
	1.50	2.4	12.0	11.9	11.8
	2.00	3.2	12.9	12.8	12.7
	3.00	4.8	14.5	14.4	14.3
	≥ 4.00	≥ 6.4	15.4	15.3	15.2

Reproduced with permission from *Highway Capacity Manual*, 2000, Transportation Research Board, National Academy of Sciences, Courtesy of the National Academies Press, Washington, DC, USA.

TABLE 4.52

Passenger-Car Equivalents for RVs (E$_R$) for Estimating Average Travel Speed on Specific Upgrades

Grade (%)	Length of Grade (mi)	Length of Grade (km)	Passenger-Car Equivalent for RVs (E$_R$) Range of Directional Flow Rates (pc/h)		
			0–300	> 300–600	> 600
≥ 3 < 3.5	0.25	0.4	1.1	1.0	1.0
	0.50	0.8	1.2	1.0	1.0
	0.75	1.2	1.2	1.0	1.0
	1.00	1.6	1.3	1.0	1.0
	1.50	2.4	1.4	1.0	1.0
	2.00	3.2	1.4	1.0	1.0
	3.00	4.8	1.5	1.0	1.0
	≥ 4.00	≥ 6.4	1.5	1.0	1.0
≥ 3.5 < 4.5	0.25	0.4	1.3	1.0	1.0
	0.50	0.8	1.3	1.0	1.0
	0.75	1.2	1.3	1.0	1.0
	1.00	1.6	1.4	1.0	1.0
	1.50	2.4	1.4	1.0	1.0
	2.00	3.2	1.4	1.0	1.0
	3.00	4.8	1.4	1.0	1.0
	≥ 4.00	≥ 6.4	1.5	1.0	1.0
≥ 4.5 < 5.5	0.25	0.4	1.5	1.0	1.0
	0.50	0.8	1.5	1.0	1.0
	0.75	1.2	1.5	1.0	1.0
	1.00	1.6	1.5	1.0	1.0
	1.50	2.4	1.5	1.0	1.0
	2.00	3.2	1.5	1.0	1.0
	3.00	4.8	1.6	1.0	1.0
	≥ 4.00	≥ 6.4	1.6	1.0	1.0
≥ 5.5 < 6.5	0.25	0.4	1.5	1.0	1.0
	0.50	0.8	1.5	1.0	1.0
	0.75	1.2	1.5	1.0	1.0
	1.00	1.6	1.6	1.0	1.0
	1.50	2.4	1.6	1.0	1.0
	2.00	3.2	1.6	1.0	1.0
	3.00	4.8	1.6	1.2	1.0
	≥ 4.00	≥ 6.4	1.6	1.5	1.2
≥ 6.5	0.25	0.4	1.6	1.0	1.0
	0.50	0.8	1.6	1.0	1.0
	0.75	1.2	1.6	1.0	1.0
	1.00	1.6	1.6	1.0	1.0
	1.50	2.4	1.6	1.0	1.0
	2.00	3.2	1.6	1.0	1.0
	3.00	4.8	1.6	1.3	1.3
	≥ 4.00	≥ 6.4	1.6	1.5	1.4

Reproduced with permission from *Highway Capacity Manual*, 2000, Transportation Research Board, National Academy of Sciences, Courtesy of the National Academies Press, Washington, DC, USA.

TABLE 4.53

Determining f_G, E_T, and E_R for the Peak Direction (ATS Method) for Problem 4.20

Item	Value	Table	Comment
f_G	1.00	Table 4.50	Using directional flow rate = 1000 > 600 pcph, and grade = 4% ≥ 3.5 < 4.5, and length of grade = 2 miles
E_T	8.1	Table 4.51	Using directional flow rate = 1000 > 600 pcph, and grade = 4% ≥ 3.5 < 4.5, and length of grade = 2 miles
E_R	1.0	Table 4.52	Using directional flow rate = 1000 > 600 pcph, and grade = 4% ≥ 3.5 < 4.5, and length of grade = 2 miles

$$f_{HV} = \frac{1}{1 + P_T(E_T - 1) + P_R(E_R - 1)}$$

$\Rightarrow$

$$f_{HV} = \frac{1}{1 + 0.04(8.1 - 1) + 0.02(1.0 - 1)} = 0.779$$

$$v = \frac{V}{(PHF)(f_G)(f_{HV})}$$

$\Rightarrow$

$$v = \frac{900}{(0.90)(1.00)(0.779)} \cong 1284 \text{ pcph}$$

The opposing flow rate is computed using the procedure shown below:

A trial initial value for the opposing flow rate can be computed using the following formula:

$$v_o = \frac{V}{PHF}$$

$$v_o = \frac{500}{0.90} \cong 556 \text{ pcph}$$

The values of f_G (adjustment factor for grade) for downgrades = 1.00

E_T and E_R for the downgrade direction is determined as if it is a directional segment in level terrain from Table 4.20, and E_{TC} is determined from Table 4.44 as shown in Table 4.54:

TABLE 4.54

Determining E_T, E_R, and E_{TC} for Downgrade Direction on Two-Lane Highway (ATS Method) for Problem 4.20

Item	Value	Table	Comment
E_T	1.2	Table 4.20	Using directional flow rate = 556 > 300–600 pcph, and level terrain
E_R	1.0	Table 4.20	Using directional flow rate = 556 > 300–600 pcph, and level terrain
E_{TC}	23.1	Table 4.44	Using directional flow rate = 556 > 300–600 pcph, and difference between FFS and truck crawl speed = 40 mph ≥ 40 mph

$$f_{HV} = \frac{1}{1 + P_{TC}P_T\left(E_{TC}-1\right) + \left(1-P_{TC}\right)P_T\left(E_T-1\right) + P_R\left(E_R-1\right)}$$

$\Rightarrow$

$$f_{HV} = \frac{1}{1 + 0.10\left(0.04\right)\left(23.1-1\right) + \left(1-0.10\right)\left(0.04\right)\left(1.2-1\right) + 0.02\left(1.0-1\right)}$$

$$f_{HV} = 0.913$$

$$v_o = \frac{V}{\left(PHF\right)\left(f_G\right)\left(f_{HV}\right)}$$

$\Rightarrow$

$$v_o = \frac{500}{\left(0.90\right)\left(1.00\right)\left(0.913\right)} \cong 609 \text{ pcph}$$

$$ATS = FFS - 0.00776\left(v + v_o\right) - f_{np}$$

The free-flow speed (FFS) was computed above as follows:

$$FFS = 60 - 1.7 - 10.0 = 48.3 \text{ mph}\left(77.7 \text{ kph}\right)$$

From Table 4.39 by interpolation between the values at opposing flow rates of 600 and 800 pcph, and using the data: no-passing zones = 10% ≤ 20%, opposing demand flow rate = 609 pcph and FFS = 50 mph, the adjustment factor (f_{np}) is calculated as shown below (see Table 4.55):

TABLE 4.55
Tabulating the Values of f_{np} for the Opposing Direction (ATS Method) for Problem 4.20

Opposing Flow Rate (pcph)	f_{np}
600	0.6
609	?
800	0.4

$\Rightarrow$

$$f_{np} = 0.4 - \left\{\left(800-609\right) \times \frac{\left(0.4-0.6\right)}{\left(800-600\right)}\right\} = 0.591$$

And again, from Table 4.39 by interpolation between the values at opposing flow rates of 600 and 800 pcph and using the data: no-passing zones =10%≤20%, opposing demand flow rate =607 pcph and FFS =45 mph, the adjustment factor (f_{np}) is calculated as shown below (see Table 4.56). Then, by interpolation between the values at FFS of 45 mph and 50 mph, the adjustment factor (f_{np}) is calculated (see Table 4.57).

TABLE 4.56

Tabulating the Values of f_{np} for the Opposing Direction (ATS Method) for Problem 4.20

Opposing Flow Rate (pcph)	f_{np}
600	0.4
609	?
800	0.3

$\Rightarrow$

$$f_{np} = 0.3 - \left\{ (800 - 609) \times \frac{(0.3 - 0.4)}{(800 - 600)} \right\} = 0.396$$

TABLE 4.57

Tabulating the Values of f_{np} for the Opposing Direction (ATS Method) for Problem 4.20

FFS (mph	f_{np}
45	0.396
48.3	?
50	0.591

$\Rightarrow$

$$f_{np} = 0.591 - \left\{ (50 - 48.3) \times \frac{(0.591 - 0.396)}{(50 - 45)} \right\} = 0.525$$

$$\text{ATS} = \text{FFS} - 0.00776(v + v_o) - f_{np}$$

$\Rightarrow$

$$\text{ATS} = 48.3 - 0.00776(1284 + 609) - 0.525 = 33.1 \, \text{mph} \, (53.3 \, \text{kph})$$

From Table 4.34 for Class I two-lane highways and using the PTSF value of 87.7% > 80%, and the ATS value of 33.1 mph ≤ 40 mph, the level of service (LOS) for the analysis direction of this highway is E.

4.21 An urban 6-lane divided highway segment in level terrain has the following characteristics:
Traffic data:
- Peak-hourly volume = 1800 vph
- PHF = 0.95
- Trucks = 12%
- RVs = 2%
- Free-flow speed (FFS) = 50 mph
- Commuter traffic

Geometric data:
- Length = 0.75 mile
- Grade = 2%
- Lane width = 11 ft
- Shoulder width = 4 ft

Determine the level of service (LOS) for the upgrade direction.

Solution:

A highway segment having a grade ≥ 3% and length > 0.5 mile or a grade < 3% and a length > 1.0 mile is considered a specific grade.

Since the given highway segment has an upgrade of 2% < 3% and length of 0.75 mile < 1.0 mile ⇒ the segment is <u>not</u> considered a specific grade, but rather it is an extended general highway segment.

Therefore, Table 4.2 for extended general highway segments is used. E_T and E_R are determined as shown in Table 4.58:

TABLE 4.58
Determining E_T and E_R Values for Multilane Highway for Problem 4.21

Item	Value	Table	Comment
E_T	1.5	Table 4.2	Using level terrain
E_R	1.2	Table 4.2	Using level terrain

$$f_{HV} = \frac{1}{1 + P_T(E_T - 1) + P_R(E_R - 1)}$$

$\Rightarrow$

$$f_{HV} = \frac{1}{1 + 0.12(1.5 - 1) + 0.02(1.2 - 1)} = 0.940$$

The adjustment factor for drivers, $f_p = 1.00$ since traffic drivers are commuters.

$$v_p = \frac{V}{(PHF)(N)(f_p)(f_{HV})}$$

$\Rightarrow$

$$v_p = \frac{1800}{(0.95)(2)(1.00)(0.940)}$$

$$v_p \cong 1008 \text{ pcphpl}$$

The average passenger car speed (S) can be determined using speed-flow curves with LOS criteria for multilane highways.

When the flow rate ≤ 1400 pcphpl, the average passenger car speed (S) is equal to the FFS; and when the flow rate > 1400 pcphpl, S is determined from the curves.

In this case; since $v_p = 1008$ pcphpl ≤ 1400 pcphpl, S = FFS = 50 mph.

$$D = \frac{v_p}{S}$$

Where:
D = density (pc/mile/lane)
S = average passenger car speed (mph)
v_p = 15-minute service flow rate (pcphpl)

$\Rightarrow$

$$D = \frac{1008}{50} \cong 21 \text{ pc/mi/lane}$$

From Table 4.59, for $18 \leq D \leq 26$ pc/mi/lane, the LOS is C.

TABLE 4.59

Level-of-Service (LOS) Criteria for Multilane Highways

Criteria	LOS				
	A	B	C	D	E
FFS = 60 mi/h (100 km/h)					
Maximum density (pc/mi/ln)	11 (7)	18 (11)	26 (16)	35 (22)	40 (25)
Maximum speed (mi/h)	60.0 (100.0)	60.0 (100.0)	59.4 (98.4)	56.7 (91.5)	55.0 (88.0)
Maximum v/c	0.32	0.50	0.72	0.92	1.00
Maximum service flow rate (pc/h/ln)	700	1100	1575	2015	2200
FFS = 55 mi/h (90 km/h)					
Maximum density (pc/mi/ln)	11 (7)	18 (11)	26 (16)	35 (22)	40 (25)
Maximum speed (mi/h)	55.0 (90.0)	55.0 (90.0)	54.9 (89.8)	52.9 (84.7)	51.2 (80.8)
Maximum v/c	0.30	0.47	0.68	0.89	1.00
Maximum service flow rate (pc/h/ln)	630	990	1435	1860	2100
FFS = 50 mi/h (80 km/h)					
Maximum density (pc/mi/ln)	11 (7)	18 (11)	26 (16)	35 (22)	40 (25)
Maximum speed (mi/h)	50.0 (80.0)	50.0 (80.0)	50.0 (80.0)	48.9 (77.6)	47.5 (74.1)
Maximum v/c	0.28	0.44	0.64	0.85	1.00
Maximum service flow rate (pc/h/ln)	560	880	1280	1705	2000
FFS = 45 mi/h (70 km/h)					
Maximum density (pc/mi/ln)	11 (7)	18 (11)	26 (16)	35 (22)	40 (25)
Maximum speed (mi/h)	45.0 (70.0)	45.0 (70.0)	45.0 (70.0)	44.4 (69.6)	42.2 (67.9)
Maximum v/c	0.26	0.41	0.59	0.81	1.00
Maximum service flow rate (pc/h/ln)	490	770	1120	1530	1900

Reproduced with permission from *Highway Capacity Manual*, 2000, Transportation Research Board, National Academy of Sciences, Courtesy of the National Academies Press, Washington, DC, USA.

4.22 Determine the level of service (LOS) for the upgrade direction (peak direction) of a rural 4-lane divided highway segment with the following characteristics:
Traffic data:
- Average Annual Daily Traffic (AADT) = 22000 veh/day
- K (proportion of the AADT in the peak hour) = 0.15
- D (proportion of the peak-hourly traffic in the peak direction) = 0.60
- PHF = 0.90
- Trucks = 5%
- RVs = 4%
- Free-flow speed (FFS) = 55 mph
- Commuter traffic

Geometric data:
- Rolling terrain
- Length = 1.00 mile
- Grade = 4%
- Lane width = 12 ft
- Shoulder width = 6 ft

Solution:

A highway segment having a grade $\geq 3\%$ and length > 0.5 mile or a grade < 3% and a length > 1.0 mile is considered a specific grade.

Since the given highway segment has an upgrade of $4\% \geq 3\%$ and length of 1.0 mile > 0.5 mile $\Rightarrow$ the segment is considered a specific grade.

Therefore, Tables 4.13 and 4.14 for upgrades are used to determine E_T and E_R, respectively as shown in Table 4.60.

TABLE 4.60

Determining Values of E_T and E_R for Upgrade Direction on Multilane Highway for Problem 4.21

Item	Value	Table	Comment
E_T	2.5	Table 4.13	Using a percentage of trucks of 5%, upgrade of 4% > 3–4%, and length of grade of 1.00 mile > 0.75–1.00 mile
E_R	2.5	Table 4.14	Using a percentage of RVs of 4%, upgrade of 4% > 3–4%, and length of grade of 1.00 mile > 0.50 mile

---

$$f_{HV} = \frac{1}{1 + P_T\left(E_T - 1\right) + P_R\left(E_R - 1\right)}$$

$\Rightarrow$

$$f_{HV} = \frac{1}{1 + 0.05\left(2.5 - 1\right) + 0.04\left(2.5 - 1\right)} = 0.881$$

The adjustment factor for drivers, $f_p = 1.00$ since traffic drivers are commuters.

$$DDHV = AADT \times K \times D$$

Where:
DDHV = directional design hourly volume
K = proportion of the AADT in the peak hour
D = proportion of the peak-hourly traffic in the peak direction

$\Rightarrow$

$$DDHV = 22000 \times 0.15 \times 0.60 = 1980 \text{ vph}$$

$$v_p = \frac{V}{(\text{PHF})(N)(f_p)(f_{\text{HV}})}$$

$\Rightarrow$

$$v_p = \frac{1980}{(0.90)(2)(1.00)(0.881)}$$

$$v_p \cong 1249 \text{ pcphpl}$$

Since the flow rate $= 1249$ pcphpl ≤ 1400 pcphpl, the average passenger car speed (S) is equal to the FFS.
   S = FFS = 55 mph.

$$D = \frac{v_p}{S}$$

$\Rightarrow$

$$D = \frac{1249}{55} \cong 23 \text{ pc/mi/lane}$$

From Table 4.59, for $18 \leq D \leq 26$ pc/mi/lane, the LOS is C.

5 Intersection Design, Operation, and Control

Chapter 5 deals with the main types of intersections in terms of design, operation, and control. Intersections can be categorized into three major categories: (1) at-grade intersections, (2) grade-separated intersections without ramps, and (3) grade-separated intersections with ramps (interchanges). Grade-separated intersections (such as interchanges) provide uninterrupted crossing of traffic flow. The use of different levels of traffic flow by means of bridges, resulting in overpass and underpass, controls traffic flow, reduces delays, improves travel speeds, and enhances safety. On the other hand, at-grade intersections, due to the fact that traffic flows from different approaches intersect in one area, require traffic control systems to reduce delays, increase highway capacity, and improve highway safety by reducing crashes. In this section, the focus will be on this type of intersection (at-grade intersections) due to the complicated situation and the need to control devices at these intersections that improve the operation and safety at these locations. Traffic control systems generally aim at facilitating traffic flow in an orderly and systematic way that ensures smooth operation, safety, and efficiency. Traffic control systems (or devices) include but are not limited to signs (stop signs, yield signs, etc.), markings (white lines, yellow lines, dotted ones, solid ones, etc.), channelization (islands, medians, barriers, etc.), and traffic signals. The "Manual on Uniform Traffic Control Devices" (MUTCD) by the Federal Highway Administration (FHWA) recommends taking into consideration factors such as design, placement, operation, maintenance, and uniformity for an effective control device. Traffic signals are considered one of the most important devices for at-grade intersections that control traffic flow smoothly and efficiently. Lane grouping, signal phasing, and signal timing are discussed in detail in this section for different types of at-grade intersections including T- (3-leg) intersections and 4-leg intersections. Signal timing of successive intersections on arterial routes is discussed. The coordination of signals of adjacent intersections in urban areas is important to ensure smooth flow of traffic and reduce the delay of vehicles on the arterial. Three methods are used to perform the coordination between successive signals: the simultaneous system, the alternate system, and the progressive system.

5.1 On a major approach of an intersection, the following vehicular volumes during 15-minute periods of the peak hour are obtained (see Table 5.1).

TABLE 5.1
Traffic Volumes During the Peak Hour for Problem 5.1

Time Period	Traffic Count
7:00–7:15am	280
7:15–7:30am	320
7:30–7:45am	400
7:45–8:00am	350

Determine the peak-hour factor (PHF) and the design hourly volume (DHV) at this intersection approach.

Solution:

$$PHF = \frac{\text{Hourly volume}}{4 \times \text{Peak 15} - \text{minute volume in the peak hour}}$$ (5.1)

Therefore:

$$PHF = \frac{280 + 320 + 400 + 350}{4 \times 400} = 0.844$$

$$DHV = \frac{\text{Peak} - \text{hourly volume}}{PHF}$$ (5.2)

$\Rightarrow$

$$DHV = \frac{1350}{0.844} = 1600 \text{ vph}$$

The screen image of the MS Excel worksheet used to perform the computations of this problem is shown in Figure 5.1.

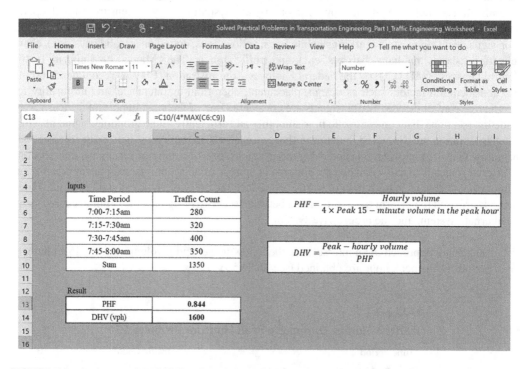

FIGURE 5.1   An image of the MS Excel worksheet used for the computations of Problem 5.1.

5.2 If the peak-hour factor (PHF) at an intersection is 0.85 and the volume during the peak 15-minute period within the peak hour is 500 vehicles, determine the volume at the intersection during the peak hour and the design hourly volume (DHV).

Solution:

$$PHF = \frac{\text{Hourly volume}}{4 \times \text{Peak 15} - \text{minute volume in the peak hour}}$$

Therefore:

$$\text{Hourly volume} = \text{PHF} \times 4 \times \text{Peak 15} - \text{minute volume in the peak hour}$$

$$\text{Hourly volume} = 0.85 \times 4 \times 500 = 1700 \text{ vph}$$

$$\text{DHV} = \frac{\text{Peak} - \text{hourly volume}}{\text{PHF}}$$

$\Rightarrow$

$$\text{HDV} = \frac{1700}{0.85} = 2000 \text{ vph}$$

A screen image of the MS Excel worksheet used to perform the computations of this problem is shown in Figure 5.2.

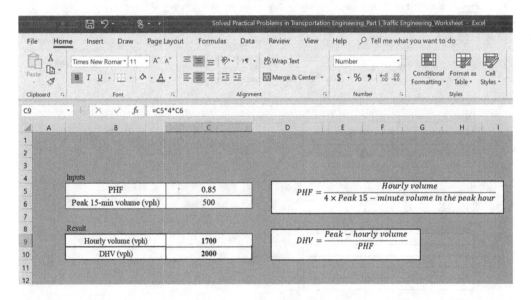

FIGURE 5.2 An image of the MS Excel worksheet used for the computations of Problem 5.2.

5.3 An intersection having a width of 50 ft (15.2 m) and the speed limit on the intersection approaches is 20 mph (32.2 kph). Compute the minimum yellow interval that will eliminate the dilemma zone at this intersection assuming that the average length of vehicle is 20 ft (6.1 m).

Solution:

The minimum yellow interval at an intersection is given by the following formula:

$$\tau_{\min} = \delta + \frac{W + L}{u_0} + \frac{u_0}{2a} \tag{5.3}$$

Where:

$\tau_{\min}$ = the minimum yellow interval (sec)

δ = the perception-reaction time (typical value = 2.5 sec)

u_0=the speed limit on approach (ft/sec)
W=the width of intersection (ft)
a=constant rate of braking deceleration (typical value=11.2 ft/sec^2)
L=the length of the vehicle (ft)

Therefore:

$$\tau_{min} = 2.5 + \frac{50+20}{20 \times \left(\dfrac{5280}{3600}\right)} + \frac{20 \times \left(\dfrac{5280}{3600}\right)}{2(11.2)} = 6.2 \text{ sec}$$

(5280/3600) is a conversion factor to convert the speed from mph to kph.
A screen image of the MS Excel worksheet used to conduct the computations of this problem is shown in Figure 5.3.

	B	C
	Inputs (US Customary Units)	
	L (ft)	20.0
	W (ft)	50.0
	Speed Limit (mph)	20.0
	δ (sec)	2.5
	a (ft/sec^2)	11.2
	Results (Customary Units)	
	τ (sec)	6.2
	Inputs (SI Units)	
	L (m)	6.1
	W (m)	15.2
	Speed (kph)	32.2
	δ (sec)	2.5
	a (m/sec^2)	3.4
	Results (SI Units)	
	τ (sec)	6.2

C12: =C8+(C6+C5)/(C7*5280/3600)+C7*(5280/3600)/(2*C9)

$$\tau_{min} = \delta + \frac{W+L}{u_0} + \frac{u_0}{2a}$$

FIGURE 5.3 An image of the MS Excel worksheet used for the computations of Problem 5.3.

5.4 Six vehicles are stopping at a signalized intersection as shown in Figure 5.4. When the signal turns to green, it takes the last vehicle (Vehicle #6) 10 seconds to cross the intersection at a speed of 15 mph (24.1 kph). If the width of the intersection is 60 ft (18.3 m), and the length of each vehicle is 20 ft (6.1 m), determine the jam density at the signal.

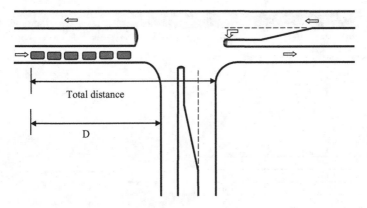

FIGURE 5.4   A signalized T-intersection for Problem 5.4.

Solution:

The total distance the last vehicle (Vehicle #6) crosses within the 10 seconds is determined:

$$\text{Total distance} = 10 \times 15 \times \frac{5280}{3600} = 220 \text{ ft (67.1 m)}$$

The total distance minus the width of the intersection provides the distance over which the six vehicles are stopping at the intersection. Therefore:

$$D = 220 - 60 = 160 \text{ ft} \left(48.8 \text{ m}\right)$$

The jam density at the signal is equal to the number of vehicles divided by the distance, D. Hence:

$$k = \frac{6}{\left(\dfrac{160}{5280}\right)} = 198 \text{ vpm} \left(123 \text{ veh /km}\right)$$

Screen images of the MS Excel worksheet used to compute the jam density at the signal in this problem are shown in Figures 5.5 and 5.6.

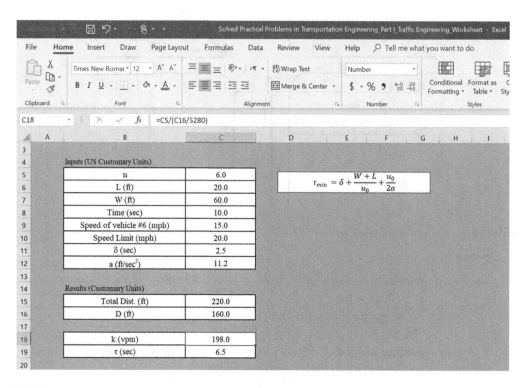

FIGURE 5.5 An image of the MS Excel worksheet used for the computations of Problem 5.4 (US customary units).

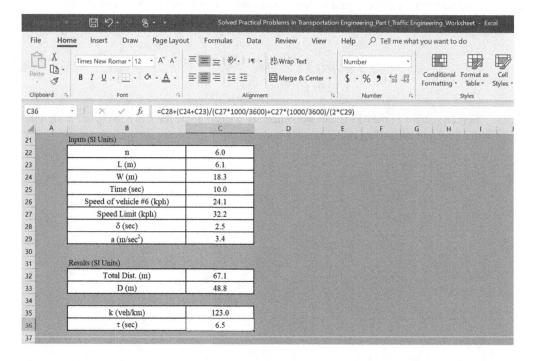

FIGURE 5.6 An image of the MS Excel worksheet used for the computations of Problem 5.4 (SI units).

5.5 In Problem 5.4, if the perception–reaction time is 2.5 seconds and the speed limit for vehicles is 20 mph (32.2 kph), determine the minimum yellow interval that should be introduced to the traffic signal to minimize the dilemma zone.

Solution:

$$\tau_{min} = \delta + \frac{W+L}{u_0} + \frac{u_0}{2a}$$

$$\tau_{min} = 2.5 + \frac{60+20}{20 \times \left(\dfrac{5280}{3600}\right)} + \frac{20 \times \left(\dfrac{5280}{3600}\right)}{2(11.2)} = 6.5 \text{ sec}$$

The screen image of the MS Excel worksheet used to compute the minimum yellow interval required in this problem is the same image shown in Figure 5.5 of the worksheet used for the computations of Problem 5.4.

5.6 Figure 5.7 illustrates a major 3-leg intersection on a highway. Establish correct lane groups for the three approaches of the intersection based on the traffic movements and geometry provided for the intersection.

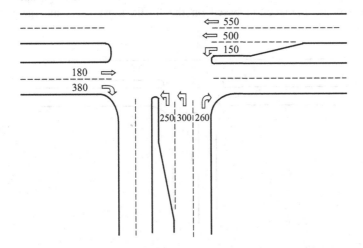

FIGURE 5.7 A 3-leg intersection for Problem 5.6.

Solution:

Before establishing lane groups, the definition for a lane group should be first presented. A lane group is defined as a group (set) of one or more lanes at an intersection approach that has the same green phase. The four guidelines shown below are used to establish a lane group:

1. Exclusive left-turn lanes are considered separate lane groups.
2. Exclusive right-turn lanes are given separate lane groups.
3. At an approach with exclusive left-turn lanes or/and right-turn lanes, all other lanes are typically given a single lane group.
4. The operation of a shared left-turn lane should be evaluated when an approach with more than one lane has a shared left-turn.

Based on the above guidelines, the lane groups are established. For the eastbound approach, the exclusive right-turn lane is a single lane group and all other lanes (the lane that is going straight) in the same approach is another single lane group. In a similar manner; for the westbound approach, the exclusive left-turn lane is a single lane group, and all other lanes (the two lanes that are moving straight) in the same approach are considered another single lane group. Finally, for the northbound direction, the exclusive left-turn lanes are a single lane group, and the exclusive right-turn lane in the same approach is another single lane group. The six resulting lane groups for this intersection are illustrated in Table 5.2.

TABLE 5.2

The Resulting Lane Groups for the 3-Leg Intersection in Problem 5.6

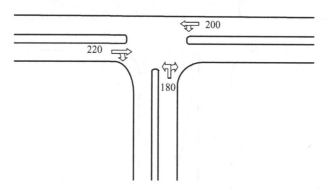

Approach	1 (Eastbound)	2 (Westbound)	3 (Northbound)
Lane Groups			

5.7 For the T-intersection shown in Figure 5.8, establish correct lane groups for the three approaches of the intersection based on the traffic movements and geometry of the intersection.

FIGURE 5.8   A T-intersection for Problem 5.7.

Solution:

The lane groups are established using the guidelines discussed above. Since there are no exclusive left-turn or right-turn lanes in any of the three approaches, only the other lanes are considered. For the eastbound approach, the shared lane (the lane with the right-turn and straight movements) is a single lane group. In a similar manner; for the westbound approach, the shared lane (the lane with the left-turn and straight movements) is a single lane group. Finally, for the northbound direction, the shared lane (the lane with the right-turn and left-turn movements) is a single lane group. The three resulting lane groups for this intersection are illustrated in Table 5.3.

TABLE 5.3

The Resulting Lane Groups for the T-Intersection in Problem 5.7

Approach	1 (Eastbound)	2 (Westbound)	3 (Northbound)
Lane Groups			

5.8   It is required to design a traffic signal for the T-intersection shown in Figure 5.9. The peak-hour volumes for each lane at the three approaches are given as in the figure. The saturation flow rate is 1800 vph for all lanes and the peak-hour factor (PHF) is 0.90 for the entire intersection. If the lost time per phase is 4.0 seconds and the yellow interval is 3.0 seconds:

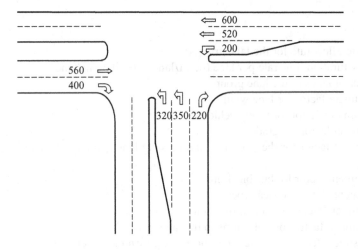

FIGURE 5.9   A T-intersection for Problem 5.8.

(a)   Establish the lane groups for this intersection to be used in the signal design.
(b)   Provide the phasing system that will offer the shortest cycle length using the Highway Capacity Method (HCM).
(c)   Provide the critical (v/s) ratio for each phase.
(d)   Determine the cycle length using the HCM that will avoid oversaturation.
(e)   Determine the actual green times for all phases.

Solution:

The following formulas are used in solving this problem:

$$v_{ij} = \frac{V_{ij}}{PHF} \qquad (5.4)$$

Where:
   v_{ij} = actual flow rate for lane group or approach i in phase j (veh/h)
   V_{ij} = traffic volume for lane group or approach i in phase j (veh/h)

PHF = peak-hour factor

$$c_{ij} = s_{ij} \left(\frac{g_{ij}}{C} \right) \tag{5.5}$$

Where:

c_{ij} = capacity of lane group or approach i in phase j (veh/h)
s_{ij} = saturation flow rate for lane group or approach i in phase j (veh/h of green time)
g_{ij} = effective green for lane group or approach i in phase j (sec)
C = cycle length for traffic signal (sec)

$$s = s_o N f_W \; f_{HV} \; f_g \; f_p \; f_{bb} \; f_a \; f_{LU} \; f_{RT} \; f_{Lpb} \; f_{Rpb} \tag{5.6}$$

Where:

S = saturation flow rate for the lane group (vph)
S_0 = ideal saturation flow rate per lane (pc/hr/lane) = 1900 pc/hr/lane
N = number of lanes in the lane group
f_W = adjustment factor for lane width
f_{HV} = adjustment factor for heavy vehicles
f_g = adjustment factor for grade
f_p = adjustment factor for the existence of a parking lane and parking activity adjacent to the lane group
f_{bb} = adjustment factor for bus blocking
f_a = adjustment factor for area type
f_{LU} = adjustment factor for lane utilization
f_{RT} = adjustment factor for right turns in the lane group
f_{Lpb} = adjustment factor for pedestrian/bicycle left-turn movements
f_{Rpb} = adjustment factor for pedestrian/bicycle right-turn movements

$$\left(\frac{v}{c} \right)_{ij} = X_{ij} = \frac{v_{ij}}{s_{ij} \left(\dfrac{g_{ij}}{C} \right)} \tag{5.7}$$

Where:

X_{ij} = $(v/c)_{ij}$ ratio for lane group or approach i in phase j

$$X_c = \left(\sum_j \left(\frac{v}{s} \right)_{cj} \right) \frac{C}{C - L} \tag{5.8}$$

Where:

X_c = critical (v/c) ratio for the intersection
$\sum_j \left(\dfrac{v}{s} \right)_{cj}$ = summation of critical (v/s) ratios for all phases

v_{ij} = actual flow rate for lane group or approach i in phase j (veh/h)
L = total lost time per cycle (sec)

$$L = R + \sum l_j \tag{5.9}$$

Where:
l_j = lost time for critical signal phase j (sec)
R = total all-red time in the cycle

$$G_{te} = C - L = C - \left(R + \sum l_j \right) \tag{5.10}$$

Where:
G_{te} = total effective green time in the cycle (sec)

$$G_{ej} = \frac{\left(\dfrac{v}{s} \right)_{cj}}{\sum_j \left(\dfrac{v}{s} \right)_{cj}} G_{te} \tag{5.11}$$

Where:
G_{ej} = effective green time for phase j (sec)

$$l_j = G_{aj} + \tau_j - G_{ej} \tag{5.12}$$

Where:
G_{aj} = actual green time for phase j (sec)
τ_j = actual green time for phase j (sec)

(a) The lane groups for this intersection are established using the guidelines discussed earlier. Two lane groups per each approach are established as shown in Table 5.4.

TABLE 5.4
The Established Lane Groups for the T-Intersection in Problem 5.8

Approach	1 (Eastbound)		2 (Westbound)		3 (Northbound)	
Lane Group #	RE[a]	TE	TW+TW	LW	RN	LN+LN
Lane Group						

[a] R, T, L refer to right-turn, through, and left-turn lanes, respectively. E, W, N, S refer to eastbound, westbound, northbound, and southbound, respectively.

(b) The phasing system that will provide the shortest cycle length using the Highway Capacity Method (HCM) is shown in Table 5.5.

TABLE 5.5

The Phasing System for the T-Intersection in Problem 5.8

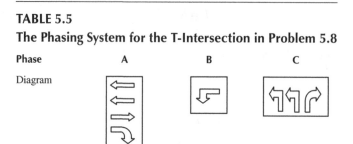

Phase	A	B	C
Diagram			

(c) The critical (v/s) ratio and computations for each phase are summarized in Table 5.6:

Phase A is composed of the following three lane groups:

The traffic volumes for these three lane groups are 1120 (600 + 520), 560, and 400 vph, respectively. Since the saturation flow rate for each lane is 1800 vph, the saturation flow rate for the three lane groups (s_j) of Phase A are 3600, 1800, and 1800 veh/h of green time, respectively. Lane group #1 has two lanes and that is why the saturation flow rate for this lane group is equal to $2 \times 1800 = 3600$ veh/h of green time. In addition, the actual flow rate for Lane Group #1 in Phase A is computed using the formula below, and the (v_{ij}/s_{ij}) ratio is computed for the lane group.

$$v_{ij} = \frac{V_{ij}}{PHF}$$

⇒

$$v_{ij} = \frac{1120}{0.90} = 1244.4 \text{ veh/h}$$

$$\frac{v_{ij}}{s_{ij}} = \frac{1244.4}{3600} = 0.346$$

In a similar manner, the traffic volumes, the saturation flow rates, and the actual flow rates for the other lane groups for all three phases are determined. The results are shown in Table 5.6.

TABLE 5.6

HCM Signal Timing Computations and Results for the Three Phases of the T-Intersection in Problem 5.8

Phase	A			B	C	
Lane Group	TW+TW[a]	RE	TE	LW	RN	LN+LN
Volume (vph)	1120	400	560	200	220	670
v_{ij} (vph)	1244.4	444.4	622.2	222.2	244.4	744.4
s_{ij} (vph)	3600	1800	1800	1800	1800	3600
v_{ij}/s_{ij}	0.346	0.247	0.346	0.123	0.136	0.207
$(v/s)_{cj}$		0.346		0.123		0.207

[a] R, T, L refer to right-turn, through, and left-turn lanes, respectively. E, W, N, S refer to eastbound, westbound, northbound, and southbound, respectively.

The critical (v/s) ratio for each phase is the maximum (v/s) ratio among the (v/s) ratios for all lane groups in that phase. Therefore, the $(v/s)_{cj}$ is equal to 0.346, 0.123, and 0.207 for phases A, B, and C, respectively. Therefore, the summation of these critical ratios for all phases is:

$$\sum_j \left(\frac{v}{s}\right)_{cj} = 0.346 + 0.123 + 0.207 = 0.676$$

(d) The cycle length that will avoid oversaturation is happening at critical (v/c) ratio, $X_c = 1.0$.

$$L = R + \Sigma l_j$$

$\Rightarrow$

$$L = 4 \times 3 = 12 \text{ sec}$$

$$X_c = \left(\sum_j \left(\frac{v}{s}\right)_{cj}\right) \frac{C}{C-L}$$

$\Rightarrow$

$$1.0 = (0.676)\frac{C}{C-12}$$

Solving for C $\Rightarrow$

$$C = 37 \text{ sec}$$

Use C = 40 sec (rounded to the nearest 5 seconds) (see Table 5.7).

$$G_{te} = C - L$$

$\Rightarrow$

$$G_{te} = 40 - 12 = 28 \text{ sec}$$

TABLE 5.7

Determination of the Cycle Length for Problem 5.8

X_c	1.0
$\Sigma(v/s)_{cj}$	0.676
L (sec)	12.0
C (sec)	37.0 (Use 40)
G_{te} (sec)	28

(e)   The actual green times for the three phases are computed and illustrated in Table 5.8:

For Phase A:

$$G_{ej} = \frac{\left(\dfrac{v}{s}\right)_{cj}}{\sum_{j}\left(\dfrac{v}{s}\right)_{cj}} G_{te}$$

$\Rightarrow$

$$G_{eA} = \frac{0.346}{0.676} \times 28 = 14.3 \text{ sec}$$

$$l_j = G_{aj} + \tau_j - G_{ej}$$

Or:

$$G_{aj} = l_j - \tau_j + G_{ej}$$

$\Rightarrow$

$$G_{aA} = 4 - 3 + 14.3 = 15.3 \text{ sec}$$

The effective green time and actual green time were computed for the three phases in the same way as shown in Table 5.8.

TABLE 5.8

The Effective and Actual Green Times for the Three Phases of the T-Intersection in Problem 5.8

Phase	A	B	C
G_{ej} (sec)	14.3	5.1	8.6
G_{aj} (sec)	15.3	6.1	9.6

The MS Excel worksheet used to solve this problem with the details, formulas, and computations is shown in Figure 5.10.

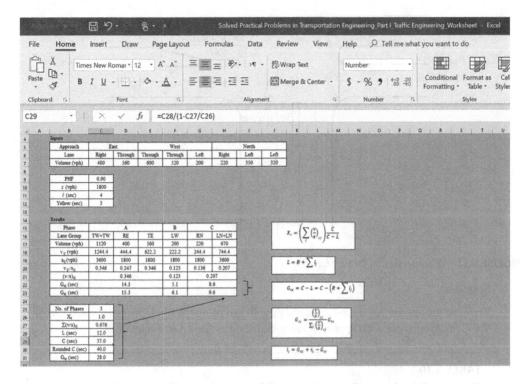

FIGURE 5.10 An image of the MS Excel worksheet used for the computations of Problem 5.8.

5.9 The T-intersection shown in Figure 5.11 with peak-hour volumes (vph) at the three approaches is having an existing cycle length of 60 seconds. Due to some delays in the major westbound approach, a traffic engineering designer is asked to investigate the existing signal timing for this intersection such that oversaturation would not occur at the intersection. The peak-hour factor (PHF) for the intersection is 0.85, the saturation flow rate for all lanes is 1600 vph, and the phasing system to be used in the design is shown in Table 5.9.

TABLE 5.9

The Phasing System for the T-Intersection in Problem 5.9

Phase	A	B	C
Diagram			

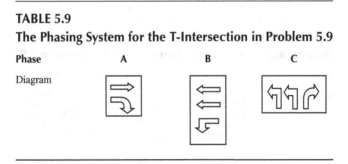

Knowing that the critical (v/s) ratio on the major westbound approach that will be used to determine the cycle length is always that for the through lanes due to the heavy volume in that direction and the total lost time in the cycle is 12 seconds, determine the maximum allowable traffic volume that the second through lane in the major westbound approach can carry to avoid oversaturation.

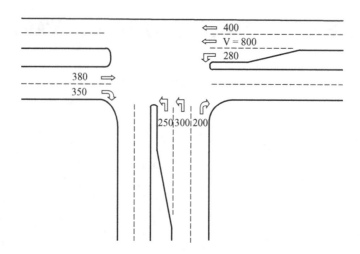

FIGURE 5.11 A T-intersection for Problem 5.9.

Solution:

The lane groups for this intersection are established using the guidelines discussed earlier. Two lane groups per each approach are established as shown in Table 5.10.

TABLE 5.10
The Established Lane Groups for the T-Intersection in Problem 5.9

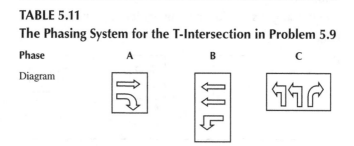

The phasing system is given in the problem as shown in Table 5.11.

TABLE 5.11
The Phasing System for the T-Intersection in Problem 5.9

Phase	A	B	C
Diagram			

The critical (v/s) ratio and computations for each phase are summarized in Table 5.12. Phase A is composed of the following lane groups:

The traffic volumes for these two lane groups are 380 and 350 vph, respectively. The actual flow rate and the (v_{ij}/s_{ij}) ratio are computed as shown below. It has to be noted that the saturation flow rate for any lane group with only one lane is equal to $s = 1600$ veh/h of green, and for any lane group with two lanes is equal to $2s = 3200$ veh/h of green.

$$v_{ij} = \frac{V_{ij}}{\text{PHF}}$$

$\Rightarrow$

$$v_{ij} = \frac{380}{0.85} = 447.1 \text{ veh/h}$$

$$\frac{v_{ij}}{s_{ij}} = \frac{447.1}{1600} = 0.279$$

In a similar manner, the traffic volumes, the saturation flow rates, and the actual flow rates for the other lane groups for all three phases are determined. The results are shown in Table 5.12.

TABLE 5.12

HCM Signal Timing Computations and Results for the Three Phases of the T-Intersection in Problem 5.9

Phase	A		B		C	
Lane Group	RE[a]	TE	TW+TW	LW	RN	LN+LN
Volume (vph)	350	380	400+800	280	200	350+320
v_{ij} (vph)	350/PHF	380/PHF	1200/PHF	280/PHF	200/PHF	550/PHF
s_{ij} (vph)	s	s	2s	s	s	2s
v_{ij}/s_{ij}	350/(PHF×s)	380/(PHF×s)	600/(PHF×s)	280/(PHF×s)	200/(PHF×s)	275/(PHF×s)
$(v/s)_{cj}$	380/(PHF×s)		600/(PHF×s)		275/(PHF×s)	

[a] R, T, L refer to right-turn, through, and left-turn lanes, respectively. E, W, N, S refer to eastbound, westbound, northbound, and southbound, respectively.

The summation of the critical (v/s) ratios for all three phases is computed:

$$\sum_j \left(\frac{v}{s}\right)_{cj} = \frac{380}{\text{PHF} \times s} + \frac{600}{\text{PHF} \times s} + \frac{275}{\text{PHF} \times s}$$

$$L = 12 \text{ sec}$$

$$X_c = \left(\sum_j \left(\frac{v}{s}\right)_{cj}\right) \frac{C}{C - L}$$

$\Rightarrow$

$$X_c = \left(\frac{380}{\text{PHF} \times s} + \frac{600}{\text{PHF} \times s} + \frac{275}{\text{PHF} \times s} \right) \frac{60}{60-12}$$

$$X_c = \left(\frac{380+600+275}{0.85 \times 1600} \right) \frac{60}{60-12} = 1.15 > 1.0$$

⇒ Oversaturation occurs at the intersection.

The MS Excel worksheet used to solve the above part of the problem is shown in the screen image in Figure 5.12.

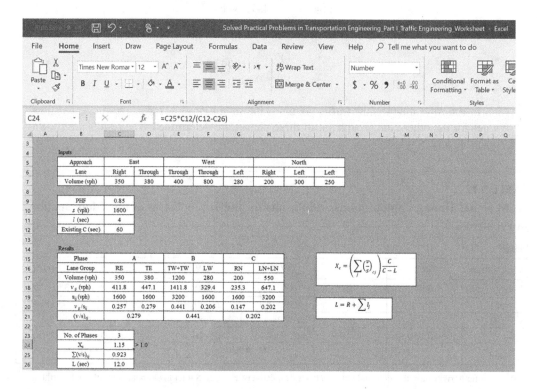

FIGURE 5.12   An image of the MS Excel worksheet used for the computations of Part 1 of Problem 5.9.

To avoid oversaturation, the critical (v/c) ratio, $X_c = 1.0$.

$$\sum_j \left(\frac{v}{s} \right)_{cj} = \frac{380}{\text{PHF} \times s} + \frac{400+V}{\text{PHF} \times 2s} + \frac{275}{\text{PHF} \times s}$$

$$X_c = \left(\sum_j \left(\frac{v}{s} \right)_{cj} \right) \frac{C}{C-L}$$

⇒

$$1.0 = \left(\frac{380}{\text{PHF} \times s} + \frac{200+V/2}{\text{PHF} \times s} + \frac{275}{\text{PHF} \times s} \right) \frac{60}{60-12}$$

$\Rightarrow$

$$1.0 = \left(\frac{380 + 200 + V/2 + 275}{0.85 \times 1600} \right) \frac{60}{60 - 12}$$

$\Rightarrow$

$$V = 466 \text{ vph}$$

In conclusion, the designer would have to reduce the traffic volume in the second through lane from 800 to 466 vph to avoid oversaturation at the intersection. Another solution would be to increase the cycle length. Of course, reducing the traffic volume on the through lane would accompany finding other alternatives for this traffic by either detouring the traffic or expanding the approach.

The MS Excel worksheet used to solve this part of the problem with its details and computations is shown in Figure 5.13.

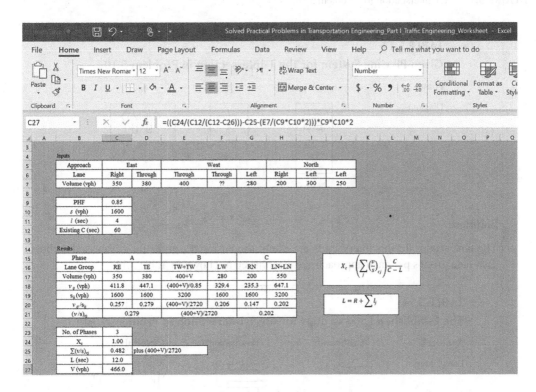

FIGURE 5.13 An image of the MS Excel worksheet used for the computations of Part 2 of Problem 5.9.

5.10 Figure 5.14 shows a T-intersection with the flow ratios (v/s) for all traffic movements of the three approaches. The saturation flow rates are 1800, 1850, and 1600 vph for right lane, through lane, and left lane, respectively. The phasing system for the intersection is given in Table 5.13 and the lost time per phase is 3 seconds. Using the Highway Capacity Method (HCM), determine the shortest cycle length (to the nearest 5 seconds) that will avoid oversaturation.

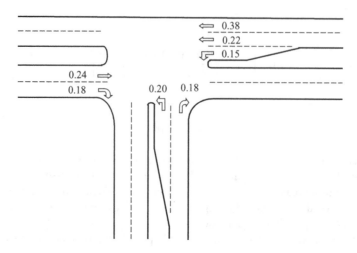

FIGURE 5.14   A T-intersection for Problem 5.10.

TABLE 5.13

Given Phasing System for the T-Intersection in Problem 5.10

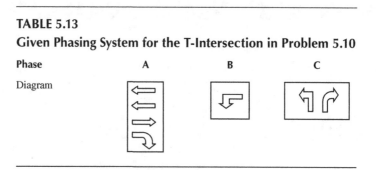

Solution:

The lane groups for this intersection are established using the guidelines discussed earlier. Two lane groups per each approach are established as shown in Table 5.14.

TABLE 5.14

The Established Lane Groups for the T-Intersection in Problem 5.10

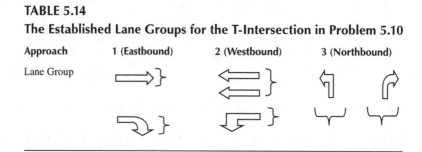

The actual flow rates for all lanes in all three phases A, B, and C are determined. The results are shown in Table 5.15.

Sample Calculation:

For the Through Lane #1 (Phase A):

$$\frac{v}{s} = \frac{v}{1850} = 0.38$$

$$\Rightarrow$$

$$v = 703 \text{ vph}$$

In a similar manner, the actual flow rates for the other lanes for all three phases are determined. The results are shown in Table 5.15. After that, the actual flow rates for the lane groups are computed. The reason behind this step is that some lane groups consist of more than one lane; and therefore, the flow rate for the lane group is equal to summation of the flow rates in all lanes.

TABLE 5.15
HCM Signal Timing Computations and Results for the Three Phases of the T-Intersection in Problem 5.10

Phase	A				B	C	
Lane	1 ⇐	2 ⇐	3 ⤵	4 ⇒	1 ⤸	1 ⤷	2 ⤶
v/s	0.380	0.220	0.180	0.240	0.150	0.180	0.200
v (vph)	703	407	324	444	240	324	320
Lane Group	1W+2W[a]		1E	2E	3W	1N	2N
v_{ij} (vph)	703+407=1110		324	444	240	324	320
s_{ij} (vph)	2×1850=3700		1800	1850	1600	1800	1600
v_{ij}/s_{ij}	0.300		0.180	0.240	0.150	0.180	0.200
$(v/s)_{cj}$	0.300				0.150	0.200	

[a] This lane group consists of Lane 1W and Lane 2W (i.e., two through lanes westbound).

The summation of the critical (v/s) ratios for the three phases is computed:

$$\sum_j \left(\frac{v}{s}\right)_{cj} = 0.300 + 0.150 + 0.200 = 0.650$$

The cycle length that will avoid oversaturation is happening at critical (v/c) ratio, $X_c = 1.0$.

$$L = 3 \times 3 = 9 \text{ sec}$$

$$\Rightarrow$$

$$X_c = \left(\sum_j \left(\frac{v}{s}\right)_{cj}\right)\frac{C}{C-L}$$

$$\Rightarrow$$

$$1.0 = (0.650)\frac{C}{C-9}$$

Solving for C ⇒

$$C = 25.7 \text{ sec}$$

Use C=30 sec (rounded to the nearest 5 seconds).

The MS Excel worksheet used to solve this problem with all details and computations is shown in Figure 5.15.

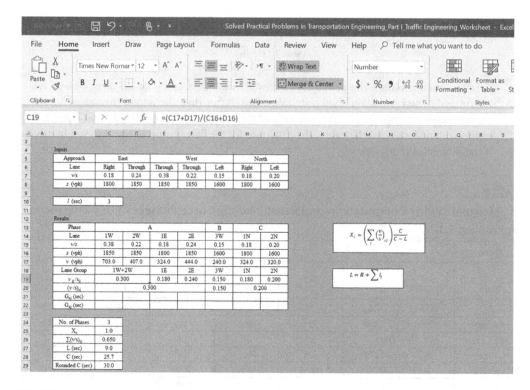

FIGURE 5.15 An image of the MS Excel worksheet used for the computations of Problem 5.10.

5.11 In Problem 5.10 above, if a cycle length of 60 seconds is used for this intersection, then compute the maximum allowable flow rate in the left lane of the major approach to avoid oversaturation.

Solution:

The left-turn lane on the major approach (westbound approach) in this case is unknown. The same (v/s) values from Problem 5.10 are used for the other lanes. Similar computations are performed as seen in Table 5.16:

TABLE 5.16

HCM Signal Timing Computations and Results for the Three Phases of the T-Intersection in Problem 5.11

Phase	A				B	C	
Lane	1⇐	2⇐	3 ⤵	4⇒	3 ⤵	1⤷	2⤴
v/s	0.38	0.22	0.18	0.24	??	0.18	0.20
v (vph)	703	407	324	444	??	324	320
Lane Group	1		2	3	1	1	2
v_{ij} (vph)	703+407=1110		324	444	??	324	320
s_{ij} (vph)	2×1850=3700		1800	1850	1600	1800	1600
v_{ij}/s_{ij}	0.30		0.18	0.24	??	0.18	0.20
$(v/s)_{cj}$	0.30				??	0.20	

$$\sum_{j}\left(\frac{v}{s}\right)_{cj} = 0.30 + \left(\frac{v}{s}\right)_{\text{critical-Phase } B} + 0.20$$

$$X_c = \left(\sum_{j}\left(\frac{v}{s}\right)_{cj}\right)\frac{C}{C-L}$$

$\Rightarrow$

$$1.0 = \left(0.30 + \left(\frac{v}{s}\right)_{\text{critical-Phase } B} + 0.20\right)\frac{60}{60-9}$$

$\Rightarrow$

$$\left(\frac{v}{s}\right)_{\text{critical-Phase } B} = 0.35$$

This value for (v/s) ratio is the critical value for Phase B; however, since this phase is composed of only one lane group, and the lane group consists of only one lane (the left-turn lane), the (v/s) value of 0.35 corresponds to the left-turn lane in Phase B.

$$\left(\frac{v}{s}\right)_{\text{critical-Phase } B} = 0.35 = \frac{v}{1600}$$

$\Rightarrow$

$$v = 560 \text{ vph}$$

5.12 In the 4-leg intersection shown in Figure 5.16, the peak-hour volumes for all traffic movements are given: (a) establish lane grouping for this intersection, (b) provide the phasing system that will provide the shortest cycle length using Webster method, and (c) provide the critical lane group volume for each phase, if the saturation flow rate and the peak-hour factor are the same for all lanes.

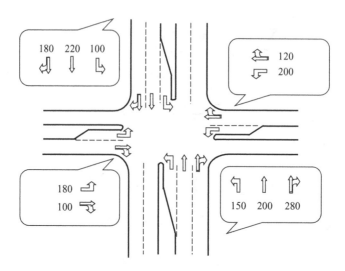

FIGURE 5.16   A 4-leg intersection for Problem 5.12.

Solution:

(a)   The lane groups for this intersection are established using the guidelines discussed earlier (see Table 5.17).

TABLE 5.17
The Established Lane Groups for the 4-Leg Intersection in Problem 5.12

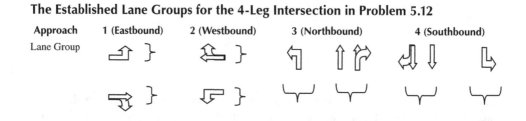

(b)   The phasing system that will provide the shortest cycle length is illustrated in Table 5.18.

TABLE 5.18
The Phasing System for the 4-Leg Intersection in Problem 5.12

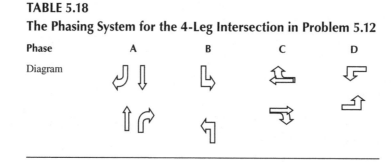

When the highest volume in one approach is combined with the highest volume in an opposite approach in one phase, this situation normally creates the optimum phasing system that will provide the shortest cycle length. In this case, the through lanes with the highest traffic volume (280 + 200) in the northbound approach are combined with the through lanes with the highest traffic volume (180+220) in the southbound approach to compose one phase. Similarly, the left-turn lane with the highest traffic volume (180) in the eastbound approach is combined with the left-turn lane with the highest traffic volume (200) in the westbound approach to provide one phase.

(c)   The lane group volumes along with the critical lane group volume for each phase are summarized in Table 5.19.

TABLE 5.19

Lane Group Volumes and Critical Lane Group Volume for Problem 5.12

Phase	A		B		C		D	
Lane Group	1N	1S	2N	2S	1E	1W	2E	2W
Volume (vph)	280+200=480	180+220=400	150	100	100	120	180	200
$V_{critical}$ (vph)		480		150		120		200

5.13 For Problem 5.12, if the saturation flow rate is 1600 vph for all lanes, the peak-hour factor for the intersection approaches is 0.90, the lost time per phase is 3 seconds, and the yellow interval is 3 seconds, determine the cycle length using Webster method and the actual green times for all phases.

Solution:

The formula for Webster method to calculate the cycle length is shown below:

$$C_o = \frac{1.5L + 5}{1 - \sum_j \left(\frac{v}{s}\right)_{cj}} \tag{5.13}$$

Where:
   C_o = optimum cycle length for the intersection (sec)
   L = total lost time in the cycle (sec)

$$\sum_j \left(\frac{v}{s}\right)_{cj} = \text{summation of critical (v/s) ratios for all phases}$$

$$\left(\frac{v}{s}\right)_{cj} = \text{critical flow ratio for phase } j$$

   v = actual flow rate for lane group (veh/h)
   s = saturation flow rate for lane group (veh/h of green time)

Sample Calculation:
For Phase A, Lane Group #1:
V = 280+200=480 vph
s = 1600×2=3200 vph (since this lane group consists of two lanes).

$$v = \frac{V}{PHF}$$

$\Rightarrow$

$$v = \frac{480}{0.90} = 533.3 \text{ vph}$$

And

$$\frac{v}{s} = \frac{533.3}{3200} = 0.167$$

In a similar manner, the actual flow rates and the (v/s) ratios are calculated for all lane groups of the four phases. The results are shown in Table 5.20.

TABLE 5.20

Computation of (v/s) and Critical (v/s) Ratios for the Lane Groups of the Four Phases for Problem 5.13

Phase	A		B		C		D	
Lane Group	1N	1S	2N	2S	1E	1W	2E	2W
Volume (vph)	480	400	150	100	100	120	180	200
$V_{critical}$ (vph)	480		150		120		200	
v (vph)	533.3	444.4	166.7	111.1	111.1	133.3	200.0	222.2
s (vph)	3200	3200	1600	1600	1600	1600	1600	1600
v/s	0.167	0.139	0.104	0.069	0.069	0.083	0.125	0.139
$(v/s)_{critical}$	0.167		0.104		0.083		0.139	
Summation				0.493				

$$C_o = \frac{1.5L + 5}{1 - \sum_j \left(\frac{v}{s}\right)_{cj}}$$

$\Rightarrow$

$$C_o = \frac{1.5(3 \times 4) + 5}{1 - 0.493} \cong 46 \text{ seconds}$$

$\Rightarrow$

$$C_o = 50 \text{ seconds (rounded to the nearest 5 seconds)}$$

$$G_{te} = C - L$$

$\Rightarrow$

$$G_{te} = 50 - 12 = 38 \text{ sec}$$

The effective green times and the actual green times for the four phases are computed and summarized in Table 5.21.

Sample Calculation:

For Phase A:

$$G_{ej} = \frac{\left(\dfrac{v}{s}\right)_{cj}}{\sum_j \left(\dfrac{v}{s}\right)_{cj}} G_{te}$$

$\Rightarrow$

$$G_{eA} = \frac{0.167}{0.493} \times 38 = 12.9 \text{ sec}$$

$$l_j = G_{aj} + \tau_j - G_{ej}$$

Or:

$$G_{aj} = l_j - \tau_j + G_{ej}$$

$\Rightarrow$

$$G_{aA} = 3 - 3 + 12.9 = 12.9 \text{ sec}$$

TABLE 5.21

The Effective and Actual Green Times for the Four Phases of the 4-Leg Intersection in Problem 5.13

Phase	A	B	C	D
$(v/s)_{critical}$	0.167	0.104	0.083	0.139
τ (sec)	3	3	3	3
l (sec)	3	3	3	3
G_e (sec)	12.9	8.0	6.4	10.7
G_a (sec)	12.9	8.0	6.4	10.7

5.14 In Problem 5.13, check the minimum green times required for pedestrians crossing if the following pedestrian-related data is given:

Number of pedestrians crossing the northbound approach = 22
Number of pedestrians crossing the southbound approach = 20
Number of pedestrians crossing the eastbound approach = 15
Number of pedestrians crossing the westbound approach = 12
Effective crosswalk width for each approach = 8 ft
Crosswalk length in N-S approach = 64 ft (4 lanes; lane width = 12 ft, median width = 16 ft)
Crosswalk length in E-W approach = 40 ft (2 lanes; lane width = 12 ft, median width = 16 ft)
Speed limit on each approach = 25 mph

Solution:

The minimum green times required for pedestrian crossing can be determined using the HCM formulas given below:

For $W_E \leq 10$ ft:

$$G_p = 3.2 + \frac{L}{S_P} + \left(0.27 N_{ped}\right) \tag{5.14}$$

For $W_E > 10$ ft:

$$G_p = 3.2 + \frac{L}{S_P} + \left(2.7 \frac{N_{ped}}{W_E}\right) \tag{5.15}$$

Where:

G_p = minimum green time (sec)
L = crosswalk length (ft)
S_p = average pedestrians speed (typical value = 4 ft/sec)
3.2 = pedestrian reaction (start-up) time (sec)
W_E = effective crosswalk width (ft)
N_{ped} = number of pedestrians crossing an approach

Since $W_E = 8$ ft ≤ 10 ft,
$\Rightarrow$

$$G_p = 3.2 + \frac{L}{S_P} + \left(0.27 N_{ped}\right)$$

For northbound approach:

$$G_{p\text{-north}} = 3.2 + \frac{64}{4} + \left(0.27 \times 22\right) = 23.2 \text{ sec}$$

For southbound approach:

$$G_{p\text{-south}} = 3.2 + \frac{64}{4} + \left(0.27 \times 20\right) = 22.4 \text{ sec}$$

For eastbound approach:

$$G_{p\text{-east}} = 3.2 + \frac{40}{4} + \left(0.27 \times 15\right) = 19.1 \text{ sec}$$

For westbound approach:

$$G_{p\text{-west}} = 3.2 + \frac{40}{4} + \left(0.27 \times 12\right) = 18.6 \text{ sec}$$

The green times available for pedestrians for each approach based on the computed cycle length and actual green times from the signal phasing are compared with the minimum green time requirements calculated above for pedestrian crossing data as shown in Table 5.22.

TABLE 5.22

Calculated Green Time versus Available Green Time for Pedestrians for Problem 5.14

Approach	N-S Approaches	E-W Approaches
Available G_p (sec)	$6.4 + 10.7 = 17.1$	$12.9 + 8.0 = 20.9$
Required G_p	$23.2 + 22.4 = 45.6$	$19.1 + 18.6 = 37.7$
Not-Satisfied		

5.15 Solve Problem 5.13 using the following phasing system (see Table 5.23).

TABLE 5.23

Given Phasing System for Problem 5.15

Phase	A	B	C	D
Diagram				

Solution:

Based on the given phasing system, the traffic volumes, the saturation flow rates, and the actual flow rates for the lane groups of the four phases are determined.

Sample Calculation:

For Phase D, Lane Group #2:

$V = 200$ vph

$s = 1600 \times 1 = 1600$ vph (since this lane group consists of only one lane).

$$v = \frac{V}{\text{PHF}}$$

$\Rightarrow$

$$v = \frac{200}{0.90} = 222.2 \text{ vph}$$

And

$$\frac{v}{s} = \frac{222.2}{1600} = 0.139$$

In a similar manner, the actual flow rates and the (v/s) ratios are calculated for all lane groups of the four phases. The results are shown in Table 5.24.

TABLE 5.24

Computation of (v/s) and Critical (v/s) Ratios for the Lane Groups of the Four Phases for Problem 5.15

Phase	A		B		C		D	
Lane Group	1N	2N	1S	2S	1E	2E	1W	2W
Volume (vph)	480	150	400	100	100	180	120	200
$V_{critical}$ (vph)	480		400		180		200	
v (vph)	533.3	166.7	444.4	111.1	111.1	200.0	133.3	222.2
s (vph)	3200	1600	3200	1600	1600	1600	1600	1600
v/s	0.167	0.104	0.139	0.069	0.069	0.125	0.083	0.139
$(v/s)_{critical}$	0.167		0.139		0.125		0.139	
Summation				0.570				

$$C_o = \frac{1.5L + 5}{1 - \sum_j \left(\frac{v}{s}\right)_{cj}}$$

$\Rightarrow$

$$C_o = \frac{1.5(3 \times 4) + 5}{1 - 0.570} \cong 53.5 \text{ seconds}$$

$\Rightarrow$

$$C_o = 55 \text{ seconds} \left(\text{rounded to the nearest 5 seconds}\right)$$

Therefore, by using the phasing system in Problem 5.15, the cycle length is longer than that when using the phasing system in Problem 5.13.

5.16 For the intersection with the information below (see Figure 5.17): (a) establish the lane groups at the intersection, (b) Select an appropriate phasing system, (c) design a signal timing for the intersection using the Webster method.
   • The numbers shown at the intersection are the peak-hour volumes for the different traffic movements.
   • Assume that the yellow interval at the intersection is 3 seconds and the lost time per phase is 3.5 seconds.
   • PHF=0.95.
   • The saturation flow rate (vph) for each lane is shown in Table 5.25.

TABLE 5.25
The Given Saturation Flow Rates for the Lanes of the 4-Leg Intersection in Problem 5.16

1700	1800	1800	1600	1600	1800	1800	1700	1600	1800	1800	1800	1700	1700	1800	1800	1600

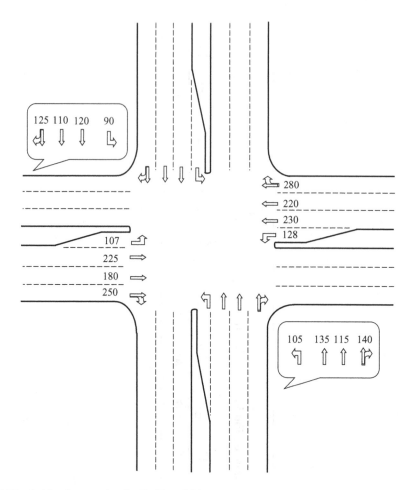

FIGURE 5.17 A 4-leg intersection for Problem 5.16.

Solution:

(a) Lane Groups: Lane groups are established as follows: per each approach, the left-turn lane on each approach is given a separate lane group, and the other lanes (the shared lane with the through lanes) are all given a separate lane group since there is no exclusive right-turn lane. The resulting lane groups are shown in Table 5.26.

TABLE 5.26

The Established Lane Groups for the Four Approaches of the Intersection in Problem 5.16

Approach	1 (Eastbound)		2 (Westbound)		3 (Northbound)		4 (Southbound)	
Lane Group No.	1E	2E	1W	2W	1N	2N	1S	2S
Lane Group	⇒⇒⇒	⤴	⇖⇐ ⇐	⤶	⇑⇑⇗	⇖	⤸⇓⇓	⤵

(b) Phasing System: The phasing system at the intersection is established based on the peak-hour volumes for the different traffic movements. In this case, two phasing systems are used and evaluated. The first one is described in Table 5.27.

TABLE 5.27

Phasing System #1 for the 4-Leg Intersection in Problem 5.16

Phase	A	B	C	D
Diagram				

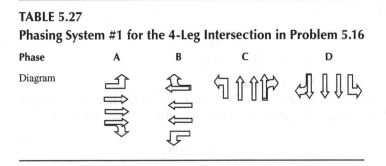

The second phasing system is shown in Table 5.28.

TABLE 5.28

Phasing System #2 for the 4-Leg Intersection in Problem 5.16

Phase	A	B	C	D
Diagram				

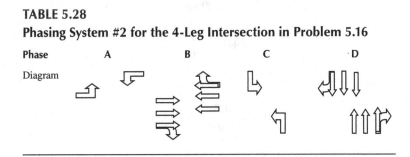

(c) Signal Timing: The design hourly volumes (DHVs) are determined by dividing the peak-hour volumes by the PHF. Afterwards, the actual flow rate is determined by dividing the design hourly volume by the saturation flow rate.

Sample Calculation for Phasing System #1:
   For Phase A, Lane Group #1:
   This lane groups consists of three lanes: one shared lane (right-turn and through movements) and two through lanes. Therefore, the traffic volume for this lane group is the summation of the traffic volumes in the three lanes:

$$V = 250 + 180 + 225 = 655 \text{ vph}$$

The saturation flow rate for this lane group is simply the addition of the saturation flow rates of the three lanes in the lane group; that is:

$$s = 1700 + 1800 + 1800 = 5300 \text{ vph}$$

$$DHV = \frac{V}{PHF}$$

$\Rightarrow$

$$DHV = \frac{655}{0.95} = 689.5 \text{ vph}$$

And

$$\frac{v}{s} = \frac{689.5}{5300} = 0.130$$

For Phase C, Lane Group #2:

This lane groups consists of only one exclusive left-turn lane. Therefore, the traffic volume for this lane group is equal to the traffic volume on that lane:

$$V = 105 \text{ vph}$$

The saturation flow rate for this lane group is simply the saturation flow rate of the left-turn lane; that is:

$$s = 1600 \text{ vph}$$

$$DHV = \frac{V}{PHF}$$

$\Rightarrow$

$$DHV = \frac{105}{0.95} = 110.5 \text{ vph}$$

And

$$\frac{v}{s} = \frac{110.5}{1600} = 0.069$$

In a similar manner, the actual flow rates and the (v/s) ratios are calculated for all the lane groups of the four phases in this phasing system. The results are shown in Table 5.29.

TABLE 5.29

Computation of (v/s) and Critical (v/s) Ratios for the Lane Groups of Phasing System #1 for Problem 5.16

Phase	A		B		C		D	
Lane Group	1E	2E	1W	2W	1N	2N	1S	2S
Volume (vph)	655	107	730	128	390	105	355	90
v (vph)	689.5	112.6	768.4	134.7	410.5	110.5	373.7	94.7
s (vph)	5300	1600	5300	1600	5300	1600	5300	1600
v/s	0.130	0.070	0.145	0.084	0.077	0.069	0.071	0.059
$(v/s)_{critical}$	0.130		0.145		0.077		0.071	
Summation				0.423				

The cycle length is calculated using the formula of Webster method. The total lost time for the traffic signal is estimated first. Since the lost time per phase is 3.5 seconds, the total lost time for the four phases is determined as follows:

$$L = R + \Sigma l_j$$

$\Rightarrow$

$$L = 0 + 3.5 \times 4 = 14 \text{ sec}$$

$$C_o = \frac{1.5L + 5}{1 - \sum_j \left(\frac{v}{s}\right)_{cj}}$$

$\Rightarrow$

$$C_o = \frac{1.5(14) + 5}{1 - 0.423} = 45.1 \text{ seconds}$$

$\Rightarrow$

$$C_o = 50 \text{ seconds (rounded to the nearest 5 seconds)}$$

$$G_{te} = C - L$$

$\Rightarrow$

$$G_{te} = 50 - 14 = 36 \text{ sec}$$

The effective green times and the actual green times for the four phases are computed and summarized in Table 5.30.

Sample Calculation:

For Phase A:

$$G_{ej} = \frac{\left(\frac{v}{s}\right)_{cj}}{\sum_j \left(\frac{v}{s}\right)_{cj}} G_{te}$$

$\Rightarrow$

$$G_{eA} = \frac{0.130}{0.423} \times 36 = 11.1 \text{ sec}$$

$$l_j = G_{aj} + \tau_j - G_{ej}$$

Or:

$$G_{aj} = l_j - \tau_j + G_{ej}$$

$\Rightarrow$

$$G_{aA} = 3.5 - 3 + 11.1 = 11.6 \text{ sec}$$

TABLE 5.30

The Effective and Actual Green Times for Phasing System #1 of the 4-Leg Intersection in Problem 5.16

Phase	A	B	C	D
$(v/s)_{critical}$	0.130	0.145	0.077	0.071
τ (sec)	3	3	3	3
l (sec)	3.5	3.5	3.5	3.5
G_e (sec)	11.1	12.3	6.6	6.0
G_a (sec)	11.6	12.8	7.1	6.5

Calculations for Phasing System #2:

The design-hourly volumes, saturation flow rates, actual flow rates, and (v/s) ratios for lane groups have been calculated in the previous step for Phasing System #1. The same values will be used in Phasing System #2 but with different phases than before. The results for Phasing System #2 are shown in Table 5.31.

TABLE 5.31

Computation of (v/s) and Critical (v/s) Ratios for the Lane Groups of Phasing System #2 for Problem 5.16

Phase	A		B		C		D	
Lane Group	2E	2W	1E	1W	2N	2S	1N	1S
Volume (vph)	107	128	655	730	105	90	390	355
v (vph)	112.5	134.7	689.5	768.4	110.5	94.7	410.5	373.7
s (vph)	1600	1600	5300	5300	1600	1600	5300	5300
v/s	0.070	0.084	0.130	0.145	0.069	0.059	0.077	0.071
$(v/s)_{critical}$	0.084		0.145		0.069		0.077	
Summation				0.375				

The total lost time for the traffic signal is estimated as follows:

$$L = R + \Sigma l_j$$

$\Rightarrow$

$$L = 0 + 3.5 \times 4 = 14 \text{ sec}$$

The cycle length is calculated using the formula of Webster method shown below:

$$C_o = \frac{1.5L + 5}{1 - \sum_j \left(\frac{v}{s}\right)_{cj}}$$

$\Rightarrow$

$$C_o = \frac{1.5(14) + 5}{1 - 0.375} = 41.6 \text{ seconds}$$

$\Rightarrow$

$$C_o = 45 \text{ seconds} \left(\text{rounded to the nearest 5 seconds}\right)$$

$$G_{te} = C - L$$

$\Rightarrow$

$$G_{te} = 45 - 14 = 31 \text{ sec}$$

The effective green times and the actual green times for the four phases are computed and summarized in Table 5.32.

Sample Calculation:

For Phase A:

$$G_{ej} = \frac{\left(\dfrac{v}{s}\right)_{cj}}{\displaystyle\sum_j \left(\dfrac{v}{s}\right)_{cj}} G_{te}$$

$\Rightarrow$

$$G_{eA} = \frac{0.084}{0.375} \times 31 = 6.9 \text{ sec}$$

$$l_j = G_{aj} + \tau_j - G_{ej}$$

Or:

$$G_{aj} = l_j - \tau_j + G_{ej}$$

$\Rightarrow$

$$G_{aA} = 3.5 - 3 + 6.9 = 7.4 \text{ sec}$$

TABLE 5.32

The Effective and Actual Green Times for Phasing System #2 of the 4-Leg Intersection in Problem 5.16

Phase	A	B	C	D
$(v/s)_{\text{critical}}$	0.084	0.145	0.069	0.077
τ (sec)	3	3	3	3
l (sec)	3.5	3.5	3.5	3.5
G_e (sec)	6.9	12.0	5.7	6.4
G_a (sec)	7.4	12.5	6.2	6.9

In conclusion, using Phasing System #2 provides a cycle length that is shorter than that provided by Phasing System #1.

5.17 It is required to design a complete signal phasing and timing plan for the 4-leg intersection shown in Figure 5.18. It is also required to check if protected left-turn phase should be provided for any of the approaches using the cross-product guideline. The geometry, number of lanes for all approaches of the intersection, and the design-hourly peak volumes (DHVs) are provided in the diagram. The following data is given:

- The yellow interval at the intersection is assumed to be 3.5 seconds for each phase, and the lost time per phase is also assumed as 3 seconds. No all-red phase will be used for the traffic signal.
- The speeds of vehicles are 30 mph (43.3 kph) and 35 mph (56.3 kph) for the N-S approach and the E-W approach, respectively.
- Number of pedestrians crossing the N-S approaches = 9.
- Number of pedestrians crossing the E-W approaches = 9.
- Effective crosswalk width for each approach = 8 ft.
- The saturation flow rates for the different lanes are given below (see Table 5.33).

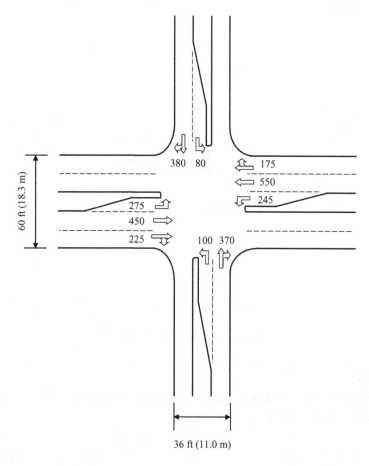

FIGURE 5.18   A 4-leg intersection for Problem 5.17.

TABLE 5.33

The Given Saturation Flow Rates for the Lanes of the 4-Leg Intersection in Problem 5.17

↙↓	↳	↑↰	↱	↰→	⇒	⇛↰	↲	⇐	↱↓
1300	900	900	1300	1200	1500	1350	1350	1500	1200

- The phasing system used for the design of a traffic signal at this intersection is shown in Table 5.34.

TABLE 5.34

The Phasing System of the 4-Leg Intersection in Problem 5.17

Phase	A	B	C	D
Diagram	↰→ ⇒ ⇛↰	⇐↰ ⇐ ↱↓	↑↰ ↱	↙↓ ↳

Solution:

Lane Groups:

Lane groups are established as follows: Per each approach, the left-turn lane on each approach is given a separate lane group, and the other lanes (the shared lane with the through lanes) are all given a separate lane group since there is no exclusive right-turn lane. The resulting lane groups are shown in Table 5.35.

TABLE 5.35

The Established Lane Groups for the 4-Leg Intersection in Problem 5.17

Approach	1 (Eastbound)		2 (Westbound)		3 (Northbound)		4 (Southbound)	
Lane Group No.	1E	2E	1W	2W	1N	2N	1S	2S
Lane Group	⇒ ⇛↰	↰→	⇐↰ ⇐	↱↓	↱	↑↰	↙↓	↳

Requirement for Any Protected Left-Turn Phase:

The determination if protected left-turn phase should be provided for any of the approaches is verified using the cross-product guideline. The *Highway Capacity Manual* offers the following criteria for the more common guidelines (using the cross product of the left-turn volume and opposing through and right-turn volumes), as follows:

The use of protected left-turn phase should be considered when the product of left-turning vehicles and opposing traffic volumes:

- Exceeds 50000 during the peak hour for one opposing lane, or
- Exceeds 90000 during the peak hour for two opposing lanes, or
- Exceeds 110000 during the peak hour for three or more opposing lanes

Accordingly,

For the E–W approaches:

- The product of eastbound left-turning vehicles and westbound opposing traffic (through and right-turn vehicles) $= 275 \ (550 + 175) = 199375 > 90000$ (requirement for two opposing lanes). Thus, this movement should be protected.
- The product of westbound left-turning vehicles and eastbound opposing traffic (through and right-turn vehicles) $= 245 \ (450 + 225) = 165375 > 90000$ (requirement for two opposing lanes). Thus, this movement should be also protected.

For the N–S approaches:

- The product of northbound left-turning vehicles and southbound opposing traffic (through and right-turn vehicles) $= 100 \ (380) = 38000 < 50000$ (requirement for one opposing lanes). Thus, this movement could be permitted (not protected)
- The product of southbound left-turning vehicles and northbound opposing traffic (through and right-turn vehicles) $= 80 \ (370) = 29600 < 50000$ (requirement for one opposing lanes). Thus, this movement could also be permitted (not protected).

Recommended Phasing Plan:

Based on the above analysis, the given phasing system is modified for permitted (not protected) left-turn phases for the N–S approaches, and a three-phase traffic- signal plan is recommended as shown in Table 5.36.

TABLE 5.36

The Phasing System for the 4-Leg Intersection in Problem 5.17

Phase	A	B	C

* Note: The dashed lines represent permitted movements and the solid lines represent protected movements.

Saturation Flow Rates:

The saturation flow rates given in the problem are determined for each lane group using the formula:

$$s = s_o N f_w f_{HV} f_g f_p f_{bb} f_a f_{LU} f_{RT} f_{Lpb} f_{Rpb}$$

and based upon the existing conditions for the intersection. The adjustment factors for the prevailing conditions at the intersection are determined to the best of knowledge and data related to the intersection. The ideal saturation flow rate is 1900 vphpl (used under ideal conditions) and adjusted for the different existing conditions. The final saturation flow rates are summarized in Table 5.36.

Flow Ratios and Critical Lane Groups:

The design-hourly volumes are determined by dividing the peak-hour volumes by the PHF. This step is already done since the DHVs are already given in the problem. The flow ratio (v/s) for each lane groups is computed for all three phases. The critical (v/s) ratio (the maximum value among the values of the different lane groups in the same phase) is determined. Below is a sample calculation for Phase B, Lane Group 1E:

Sample Calculation for Phase B, Lane Group 1E:

This lane groups consists of two lanes: one shared lane (right-turn and through movements) and one through lane. Therefore, the design-hourly volume for this lane group is the summation of the design-hourly volumes in the two lanes:

$$DHV = 225 + 450 = 675 \text{ vph}$$

The saturation flow rate for this lane group is simply the addition of the saturation flow rates of the two lanes in the lane group; that is:

$$s = 1350 + 1500 = 2850 \text{ vph}$$

$\Rightarrow$

$$\frac{v}{s} = \frac{675}{2850} = 0.237$$

The results are shown in Table 5.37.

TABLE 5.37

Computation of (v/s) and Critical (v/s) Ratios for the Lane Groups of the Three Phases for Problem 5.17

Phase	A		B		C			
Lane Group	2E	2W	1E	1W	1N	1S	2N	2S
v, DHV (vph)	275	245	675	795	370	380	100	80
s (vph)	1200	1200	2850	2850	1300	1300	900	900
v/s	0.229	0.204	0.237	0.279	0.285	0.292	0.111	0.089
$(v/s)_{critical}$		0.229		0.279			0.292	
Summation				0.800				

Notes:

(1)   The maximum value of the flow ratio (v/s) for all lane groups in each phase is determined as shown above, considering that the Critical Lane Group is the lane group that requires the longest green time in a phase. And therefore, this lane group determines the green time that is assigned to that phase.

(2)   The summation of the total (v/s) ratios provides an idea of the required timing for each lane group based on the traffic volume for that lane group and the saturation flow rate of the lane group. In other words, higher traffic volume needs higher green time, but also the saturation flow rate of the lane group affects this timing. For instance, two lane groups with the same traffic volume may have different green times if they are in different phases depending on their saturation flow rates. A lower saturation flow rate would provide a higher green time and vice versa because prevailing conditions that reduce the saturation flow rate would also impact the movement of the vehicles at the intersection when the green turns on.

Calculation of the Total Lost Time:

It has to be noted that the lost time per cycle is the total of lost times for all phases plus the all-red time for the cycle. The lost time per phase is the time wasted at the beginning of the green phase when the signal turns on green and at the end of the yellow phase due to the reaction times of drivers. Therefore; the total lost time is determined:

$$L = R + \Sigma l_j$$

Since there is no all-red phase, $R = 0$, and lost time per phase $= 3$ seconds

$$L = 0 + 3 \times 4 = 12 \text{ sec}$$

Determination of the Optimum Cycle Length:

The optimum cycle length is determined using the following equation (Webster Method for pre-timed (fixed) signals):

$$C_o = \frac{1.5L + 5}{1 - \sum_j \left(\frac{v}{s} \right)_{cj}}$$

$\Rightarrow$

$$C_o = \frac{1.5(12) + 5}{1 - 0.800} = 115 \text{ seconds}$$

This cycle length is accepted for this type of intersection since traffic volumes are relatively high compared to the number of lanes in the intersection approaches.

The cycle length for isolated intersections should be kept below 120 seconds to avoid long delays. Cycle lengths are typically between 35 and 60 seconds. However, it is necessary sometimes to use longer cycle lengths when traffic volumes are high.

Allocation of the Green Time:

The total effective green time is determined using the formula shown below:

$$G_{te} = C - L$$

$\Rightarrow$

$$G_{te} = 115 - 12 = 103 \text{ sec}$$

To determine the effective green time for each phase, the following formula is used:

$$G_{ej} = \frac{\left(\frac{v}{s} \right)_{cj}}{\sum_j \left(\frac{v}{s} \right)_{cj}} G_{te}$$

For instance; for Phase A:

$$G_{eA} = \frac{0.229}{0.800} \times 103 = 29.5 \text{ sec}$$

Note that the number of vehicles that cross the intersection divided by the saturation flow rate provides the effective green time. The effective green time is less than the sum of the green and yellow times. This difference is considered lost time; and therefore, the actual green time for each phase is determined using the following formula:

$$l_j = G_{aj} + \tau_j - G_{ej}$$

$$\Rightarrow$$

$$G_{aj} = l_j - \tau_j + G_{ej}$$

For Phase A:

$$G_{aA} = 3 - 3.5 + 29.5 = 29.0 \text{ sec}$$

Table 5.38 summarizes the effective green times and actual green times for the three phases.

TABLE 5.38

The Effective and Actual Green Times for the Three Phases of the 4-Leg Intersection in Problem 5.17

Phase	A	B	C
$(v/s)_{\text{critical}}$	0.229	0.279	0.292
τ (sec)	3.5	3.5	3.5
l (sec)	3.0	3.0	3.0
G_e (sec)	29.5	35.9	37.6
G_a (sec)	29.0	35.4	37.1

<u>Determination of the Minimum Green Time for Safe Pedestrian Crossing:</u>

The minimum green time at the intersection that will allow pedestrians to safely cross the intersection is given by the following equations:

For $W_E \leq 10$ ft:

$$G_p = 3.2 + \frac{L}{S_P} + (0.27 N_{\text{ped}})$$

For $W_E > 10$ ft:

$$G_p = 3.2 + \frac{L}{S_P} + \left(2.7 \frac{N_{\text{ped}}}{W_E}\right)$$

Please note that the number of those pedestrians crossing during an interval are counted by visiting the intersection and seeing how many pedestrians are indeed trying to cross a certain approach within this time, considering that the interval is the total red time for the vehicles moving in the crossing direction when pedestrians can really cross a certain approach, or the green time for vehicles moving parallel to the pedestrians.

The available green time for pedestrians should be verified against the values provided by these equations.

The width of the crosswalk that is available for pedestrians at all approaches is only 8 ft, which is less than 10 ft. Therefore, the first formula above will be used to determine the minimum green time that will allow pedestrians to cross the E–W approaches during the phase for parallel vehicles (in this case Phase C) is computed as follows:

For $W_E \leq 10$ ft:

$$G_p = 3.2 + \frac{L}{S_P} + \left(0.27N_{ped}\right)$$

$L = 60$ ft
$S_p = 4$ ft/sec
$N_{ped} = 9$ pedestrians
$\Rightarrow$

$$G_p = 3.2 + \frac{60}{4} + \left(0.27 \times 9\right) = 20.6 \text{ sec}$$

The green time for Phase C is equal to 37.1 seconds > the minimum required green time for pedestrians crossing = 20.6 seconds. Consequently, the green time for the phase is enough to enable pedestrians to cross the E–W approach(s) at the intersection safely.

In a similar manner, the minimum green time that will allow pedestrians cross the N–S approach(s) during the phase for parallel vehicles (in this case Phase A or B) is computed as follows:

For $W_E \leq 10$ ft:

$$G_p = 3.2 + \frac{L}{S_P} + \left(0.27N_{ped}\right)$$

$L = 36$ ft
$S_p = 4$ ft/sec
$N_{ped} = 9$ pedestrians

$\Rightarrow$

$$G_p = 3.2 + \frac{36}{4} + \left(0.27 \times 9\right) = 14.6 \text{ sec}$$

The green times for Phases A and B are 29.0 and 35.4 seconds, respectively. Both are higher than the minimum required green time for pedestrians crossing = 14.6 seconds. Therefore, the green time for the phase is enough to enable pedestrians to cross the N–S approaches at the intersection safely.

5.18 Five successive intersections are located on urban arterial with the spacings as shown in Figure 5.19. The cycle lengths at the intersections are: 70, 65, 75, 65, and 70 seconds, respectively. A simultaneous traffic signal system is to be designed for the five intersections given that the progression speed on the arterial is 40 mph (64.4 kph). What cycle length is recommended in this case, and why?

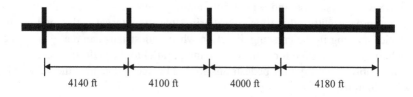

4140 ft   4100 ft   4000 ft   4180 ft

FIGURE 5.19 Five successive intersections on urban arterial for Problem 5.18.

Solution:

The following mathematical expression for the relationship between the progression speed on an arterial, the spacing between successive intersections on the arterial, and the required cycle length of the signal light for proper coordination is used:

$$u = \frac{X}{C} \tag{5.16}$$

Where:

u = progression speed (converted to ft/sec)
X = average spacing between intersections (ft)
C = cycle length (sec)

Therefore, the recommended cycle length based on the given data is computed using:

$$C = \frac{X}{u} \tag{5.17}$$

$$X = \text{average spacing} = \frac{4140 + 4100 + 4000 + 4180}{4}$$

$$= 4105 \text{ ft} \left(1251.2 \text{ m}\right)$$

$$\Rightarrow$$

$$C = \frac{4105}{40 \times \dfrac{5280}{3600}} \cong 70 \text{ sec}$$

Table 5.39 summarizes the given data along with the average spacing and recommended cycle length for simultaneous traffic signal system.

TABLE 5.39

The Average Spacing and Recommended Cycle Length for Simultaneous Traffic Signal System in Problem 5.18

Intersection	1	2	3	4	5
Cycle Length (sec)	70	65	75	65	70
Spacing (ft)		4140	4100	4000	4180
Average Spacing (ft)			4105		
Recommended C (sec)			70		

5.19 A traffic engineering designer has been asked to coordinate the traffic signals of consecutive intersections spaced at 750 ft (228.6 m) intervals on an urban arterial. If the progression speed of vehicles on the arterial is 30 mph (48.3 kph), what is the cycle length recommended for the coordination when single-, double-, and triple-alternate system is used?

Solution:

The mathematical formula given below is used to relate the progression speed, the spacing, and the required cycle length for proper coordination using the alternate system:

$$u = \frac{nX}{C} \tag{5.18}$$

Where:
u = progression speed (converted to ft/sec)
X = average spacing between intersections (ft)
C = cycle length (sec)
N = 2, 4, or 6 for single-, double-, or triple-alternate system, respectively

Thus, the recommended cycle length is given by:

$$C = \frac{nX}{u} \tag{5.19}$$

For single-alternate system (when n = 4),
⇒

$$C = \frac{2 \times 750}{30 \times \frac{5280}{3600}} = 34.1 \text{ sec}$$

The cycle length is computed using the same formula for the double- and triple-alternate systems. The results are summarized in Table 5.40.

TABLE 5.40

The Cycle Length for Single-, Double-, and Triple-Alternate Systems for Problem 5.19

Alternate System	Single	Double	Triple
n	2	4	6
X (ft)		750	
u (ft/sec)		$30 \times \frac{5280}{3600} = 44$	
Cycle Length (sec)	34.1	68.2	102.3
C (sec)[a]	35	70	105

[a] Rounded to the nearest 5 seconds.

Multiple-Choice Questions and Answers for Chapter 1
Terminology, Concepts, and Theory

1. One of the following is <u>not</u> a transportation mode:
   a. Airways
   b. Railroads
   c. Waterways
   d. Highways
   e. **None of the above**
2. The highway mode consists of:
   a. **Highway facilities and traffic systems**
   b. Highway facilities only
   c. Traffic systems only
   d. Freeways only
   e. Highways only
3. Highway facilities include:
   a. Rural roads only
   b. Urban roads only
   c. **a and b**
   d. Local streets only
   e. Traffic signals
4. One of the following is <u>not</u> among traffic systems:
   a. Signalized intersections
   b. Unsignalized intersections
   c. Roundabouts
   d. Intelligent Transportation Systems
   e. **Parking facilities**
5. The design, operation, and control of traffic systems impact one or more of the following components of traffic systems:
   a. Capacity
   b. Level of service
   c. Safety
   d. Mobility
   e. **All of the above**
6. The number of merging conflict points at a 4-leg intersection is:
   a. 4
   b. 6
   c. **8**
   d. 12
   e. 32
7. The number of diverging conflict points at a 4-leg intersection is:
   a. 4
   b. 6

   c. **8**

   d. 12

   e. 32

8. The total number of conflict points at a 4-leg intersection is equal to:

   a. 4

   b. 6

   c. 8

   d. 12

   e. **32**

9. The total number of conflict points at a 3-leg intersection is:

   a. 3

   b. 6

   c. **9**

   d. 12

   e. 15

10. One of the following is <u>not</u> among the uses and importance of traffic engineering:

   a. Safety

   b. **Geometric design of roads**

   c. Capacity of roads

   d. Transportation planning

   e. Mobility

11. A two-lane two-way highway is a highway having:

   a. A total of four lanes for both directions (two lanes per direction)

   b. **A total of two lanes for both directions (one lane per direction)**

   c. A total of eight lanes for both directions (four lanes per direction)

   d. A total of two lanes (moving in only one direction)

   e. None of the above

12. A multi-lane highway consists of:

   a. **A minimum of two lanes per direction**

   b. A total of two lanes for both directions

   c. A total of four lanes for both directions

   d. Four lanes in each direction

   e. A minimum of six lanes per direction

13. The interchange shown in the diagram below is called:

   a. **Trumpet**

   b. Cloverleaf

   c. Partial cloverleaf

   d. Full directional

   e. Directional-Y

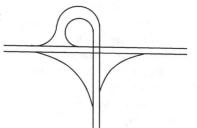

14. The interchange shown in the diagram below is called:

   a. Trumpet

   b. Cloverleaf

   c. Diamond

   d. **Full directional**

   e. Directional-Y

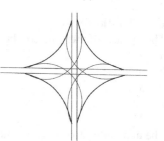

15. Traffic jams are defined in traffic engineering as:

   a. Bottleneck conditions

   b. Shock waves

   c. **Traffic congestions**
   d. Weaving and merging
   e. Traffic behaviors in off-peak hours

16. Visual perception is one of the driver characteristics in traffic engineering; one of the following characteristics is <u>not</u> considered a visual perception:
   a. Visual acuity
   b. Peripheral vision
   c. Color vision
   d. Depth perception
   e. **Reaction**

17. The ability to see the fine details of an object on a roadway is called:
   a. Brightness
   b. Contrast
   c. Peripheral vision
   d. **Visual acuity**
   e. Glare vision

18. The ability of the driver to detect moving objects on a roadway is called:
   a. **Dynamic visual acuity**
   b. Static visual acuity
   c. Peripheral vision
   d. Glare vision
   e. Depth perception

19. One of the following is <u>not</u> among the factors that impact static visual acuity of drivers:
   a. Background
   b. Brightness
   c. Contrast
   d. Time
   e. **Speed**

20. The conical angle of clear vision for drivers is equal to:
   a. 6–$8°$
   b. 10–$12°$
   c. **3–$5°$**
   d. 0–$90°$
   e. 0–$180°$

21. The conical angle for moderately clear vision is:
   a. 6–$8°$
   b. **10–$12°$**
   c. 3–$5°$
   d. 0–$90°$
   e. 0–$180°$

22. The ability of a driver to see objects beyond the conical angle of clear vision is called:
   a. Visual acuity
   b. **Peripheral vision**
   c. Color vision
   d. Glare vision and recovery
   e. Depth perception

23. The angle for peripheral vision can go up to:
   a. $5°$
   b. $8°$
   c. $90°$

    d. **160°**
    e. 180°

24. For color vision, the eye is typically most sensitive to a combination of the following two colors:
    a. Black and red
    b. White and blue
    c. Black and white
    d. Black and yellow
    e. **c and d**

25. There are two types of glare vision, which are:
    a. Simple and fussy
    b. **Direct and specular**
    c. Slow and fast
    d. Recovered and unrecovered
    e. None of the above

26. Glare recovery for drivers when moving from dark to light is approximately:
    a. **3 seconds**
    b. ≥ 6 seconds
    c. ≤ 6 seconds
    d. 8 seconds
    e. 10 seconds

27. Glare recovery for drivers when moving from light to dark is about:
    a. 3 seconds
    b. **≥ 6 seconds**
    c. ≤ 6 seconds
    d. 8 seconds
    e. 10 seconds

28. The ability of a driver to estimate speed and distance is called:
    a. Visual acuity
    b. Peripheral vision
    c. Color vision
    d. Glare vision and recovery
    e. **Depth perception**

29. Traffic control device are made standard in size, shape, and color in order to:
    a. Help color-blind drivers in identifying signs
    b. Minimize the effect of visual perception
    c. Minimize the effect of hearing perception
    d. Minimize the driver's perception–reaction time
    e. **a and b**

30. The process through which a driver estimates and reacts to an event is called:
    a. Reaction process
    b. Perception process
    c. **Perception–reaction process**
    d. Weaving process
    e. Merging process

31. The driver's perception–reaction time does <u>not</u> impact one of the following:
    a. Braking distance
    b. Sight distance
    c. Yellow phase at a signalized intersection
    d. **Green phase at a signalized intersection**
    e. All of the above

32. One of the following is <u>not</u> among the factors affecting the perception–reaction time:
    a. Driver's age
    b. Conditions under which driver is driving
    c. Complications of the event
    d. Existing environmental and road conditions
    e. **Speed**
33. One of the following is <u>not</u> among kinematic characteristics of vehicles:
    a. Acceleration
    b. Speed
    c. **Size**
    d. All of the above
    e. None of the above
34. One of the following is <u>not</u> among dynamic characteristics of vehicles:
    a. Air resistance
    b. Grade resistance
    c. Rolling resistance
    d. Curve resistance
    e. **Weight**
35. The size of a vehicle is an important characteristic in determining the following <u>except</u> one, that is:
    a. Parking space
    b. Lane width
    c. Shoulder width
    d. **Pavement structural design**
    e. Length of vertical curve
36. The weight of a vehicle is an important characteristic considered in the design of:
    a. Parking space
    b. Lane width
    c. Pavement structure
    d. Maximum road grade
    e. **c and d**
37. The American Association of State Highway and Transportation Officials (AASHTO) classify vehicles into three main categories, which are:
    a. Motorcycles, bicycles, and passenger cars
    b. Motorcycles, passenger cars, and trucks
    c. Passenger cars, light-weight trucks, and heavy-weight trucks
    d. **Passenger cars, buses/recreational vehicles, and trucks**
    e. Motorcycles, passenger cars, and buses
38. The equivalent hourly rate at which vehicles are passing a point on a roadway during a time period less than one hour is called:
    a. Volume
    b. Density
    c. **Flow**
    d. Speed
    e. Capacity
39. The number of vehicles per unit length of roadway at a specific time is called:
    a. Volume
    b. **Density**
    c. Flow
    d. Speed
    e. Capacity

40. The graph that illustrates the relationship between the position of vehicles in traffic stream and time is called:
    a. Speed–flow diagram
    b. Density–speed diagram
    c. Flow–density diagram
    d. **Time–space diagram**
    e. Time–speed diagram
41. The arithmetic mean of speeds of vehicles on a highway at a specific time is called:
    a. **Time mean speed**
    b. Space mean speed
    c. Design speed
    d. Speed limit
    e. 90th percentile of speed
42. The harmonic mean of speeds of vehicles during a specific time interval on a highway is called:
    a. Time mean speed
    b. **Space mean speed**
    c. Design speed
    d. Speed limit
    e. 90th percentile of speed
43. Time speed compared to space mean speed is always:
    a. Lower
    b. Equivalent
    c. **Higher**
    d. 50%
    e. Double
44. The difference between the time arrivals of two successive vehicles at a point on a roadway is called:
    a. Passing time
    b. Travel time
    c. Reaction time
    d. Space headway
    e. **Time headway**
45. Greenshields model can be used for the following traffic conditions:
    a. Light traffic only
    b. Dense traffic only
    c. **a or b**
    d. Shockwaves
    e. Bottleneck conditions
46. Greenberg's model can be used for the following traffic conditions:
    a. Light traffic only
    b. **Dense traffic only**
    c. a or b
    d. Shockwaves
    e. Bottleneck conditions
47. A sudden reduction of the capacity of a highway due to a certain event is called:
    a. Jammed condition
    b. Shockwave in traffic stream
    c. Bottleneck conditions
    d. **b or c**
    e. All of the above

48. The process in which a vehicle in a traffic stream joins another traffic stream traveling in the same direction is called:
    a. **Merging**
    b. Diverging
    c. Weaving
    d. Gap acceptance
    e. Gap

49. The process in which a vehicle in a traffic stream leaves the traffic stream is called:
    a. Merging
    b. **Diverging**
    c. Weaving
    d. Gap acceptance
    e. Gap

50. The process in which a vehicle first joins a traffic stream, traverses the stream, and then merges into another stream moving in the same direction is called:
    a. **Merging**
    b. Diverging
    c. Weaving
    d. Gap acceptance
    e. Gap

51. The difference between the distance a merging vehicle is away from a reference point in the area of merging and the distance a vehicle in the main stream is away from the same point at a specific time is called:
    a. Time gap
    b. Critical gap
    c. **Space gap**
    d. Clearance
    e. None of the above

52. According to Raff, the definition "the number of accepted gaps shorter than it is equal to the number of rejected gaps longer than it" refers to:
    a. Time gap
    b. **Critical gap**
    c. Space gap
    d. Clearance
    e. None of the above

53. The maximum number of vehicles a roadway can carry at a specific time is called:
    a. Density
    b. Flow
    c. Volume
    d. **Capacity**
    e. Flow rate

54. The number of vehicles passing a point on a highway during a time period less than one hour (represented as equivalent hourly rate) is called:
    a. Density
    b. Flow
    c. Volume
    d. Capacity
    e. **Flow rate**

55. The ratio between the traffic volume and the maximum equivalent hourly rate within a given one-hour period is called:
    a. **Peak-hour factor (PHF)**

b. Flow rate

c. Level of service (LOS)

d. Peak-hour volume

e. Volume-to-capacity ratio

56. Traffic conditions as well as roadway conditions impact the level of service (LOS) of a transportation facility; one of the following roadway factors does <u>not</u> affect the LOS:

a. Lane width

b. Lateral obstruction

c. Grade

d. Number of lanes

e. **Number of horizontal curves**

57. One of the following traffic factors does <u>not</u> impact the LOS of a highway:

a. Traffic composition (mix)

b. Speed

c. Drivers familiarity with the highway

d. Heavy vehicles

e. **None of the above**

58. A divided roadway with complete access control and a minimum of two lanes in each direction is called:

a. Divided highway

b. Multi-lane highway

c. **Freeway**

d. Two-lane two-way highway

e. Four-lane highway

59. A two-lane two-way is a highway with:

a. A total of four lanes for the two directions

b. **A total of two lanes (one lane in each direction)**

c. A total of two lanes in one direction

d. A total of four lanes in one direction

e. None of the above

60. Intersections having traffic flow crossing with interruption is called:

a. Signalized intersections

b. Unsignalized intersections

c. Roundabouts

d. Interchanges

e. **a or b or c**

61. Intersections having traffic flow crossing <u>without</u> interruption is called:

a. Signalized intersections

b. Unsignalized intersections

c. Roundabouts

d. **Interchanges**

e. At-grade intersections

62. Traffic control devices are mainly needed for:

a. **At-grade intersections**

b. Interchanges

c. Grade-separated intersections without ramps

d. High-volume intersections

e. All of the above

63. The most desired two goals that a traffic control device should achieve at intersections are:

a. Safety and uniformity

b. **Safety and efficiency**

   c. Efficiency and uniformity
   d. Operation and uniformity
   e. Safety and operation
64. MUTCD stands for:
   a. Manual of Uniform Traffic Control Devices
   b. **Manual on Uniform Traffic Control Devices**
   c. Manual of Uninterrupted Traffic Control Devices
   d. Manual on Uninterrupted Traffic Control Devices
   e. Manual on Uniform Traffic Capacity and Density
65. One of the following is <u>not</u> among intersection traffic control systems:
   a. Stop signs
   b. Yield signs
   c. Traffic signals
   d. Raised barriers
   e. **Electronic camera**

Part II

Pavement Materials,
Analysis, and Design

6 Terminology

Chapter 6 includes questions and problems that cover the terminology, concepts, and theory used in pavement engineering. A pavement engineering practitioner should have the minimum understanding and grasp of the terms used in pavement engineering as well as knowledge and perception of the concepts and theory behind this type of engineering. The questions will shed light on this aspect, which will comprise the major themes of pavement engineering in this book. The questions in multiple choice format and their answers are available at the end of the part.

7 Important Properties of Subgrade, Subbase, and Base Materials

Chapter 7 includes questions and practical problems that cover the key properties of subgrade, subbase, and base materials including the California bearing ratio (CBR), resilient modulus, stabilometer R-value, dry density, and moisture content. The unbound materials are used in the underlying layers of the pavement structure, which provide strength and support to the pavement, serve as drainage layers and prevent frost protection, and minimize settlements in the pavement. The underlying layers of the pavement increase the structural capacity of the pavement to carry the traffic loads during the design life of the pavement structure and distribute the traffic loading to the subgrade soil. Therefore, the properties of the unbound materials used in these layers are of high significance to pavement engineers. In this chapter, a proper understanding of these properties will be achieved through presenting practical problems on this aspect.

7.1 If the R-value (from the Hveem stabilometer test) of a soil material is 15, determine the California Bearing Ratio (CBR) of the material.

Solution:

$$M_R = 1155 + 555R \qquad (7.1)$$

Where:
M_R = resilient modulus
R = Hveem stabilometer resistance value

⇒

$$M_R = 1155 + 555(15) = 9840 \text{ psi } (1427 \text{ kPa})$$

$$M_R = 1500(\text{CBR}) \qquad (7.2)$$

Where:
M_R = resilient modulus
CBR = California bearing ratio

⇒

$$\text{CBR} = \frac{M_R}{1500}$$

$$\text{CBR} = \frac{9840}{1500} = 6.3$$

The computations of this problem are conducted using an MS Excel worksheet. An image of the MS Excel worksheet used is shown in Figure 7.1.

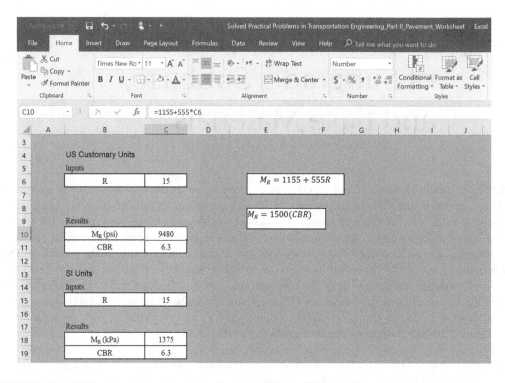

FIGURE 7.1   MS Excel worksheet image for the computations of Problem 7.1.

7.2   In a triaxial test for resilient modulus of a granular material, the confining pressure was 15
psi (103.4 kPa) and the bulk stress was 47 psi (324.1 kPa), determine the deviator stress.

Solution:

Given:
   $\theta = 47$ psi, and $\sigma_3 = 15$ psi

$$\theta = \sigma_1 + 2\sigma_3 \tag{7.3}$$

$$\sigma_d = \sigma_1 - \sigma_3 \tag{7.4}$$

$\Rightarrow$

$$\theta = \sigma_d + 3\sigma_3$$

Or:

$$\sigma_d = \theta - 3\sigma_3$$

And therefore,

$$\sigma_d = 47 - 3(15) = 2 \text{ psi } (13.8 \text{ kPa})$$

7.3   In Problem 7.2 above, determine the axial stress (σ_1).

Solution:

$$\theta = \sigma_1 + 2\sigma_3$$

$\Rightarrow$

$$\sigma_1 = \theta - 2\sigma_3$$

$$\sigma_1 = 47 - 2(15) = 17 \text{ psi } (117.2 \text{ kPa})$$

The computations and results of the MS Excel worksheet used for Problems 7.2 and 7.3 are shown in Figure 7.2.

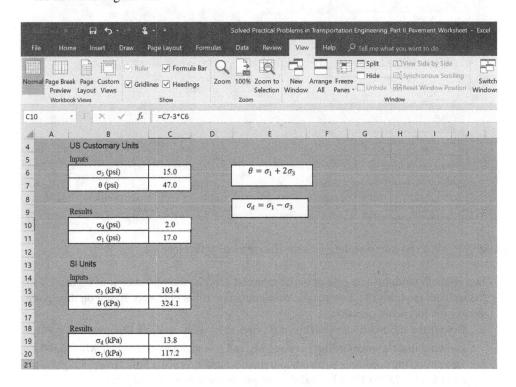

FIGURE 7.2 MS Excel worksheet image for the computations of Problems 7.2 and 7.3.

7.4 In Problem 7.2 above, if the recoverable strain for the specimen is 140 microstrains, then the resilient modulus of the specimen is:

Solution:

$$M_R = \frac{\sigma_d}{\varepsilon_r} \tag{7.5}$$

Where:
M_R = resilient modulus
ε_r = resilient (recoverable) strain

$\Rightarrow$

$$M_R = \frac{2}{140 \times 10^{-6}} = 14286 \text{ psi } (98500 \text{ kPa})$$

The MS Excel worksheet used to compute the results of this problem is shown in Figure 7.3.

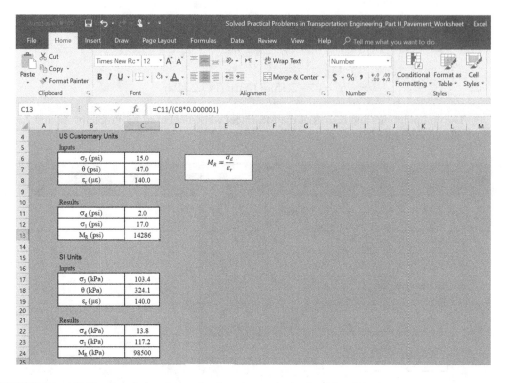

FIGURE 7.3 MS Excel worksheet image for the computations of Problem 7.4.

7.5 In a resilient modulus test of a granular material sample, the plot shown in Figure 7.4 for
the resilient modulus (M_R) versus the stress invariant (θ) is obtained. Based on this figure,
answer the following questions:
   a. Determine the nonlinear coefficient k_1 and the exponent k_2 of the granular material
knowing that the relationship between M_R and θ is given by the model $M_R = k_1\theta^{k_2}$.
   b. Determine the confining pressure and the deviator stress at a bulk stress of 8 psi if the
axial load is 75 lb (333.8 N) and the sample diameter = 4.0 in (10.2 cm).

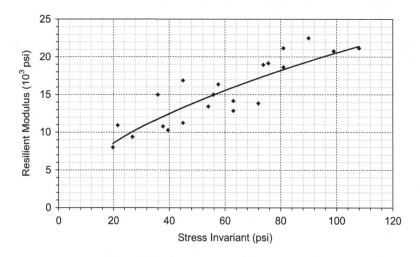

FIGURE 7.4 Stress invariant versus resilient modulus (Problem 7.5).

Solution:

a. By taking any two arbitrary points (the points with a circle legend) on the curve and computing the slope as the difference between log (resilient modulus value) divided by the difference in log (stress invariant), the value of k_2 is obtained (see Figure 7.5):

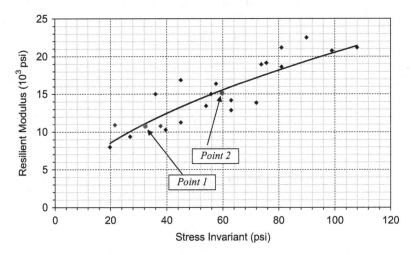

FIGURE 7.5 The selected two points on the curve (Problem 7.5).

Point 1 (32, 11000), Point 2 (60, 15500)
k_2 = slope of log-log relationship

$$k_2 = \frac{\log M_{R2} - \log M_{R1}}{\log \theta_2 - \log \theta_1} \tag{7.6}$$

Where:
M_{R1} and M_{R2} = resilient modulus of points 1 and 2, respectively.
θ_1 and θ_2 = bulk stress (stress invariant) of points 1 and 2, respectively.

$\Rightarrow$

$$k_2 = \frac{\log(15500) - \log(11000)}{\log(60) - \log(32)} = 0.546$$

k_1 can be determined by substituting k_2 in the given model and using any of the two points as shown below:

$$M_R = k_1 \theta^{k_2} \tag{7.7}$$

$\Rightarrow$

$$15500 = k_1(60)^{0.546}$$

$\Rightarrow$

$$k_1 = 1661$$

b. $\sigma_1 = \dfrac{\text{Axial Load}}{\text{Circular Area}}$

$$\sigma_1 = \frac{\text{Axial Load}}{\text{Circular Contact Area}} \tag{7.8}$$

⇒

$$\sigma_1 = \frac{75}{\left(\dfrac{\pi(4)^2}{4}\right)} = 6.0 \text{ psi (41.2 kPa)}$$

$$\theta = \sigma_1 + 2\sigma_3$$

⇒

$$\sigma_3 = \frac{(\theta - \sigma_1)}{2}$$

⇒

$$\sigma_3 = \frac{(8-6)}{2} = 1.0 \text{ psi (6.9 kPa)}$$

$$\sigma_d = \sigma_1 - \sigma_3$$

⇒

$$\sigma_d = 6.0 - 1.0 = 5.0 \text{ psi (34.1 kPa)}$$

The MS Excel worksheet used for rapid and efficient solution is shown in Figure 7.6.

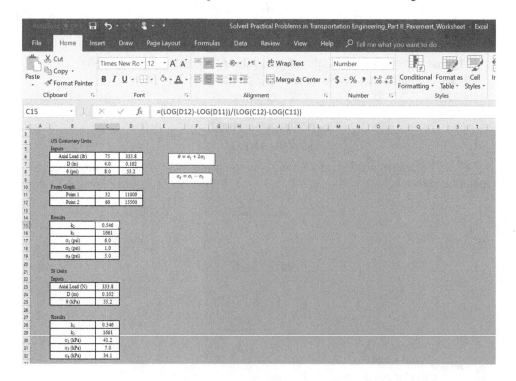

FIGURE 7.6   MS Excel worksheet image for the computations of Problem 7.5.

7.6 The results from a confined resilient modulus test of a granular soil material are given in Table 7.1.

TABLE 7.1

Bulk Stress Versus Resilient Modulus of a Granular Soil Material

M_R (10^3 psi)	Bulk Stress, θ (psi)
18.9	73.8
19.1	75.6
21.2	81.0
22.5	90.0
20.8	99.0
21.2	108.0
15.0	55.8
16.4	57.6
14.2	63.0
13.8	72.0
18.6	81.0
10.8	37.8
10.3	39.6
11.3	45.0
13.4	54.0
12.9	63.0
8.0	19.8
10.9	21.6
9.4	27.0
15.0	36.0
16.9	45.0

Plot the data to determine the nonlinear coefficient K_1 and the exponent K_2 of the granular material knowing that the relationship between M_R and θ is given by the model $M_R = k_1 \theta^{k_2}$.

Solution:

The data of the resilient modulus (M_R) and the bulk stress (θ) is plotted using MS Excel and a trend regression line is added to the scatter plot of the power type as shown in Figure 7.7.

By displaying the equation and the coefficient of determination (r^2) in MS Excel worksheet, the following model, which is also shown in the figure, is obtained:

$$M_R = 1.6691(\theta)^{0.545}$$

Note that the resilient modulus (M_R) values are multiplied by 10^3.

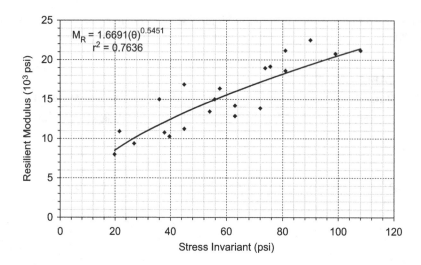

FIGURE 7.7 Stress invariant versus resilient modulus for a granular soil.

7.7 In a triaxial compression test of a granular material specimen of 4-inch diameter to determine the resilient modulus, the relationship between M_R and the bulk stress (θ) is given as $\log M_R = 4.00 + 0.5 \log \theta$. At a deviator stress and bulk stress of 10 psi (69.0 kPa) and 55 psi (379.2 kPa), respectively, determine the following:
   a. The axial compressive stress
   b. The axial load
   c. The resilient modulus
   d. The confining pressure
   e. The recoverable strain in the specimen

Solution:

Given:
   $D = 4$ inches (0.102 m), $\log M_R = 4.00 + 0.5 \log \theta$, $\sigma_d = 10$ psi (69 kPa),   and
   $\theta = 55$ psi (379.2 kPa)

a. The axial compressive stress (σ_1):

$$\theta = \sigma_1 + 2\sigma_3$$

$$\sigma_d = \sigma_1 - \sigma_3$$

$\Rightarrow$

$$\theta = 2\sigma_1 - \sigma_d$$

$\Rightarrow$

$$\sigma_1 = \frac{\theta + 2\sigma_d}{3}$$

Therefore:

$$\sigma_1 = \frac{55 + 2(10)}{3} = 25 \text{ psi (172.4 kPa)}$$

b. The axial load:

$$\sigma_1 = \frac{\text{Axial Load}}{\text{Circular Contact Area}}$$

$\Rightarrow$

$$\text{Axial Load} = \sigma_1(\text{Contact Area})$$

$\Rightarrow$

$$\text{Axial Load} = 25\left(\pi \frac{(4)^2}{4}\right) = 314.2 \text{ lb (1398 N)}$$

c. The resilient modulus (M_R):

$$\log M_R = 4.00 + 0.5 \log \theta$$

$\Rightarrow$

$$M_R = 10^{(4.00 + \log(55))} = 74162 \text{ psi } \left(511347 \text{ kPa} \cong 511.3 \text{ MPa}\right)$$

d. The confining pressure (σ_3):

$$\theta = \sigma_1 + 2\sigma_3$$

$\Rightarrow$

$$\sigma_3 = \frac{\theta - \sigma_1}{2}$$

$\Rightarrow$

$$\sigma_3 = \frac{55 - 25}{2} = 15 \text{ psi (103.4 kPa)}$$

e. The recoverable strain in the specimen (ε_r):

$$M_R = \frac{\sigma_d}{\varepsilon_r}$$

$\Rightarrow$

$$\epsilon_r = \frac{\sigma_d}{M_R}$$

$\Rightarrow$

$$\epsilon_r = \frac{10}{74162} = 135 \times 10^{-6} \text{ in/in} = 135 \ \mu\varepsilon$$

The solution of this problem is also done using the MS Excel worksheet shown in Figure 7.8.

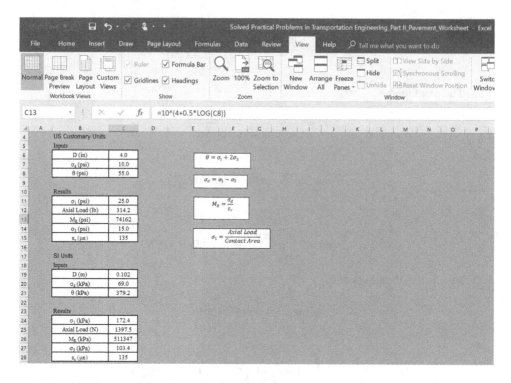

FIGURE 7.8 MS Excel worksheet image for the computations of Problem 7.7.

7.8 The results of a triaxial test of a granular base material to determine the resilient modulus are plotted in Figure 7.9. Based on these results, determine the following:

a. The nonlinear coefficient k_1 of the material if the model that describes the behavior of the material is $M_R = k_1\theta^{k_2}$.

b. The exponent k_2 of the material in the model $M_R = k_1\theta^{k_2}$.

c. If the axial load applied on the specimen is 250 lb (1112.5 N) and the specimen diameter is 4 inches (10.2 cm), then compute the confining pressure and the deviator stress at a stress invariant of 50 psi (344.8 kPa).

d. The resilient modulus in Part c above.

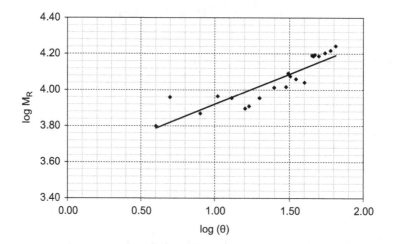

FIGURE 7.9 Bulk stress versus resilient modulus for a granular base material (log–log scale).

Solution:

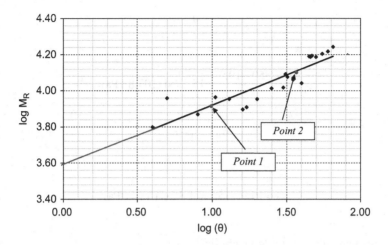

FIGURE 7.10 The two points selected on the line (Problem 7.8).

a. The nonlinear coefficient k_1 of the material:

The nonlinear coefficient k_1 of the material can be determined from the intercept of the line that represents the relationship between $\log \theta$ and $\log M_R$. By extending the line as shown in Figure 7.10, the line intersects the y-axis at 3.58. Hence,

$$k_1 = 10^{3.58} = 3802.$$

b. The exponent k_2 of the material:

The exponent k_2 is determined from the slope of the relationship between $\log \theta$ and $\log M_R$ since the original model is a power model $M_R = k_1\theta^{k_2}$. Therefore, by taking two arbitrary points on the line (Point 1 and Point 2) in Figure 7.10, the slope can be computed as shown below:

Point 1 (1, 3.92)
Point 2 (1.5, 4.12)

$\Rightarrow$

$$k_2 = \text{Slope} = \frac{4.12 - 3.92}{1.5 - 1.0} = 0.40.$$

In other words, the model between M_R and θ is:

$$M_R = 3802(\theta)^{0.40}$$

c. If the axial load applied on the specimen is 250 lb and the specimen diameter is 4 inches, then compute the confining pressure and the deviator stress at a stress invariant of 50 psi:

$$\sigma_1 = \frac{\text{Axial Load}}{\text{Contact Area}}$$

$\Rightarrow$

$$\sigma_1 = \frac{250}{\left(\pi \dfrac{(4)^2}{4}\right)} = 19.9 \text{ psi } (137.2 \text{ kPa})$$

$$\theta = \sigma_1 + 2\sigma_3$$

⇒

$$\sigma_3 = \frac{\theta - \sigma_1}{2}$$

$$\sigma_3 = \frac{50 - 19.9}{2} = 15.2 \text{ psi } (103.8 \text{ kPa})$$

$$\sigma_d = \sigma_1 - \sigma_3$$

⇒

$$\sigma_d = 19.9 - 15.1 = 4.8 \text{ psi } (33.4 \text{ kPa})$$

d. The resilient modulus in Part c above:

$$M_R = 3802(\theta)^{0.40}$$

⇒

$$M_R = 3802(50)^{0.40} = 18180 \text{ psi } (125353 \text{ kPa} \cong 125.4 \text{ MPa})$$

The MS Excel worksheet shown in Figure 7.11 is used to solve the problem for rapid solution.

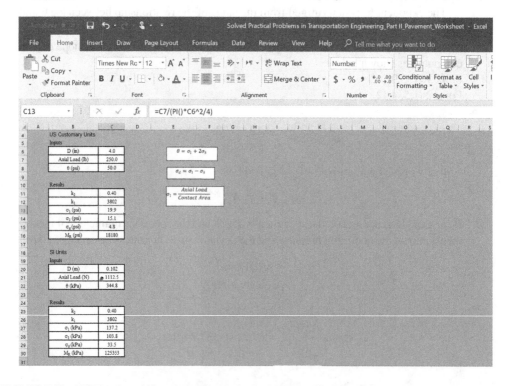

FIGURE 7.11   MS Excel worksheet image for the computations of Problem 7.8.

7.9 In a confined triaxial compression test of a granular material to determine the resilient modulus, the relationship between M_R and the bulk stress (θ) is described by the k_1-k_3 universal model being used in the Mechanistic-Empirical Pavement Design Guide (M-EPDG) $M_R = k_1(\theta)^{k_2}(\sigma_d)^{k_3}$. If the test data are given (see Table 7.2), use the multiple non-linear regression analysis techniques to:
a. Develop the non-linear model
b. Determine the material-related regression coefficients k_1, k_2, and k_3
c. Compute the coefficient of determination (r^2) of the model
d. Plot the relationship between resilient modulus (M_R) and total stress (θ)

TABLE 7.2
Resilient Modulus Test Data for a Granular Material

Confining Pressure, σ_3 (psi)	Deviator Stress, σ_d (psi)	Elastic Strain, ε_r ($\mu\varepsilon$)	M_R (psi)	Bulk Stress, θ (psi)
20	1.0	114.0	9115	73.8
	2.0	225.6	18615	75.6
	5.0	510.0	20120	81.0
	10.0	960.0	21915	90.0
	15.0	1560.0	22870	99.0
	20.0	2040.0	21435	108.0
15	1.0	144.0	12315	55.8
	2.0	264.0	14360	57.6
	5.0	762.0	12175	63.0
	10.0	1560.0	12385	72.0
	15.0	1740.0	20645	81.0
	20.0	1920.0	22175	90.0
10	1.0	200.4	10780	37.8
	2.0	420.0	9285	39.6
	5.0	960.0	11245	45.0
	10.0	1608.0	12450	54.0
	15.0	2520.0	12860	63.0
5	1.0	270.0	11575	19.8
	2.0	396.0	7615	21.6
	5.0	1152.0	8665	27.0
	10.0	1440.0	15425	36.0
	15.0	1920.0	16625	45.0
1	1.0	270.0	9870	19.8
	2.0	396.0	12980	21.6
	5.0	1152.0	9315	27.0
	7.5	1440.0	14560	36.0
	10.0	1920.0	17745	45.0

Solution:

a. Develop the non-linear model:

The MS Excel solver tool is utilized to develop the model and to determine the regression coefficients k_1, k_2, and k_3. In order to do that, initial values will have to be used for the three coefficients (k_1, k_2, and k_3) as shown in the Excel sheet shown in Figure 7.12. Excel solver uses numerical iterative approach to minimize the error associated with the developed model and provide optimum values for these coefficients. The error is simply defined as:

$$\text{Error} = \left(\left(\log(M_R) \right) - \log \left(M_{\text{R-Predicted}} \right) \right)^2 \tag{7.9}$$

Where:

$$M_{\text{R-Predicted}} = k_1 (\theta)^{k_2} (\sigma_d)^{k_3}$$

The sum of the errors is obtained for all test data points (see Table 7.3 and Figure 7.12).

TABLE 7.3
The Predicted Resilient Modulus Values with the Associated Errors

Confining Pressure, σ_3 (psi)	Deviator Stress, σ_d (psi)	Elastic Strain, ε_r ($\mu\varepsilon$)	M_R (psi)	Bulk Stress, θ (psi)	Predicted M_R (psi)	Error
20	1.0	114.0	9115	73.8	13162	2.5458E-02
	2.0	225.6	18615	75.6	14334	1.2883E-02
	5.0	510.0	20120	81.0	16238	8.6679E-03
	10.0	960.0	21915	90.0	18153	6.6918E-03
	15.0	1560.0	22870	99.0	19587	4.5292E-03
	20.0	2040.0	21435	108.0	20802	1.6929E-04
15	1.0	144.0	12315	55.8	12029	1.0448E-04
	2.0	264.0	14360	57.6	13132	1.5073E-03
	5.0	762.0	12175	63.0	14975	8.0842E-03
	10.0	1560.0	12385	72.0	16894	1.8181E-02
	15.0	1740.0	20645	81.0	18361	2.5919E-03
	20.0	1920.0	22175	90.0	19616	2.8354E-03
10	1.0	200.4	10780	37.8	10611	4.7207E-05
	2.0	420.0	9285	39.6	11639	9.6332E-03
	5.0	960.0	11245	45.0	13438	5.9857E-03
	10.0	1608.0	12450	54.0	15399	8.5252E-03
	15.0	2520.0	12860	63.0	16934	1.4284E-02
5	1.0	270.0	11575	19.8	8616	1.6436E-02
	2.0	396.0	7615	21.6	9576	9.8993E-03
	5.0	1152.0	8665	27.0	11400	1.4190E-02
	10.0	1440.0	15425	36.0	13514	3.2978E-03
	15.0	1920.0	16625	45.0	15195	1.5255E-03
1	1.0	270.0	9870	19.8	8616	3.4807E-03
	2.0	396.0	12980	21.6	9576	1.7453E-02
	5.0	1152.0	9315	27.0	11400	7.6925E-03
	7.5	1440.0	14560	36.0	13086	2.1474E-03
	10.0	1920.0	17745	45.0	14521	7.5815E-03
	Sum of Errors					**2.1388E-01**

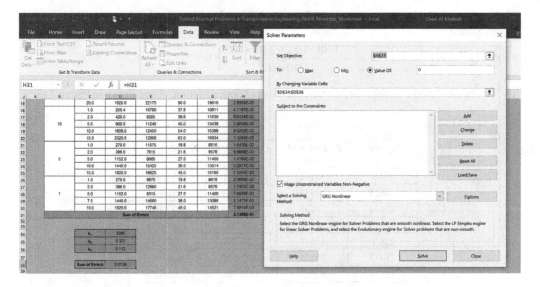

FIGURE 7.12   An Image of the MS Excel Solver and computations of Problem 7.9.

b. Determine the material-related regression coefficients k_1, k_2, and k_3:

The values of k_1, k_2, and k_3 are determined using MS Excel solver tool as shown in Figure 7.12. Numerical iterative approach is followed in the MS Excel solver tool to solve for the three regression coefficients by minimizing the sum of the errors associated with the model.

In this case,

$k_1 = 3295$

$k_2 = 0.322$

$k_3 = 0.112$

c. Compute the coefficient of determination, r^2 of the model.

The final model that fits the given resilient modulus data is:

$$M_{R\text{-Predicted}} = 3295(\theta)^{0.322}(\sigma_d)^{0.112}$$

The coefficient of determination, r^2 of the model is computed using the following formula:

$$r^2 = \frac{S_t - S_r}{S_t} \tag{7.10}$$

Where:

S_t = total sum of squares around the mean $(\bar{y})$

$$S_t = \sum_{i=1}^{n}(y_i - \bar{y})^2 \tag{7.11}$$

S_r = sum of squares of residuals (errors) around the regression model.

y_i = given (measured) resilient modulus data point.

$$S_r = \sum_{i=1}^{n}(y_i - y_{i\text{-Predicted}})^2 \tag{7.12}$$

$y_{i\text{-Predicted}}$ = calculated (predicted) resilient modulus value from the developed model.

Hence,

$$r^2 = \frac{1.40 \times 10^{11} - 2.08 \times 10^8}{1.40 \times 10^{11}} = 0.999$$

The MS Excel computations for the coefficient of determination, r^2, are shown in Figure 7.13.

	M_R (psi)	Predicted M_R (psi)	$(M_R\text{-}M_{R\text{-Avg.}})^2$	$(M_R\text{-}M_{R\text{-Predicted}})^2$
	9115	13162	28056640	16375479
	18615	14334	17666454	18329604
	20120	16238	32582955	15072010
	21915	18153	56297232	14156323
	22870	19587	71540270	10778894
	21435	20802	49324610	400359
	12315	12029	4396788	82062
	14360	13132	2689	1508049
	12175	14975	5003506	7842690
	12385	16894	4108128	20330777
	20645	18361	38852136	5215408
	22175	19616	60266469	6547632
	10780	10611	13190348	28623
	9285	11639	26284610	5543339
	11245	13438	10028951	4808272
	12450	15399	3848863	8698431
	12860	16934	2408244	16597047
	11575	8616	8047728	8753715
	7615	9576	46197195	3844108
	8665	11400	33026306	7478380
	15425	13514	1026469	3650035
	16625	15195	4898025	2044482
	9870	8616	20628418	1571683
	12980	9576	2050200	11589670
	9315	11400	25977899	4345819
	14560	13086	21948	2171302
	17745	14521	11109877	10392380
Sum	389120		1.40E+11	2.08E+08
Average	14412			

S_t	1.40E+11
S_r	2.08E+08
r^2	0.999

$$r^2 = \frac{S_t - S_r}{S_t}$$

FIGURE 7.13 MS Excel computations to determine the coefficient of determination for Problem 7.9.

d. Plot the relationship between resilient modulus (M_R) and total stress (θ):

The relationship between the given resilient modulus values and the total (bulk) stress values is plotted using arithmetic scale as shown in Figure 7.14.

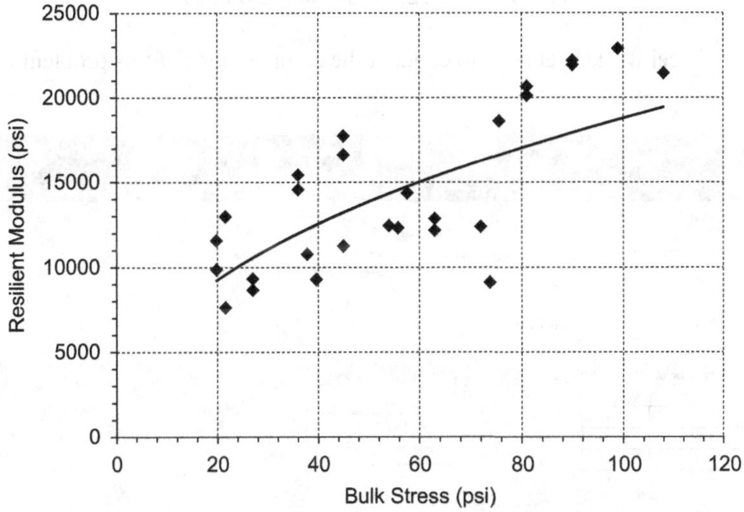

FIGURE 7.14 Bulk stress versus resilient modulus relationship for Problem 7.9.

7.10 A granular material specimen of 6-inch (152.4-mm) diameter is tested in a triaxial test to determine the resilient modulus. If the axial compressive load applied on the specimen is 276 lb (1228.2 N), and the stress invariant (bulk stress) is 20 psi (137.9 kPa), compute the axial stress, the confining pressure, and the deviator stress.

Solution:

$$\sigma_1 = \frac{\text{Axial Load}}{\text{Contact Area}}$$

$\Rightarrow$

$$\sigma_1 = \frac{276}{\left(\pi \dfrac{(6)^2}{4}\right)} = 9.76 \text{ psi (67.3 kPa)}$$

$$\theta = \sigma_1 + 2\sigma_3$$

$\Rightarrow$

$$\sigma_3 = \frac{\theta - \sigma_1}{2}$$

$$\sigma_3 = \frac{20 - 9.76}{2} = 5.12 \text{ psi (35.3 kPa)}$$

$$\sigma_d = \sigma_1 - \sigma_3$$

$\Rightarrow$

$$\sigma_d = 9.76 - 5.12 = 4.64 \text{ psi } (32.0 \text{ kPa})$$

The MS Excel worksheet used to conduct the computations of this problem is shown in Figure 7.15.

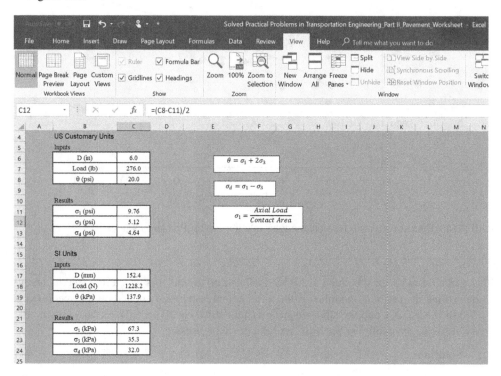

FIGURE 7.15 MS Excel worksheet image for the computations of Problem 7.10.

7.11 In a resilient modulus test of a subgrade soil material, the confining pressure is 12 psi (82.7 kPa), the axial stress is 30 psi (206.9 kPa), and the resilient strain is 715 microstrains, determine the resilient modulus of the material.

Solution:

$$\sigma_d = \sigma_1 - \sigma_3$$

$\Rightarrow$

$$\sigma_d = 30 - 12 = 18 \text{ psi } (124.1 \text{ kPa})$$

$$M_R = \frac{\sigma_d}{\varepsilon_r}$$

$\Rightarrow$

$$M_R = \frac{18}{715 \times 10^{-6}} = 25175 \text{ psi} \left(173580 \text{ kPa} \cong 173.6 \text{ MPa}\right)$$

The image in Figure 7.16 shows the MS Excel worksheet computations for the results of this problem.

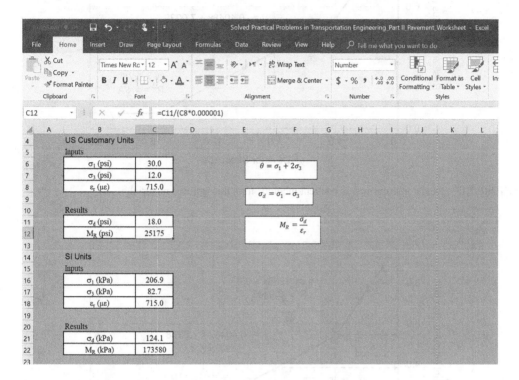

FIGURE 7.16 MS Excel worksheet image for the computations of Problem 7.11.

7.12 In a triaxial compression test of a fine-grained material to determine the resilient modulus, the results plotted in Figure 7.17 were obtained. Based on these results, determine the following:
   a. The coefficient k_1 of the material
   b. The coefficient k_2 of the material
   c. The coefficient k_3 of the material
   d. The coefficient k_4 of the material
   e. The recoverable strain at a deviator stress of 8 psi.

Solution:

   a. The coefficient k_1 of the material (see Figure 7.18):
      $k_1 = 5400$ from the figure; k_1 is the distance between the intersection point of the two linear lines and the x-axis.
   b. The coefficient k_2 of the material:
      $k_2 = 4$ from the figure; k_2 is the distance between the intersection point of the two linear lines and the y-axis.

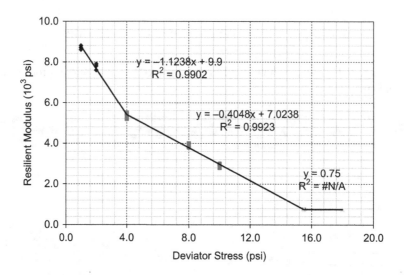

FIGURE 7.17 Deviator stress versus resilient modulus for a fine-grained material.

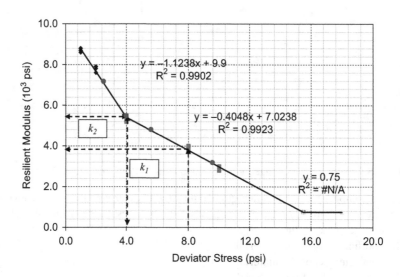

FIGURE 7.18 Determination of k1 and k2 graphically using the deviator stress versus resilient modulus relationship.

c. The coefficient k_3 of the material:

Again, by choosing two arbitrary points on the first upper portion of the curve, the slope k_3 can be estimated from these points:

Point 1 (2.4, 7200), Point 2 (4.0, 5400)

$$k_3 = \frac{(7200 - 5400)}{(4.0 - 2.4)} = 1124$$

d. The coefficient k_4 of the material:

By choosing two arbitrary points on the second lower portion of the curve, the slope k_4 can be estimated from these points:

Point 3 (5.6, 4800), Point 4 (9.6, 3180)

$$k_4 = \frac{(4800 - 3180)}{(9.6 - 5.6)} = 405$$

e. The recoverable strain at a deviator stress of 8 psi:

From the figure; when $\sigma_d = 8$ psi, $M_R = 3800$ psi.

$$M_R = \frac{\sigma_d}{\varepsilon_r}$$

$\Rightarrow$

$$\epsilon_r = \frac{\sigma_d}{M_R}$$

$$\epsilon_r = \frac{8}{3800} = 2105 \times 10^{-6} \text{ in/in} = 2105 \ \mu\epsilon$$

$\Rightarrow$

$$\varepsilon_r = \frac{\sigma_d}{M_R}$$

$$\varepsilon_r = \frac{8}{3800} = 2105 \times 10^{-6} \text{ in/in} = 2105 \ \mu\varepsilon$$

The MS Excel worksheet in Figure 7.19 shows the solution of this problem.

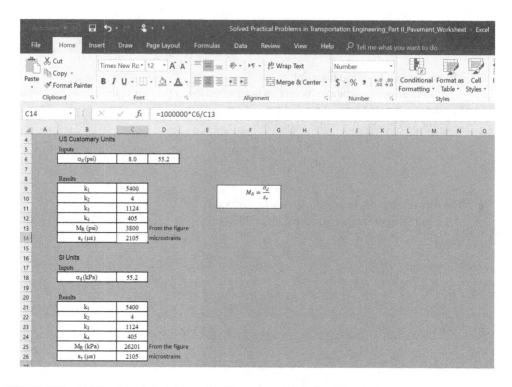

FIGURE 7.19 MS Excel worksheet image for the computations of Problem 7.12.

7.13 The CBR test results for a sand material used for the subgrade of a highway asphalt pavement are plotted in Figure 7.20. If the diameter of the piston applying the load on the sand sample is 5.0 cm (about 2 in), determine the CBR value of the sand subgrade (see Figure 7.20).

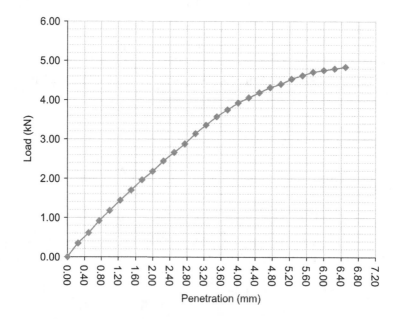

FIGURE 7.20 A Plot for the CBR test data for a subgrade sand material.

Solution:

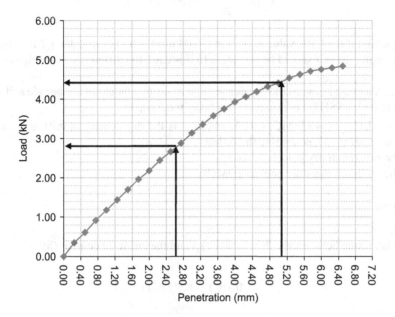

FIGURE 7.21 Determination of the CBR value using the CBR test penetration versus load relationship.

At a penetration of 0.10 inch (2.54 mm), the value of load is determined. From Figure 7.21, it is equal to 2.8 kN (629.2 lb).

$$CBR = \frac{\text{Pressure at } 0.10 \text{ in}}{\text{Standard Pressure at } 0.10 \text{ in} = 1000 \text{ psi}} \times 100 \qquad (7.13)$$

Or:

$$CBR = \frac{\text{Load at } 0.10 \text{ in}}{\text{Standard Load at } 0.10 \text{ in} = 13.2 \text{ kN}} \times 100 \qquad (7.14)$$

$\Rightarrow$

$$CBR = \frac{2.8}{13.2} \times 100 = 21.2\%$$

At a penetration of 0.20 inch (5.08 mm), the value of load is determined. From Figure 7.21, it is equal to 4.6 kN (1033.7 lb).

$$CBR = \frac{\text{Pressure at } 0.20 \text{ in}}{\text{Standard Pressure at } 0.20 \text{ in} = 1500 \text{ psi}} \times 100 \qquad (7.15)$$

Or:

$$CBR = \frac{\text{Load at } 0.20 \text{ in}}{\text{Standard Load at } 0.20 \text{ in} = 20.06 \text{ kN}} \times 100 \qquad (7.16)$$

$\Rightarrow$

$$CBR = \frac{4.4}{20.06} \times 100 = 21.9\%$$

Typically; if the CBR value at 0.20 in (50.08 mm) is higher than the CBR value at 0.10 in (25.4 mm), the test should be repeated. In this case, the difference is not significant.

7.14 If the relationship between the dry density and moisture content of a subgrade soil material in a standard AASHTO compaction test takes the form

$\gamma_d = -0.4(MC)^2 + 8.9(MC) + 94.3$, where γ_d is the dry density in lb/ft³ and MC is the moisture content (%), determine the following:
a.   The maximum dry density (max γ_d)
b.   The optimum moisture content (OMC)

Solution:

Given:
The relationship between the dry density (γ_d) and the moisture content (MC):

$$\gamma_d = -0.4(MC)^2 + 8.9(MC) + 94.3$$

a.   The maximum dry density (max γ_d):
      The relationship is plotted on arithmetic scale as shown in the graph in Figure 7.22. The maximum dry density (γ_d) is determined from the graph.
b.   The optimum moisture content (OMC):
      The optimum moisture content (OMC) is also determined from the graph, which is the moisture content on the x-axis at the maximum dry density.
      Another approach:
      The first derivative of the relationship is obtained, and the OMC is the value of MC when the derivative is zero:

$$-0.8(MC) + 8.9 = 0$$

$\Rightarrow$

$$OMC = \frac{8.9}{0.8} = 11.1\%$$

The maximum dry density is the dry density at the OMC, which is determined using the relationship:

$$\gamma_d = -0.4(MC)^2 + 8.9(MC) + 94.3$$

$\Rightarrow$

$$\gamma_d = -0.4(11.1)^2 + 8.9(11.1) + 94.3 = 143.8 \text{ lb/ft}^3 \text{ (2302.6 kg/m}^3)$$

The relationship between the moisture content and the dry density of the soil material is shown in Figure 7.22.

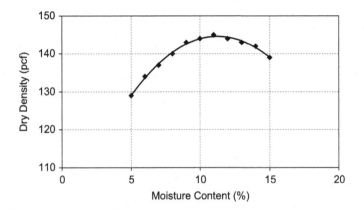

FIGURE 7.22 Moisture content versus dry density for a subgrade soil material using a standard compaction test.

The MS Excel worksheet used to solve this problem is shown in Figure 7.23.

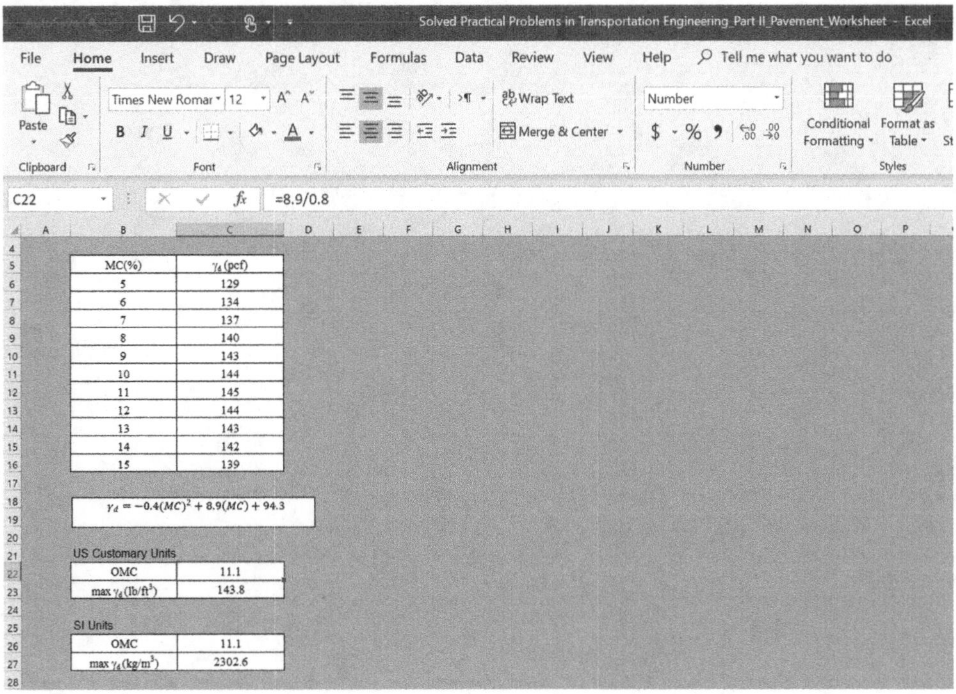

FIGURE 7.23 MS Excel worksheet image for the computations of Problem 7.14.

8 Superpave Aggregate Properties and Criteria

Chapter 8 includes questions and problems that cover the important aggregate properties, particularly the Superpave properties, minimum requirements, and criteria. Aggregate properties concern highway engineers as aggregate compose approximately 95% of the design of asphalt mixtures used in the construction of asphalt pavements in highways. Consequently, the control of these properties and the assurance of high-quality aggregate properties are crucial in order to obtain the best results in mixture design and construction. The questions and problems in the following section will encompass the key properties of aggregate used for highway construction.

8.1 In a flat and elongated (F&E) particles test for a coarse aggregate sample, the results shown in Table 8.1 are obtained. Determine the F&E value for this coarse aggregate sample, if this aggregate is to be utilized in the design of asphalt mixture that will be used in the construction of a road with 2 million equivalent single-axle loads (ESALs).

TABLE 8.1

Flat and Elongated Particles Test Data for a Coarse Aggregate Sample

Category	Flat	Elongated	Flat and Elongated	Neither Flat nor Elongated
Count of Aggregate Particles	6	4	8	82

Solution:

The percentage of F&E particles is calculated using the formula shown below:

$$\%F \& E = \frac{FE}{FE+N} \times 100\% \tag{8.1}$$

Where:
    FE = count or weight of particles that are flat, elongated, or flat and elongated,
    N = count or weight of particles that are neither flat, elongated, nor flat and elongated.

Therefore,

$$\%F \& E = \frac{6+4+8}{6+4+8+82} \times 100\% = 18\%$$

Since the traffic loading of the road is 2 million ESALs ≥ 1 million ESALs, according to Table 8.2 (the Superpave F&E Criteria), the maximum F&E value is 10%. The F&E value obtained in this problem (18%) is higher than 10%; and therefore, the coarse aggregate is rejected based on the F&E criteria.

TABLE 8.2

Superpave F&E Particles Criteria

Traffic (million ESALs)	Maximum, %
< 1	--
≥ 1	10

Reproduced with permission from *Superpave Mix Design SP-2*, 2001, by Asphalt Institute, Lexington, Kentucky, USA.

The MS Excel worksheet used to solve this problem is shown in Figure 8.1.

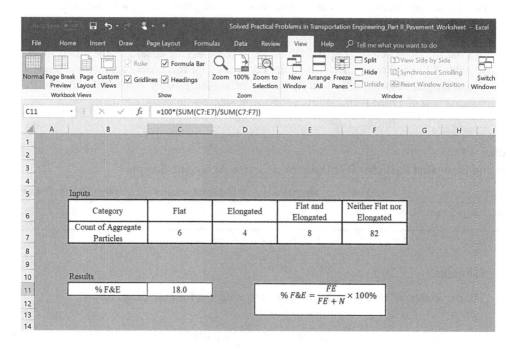

FIGURE 8.1 MS Excel worksheet image for the computations of Problem 8.1.

8.2 If the coarse aggregate angularity (CAA) of a coarse aggregate sample is recorded as 90/84, determine the following:
   a. The percentage of aggregate particles with zero fractured faces
   b. The percentage of aggregate particles with only one fractured face

Solution:

a. Since CAA=90/84, 90% of the coarse aggregate particles has one or more fractured faces, and 84% of coarse aggregate particles has two or more fractured faces. This concludes that 100−90=10% of aggregate particles has zero fractured faces.

b. In a similar manner: since 90% of the coarse aggregate particles has one or more fractured faces and 84% of coarse aggregate particles has two or more fractured faces, then 90−84=6% of aggregate particles has only one fractured face.

The MS Excel worksheet used to solve this problem is shown in Figure 8.2.

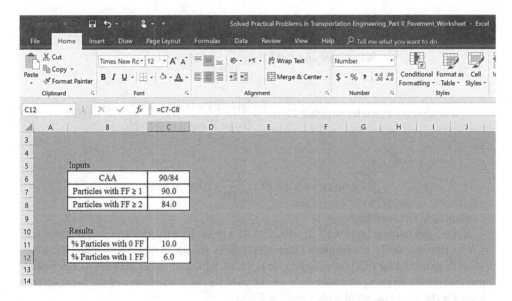

FIGURE 8.2 MS Excel worksheet image for the computations of Problem 8.2.

8.3   In a coarse aggregate angularity (CAA) test for a coarse aggregate sample, the results shown in Table 8.3 are obtained. Determine the CAA value for this coarse aggregate sample. If this coarse aggregate is to be used in the design of an asphalt mixture, can this mixture be used in the construction of a 5-cm surface course for a road pavement with 5 million ESALs?

TABLE 8.3

Coarse Aggregate Angularity (CAA) Test Data for a Coarse Aggregate Sample

FF*	0 FF	1 FF	2 FF	3 FF	4 FF
Weight of Aggregate Particles (g)	30	50	70	100	250

* **Note:** FF=Fractured Faces.

Solution:

The CAA value of aggregate is determined by calculating the percentage of aggregate particles having one or more fractured faces and the percentage of aggregate particles with two or more fractures faces. Hence, according to given data in the table, the percentage of particles with one or more fractured faces is calculated as below:

$$\text{Weight of Particles with} \geq 1 \text{ FF} = 50+70+100+250 = 470 \text{ g}$$

$$\text{Weight of Particles with} \geq 2 \text{ FF} = 70+100+250 = 420 \text{ g}$$

$$\text{Total Weight of Sample} = 30+50+70+100+250 = 500 \text{ g}$$

$$\%\text{Aggregate Particles with} \geq 1 \text{ FF} = \frac{\text{Weight of Particles with} \geq 1 \text{ FF}}{\text{Total Weight of Sample}} \times 100\% \qquad (8.2)$$

$\Rightarrow$

$$\%\text{Particles with} \geq 1 \text{ FF} = \frac{470}{500} \times 100\% = 94\%$$

$$\%\text{Aggregate Particles with} \geq 2 \text{ FF} = \frac{\text{Weight of Particles with} \geq 2 \text{ FF}}{\text{Total Weight of Sample}} \times 100\% \quad (8.3)$$

$\Rightarrow$

$$\%\text{Particles with} \geq 2 \text{ FF} = \frac{420}{500} \times 100\% = 84\%$$

Hence, % CAA = 94/84

For a 5-cm surface layer with 5 million ESALs (3–< 10 million ESALs), the Superpave CAA criteria specify a minimum value of 85/80 as shown in Table 8.4. Since the CAA value determined for this coarse aggregate is 94/84, which is higher than 85/80, the asphalt mixture produced using this aggregate can be used for the construction of the 5-cm surface layer for the road pavement with 5 million ESALs.

The MS Excel worksheet used to solve this problem is shown in Figure 8.3.

TABLE 8.4

Superpave CAA Criteria

	Depth from Surface	
Traffic (million ESALs)	< 100 mm	> 100 mm
< 0.3	55/--	--/--
0.3 – < 1	65/--	--/--
1 – < 3	75/--	50/--
3 – < 10	85/80	60/--
10 – < 30	95/90	80/75
30 – < 100	100/100	95/90
≥ 100	100/100	100/100

Reproduced with permission from *Superpave Mix Design SP-2*, 2001, by Asphalt Institute, Lexington, Kentucky, USA.

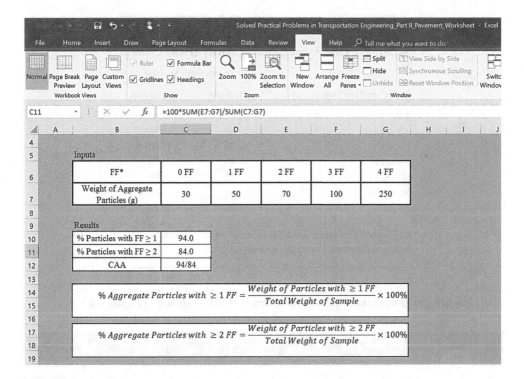

FIGURE 8.3 MS Excel worksheet image for the computations of Problem 8.3.

8.4 In a fine aggregate angularity (FAA) test, the mass of the fine aggregate sample is measured to be 145.0 g. If the bulk specific gravity of the fine aggregate is 2.452, then calculate the FAA (%) value of the aggregate. Based on this result, determine if this aggregate can be used for the same surface layer in Problem 3.3 above.

Solution:

$$\text{UVC} = \frac{V - \dfrac{M}{G_{sb}}}{V} \times 100\% \tag{8.4}$$

Where:
   UVC = uncompacted void content or FAA of fine aggregate sample (%),
   V = volume of cylinder used in the FAA test (standard volume = 100 cm³),
   M = mass of fine aggregate sample,
   G_{sb} = specific gravity of fine aggregate.

Therefore:

$$\text{UVC} = \frac{100 - \dfrac{145}{2.452}}{100} \times 100\% = 40.9\%$$

For a 5-cm surface layer with 5 million ESALs (3–< 10 million ESALs), the Superpave FAA criteria specify a minimum value of 45% as shown in Table 8.5. Since the FAA value determined for the fine aggregate in this problem is 40.9%, which is lower than the Superpave minimum requirement, the asphalt mixture produced using this aggregate is not suitable for the construction of this pavement layer.

TABLE 8.5

Superpave Fine Aggregate Angularity (FAA) Criteria

	Depth from Surface	
Traffic (million ESALs)	< 100 mm	> 100 mm
< 0.3	--	--
0.3 – < 1	40	--
1 – < 3	40	40
3 – < 10	45	40
10 – < 30	45	40
30 – < 100	45	45
≥ 100	45	45

Reproduced with permission from *Superpave Mix Design SP-2*, 2001, by Asphalt Institute, Lexington, Kentucky, USA.

The MS Excel worksheet used to solve this problem is shown in Figure 8.4.

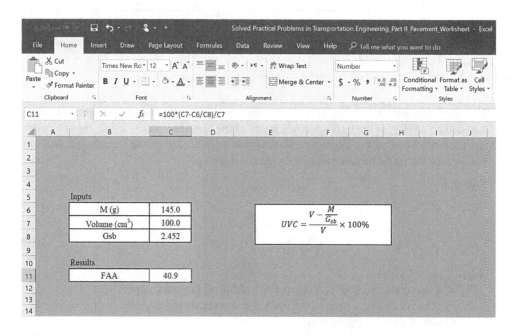

FIGURE 8.4 MS Excel worksheet image for the computations of Problem 8.4.

8.5 In a fine aggregate angularity (FAA) test, if the FAA is 50%, the volume of the cylinder is 100 cm³, and the bulk specific gravity of the aggregate is 2.590, determine the measured mass of the fine aggregate sample in the cylinder in grams.

Solution:

Using the same formula as above,

$$UVC = \frac{V - \dfrac{M}{G_{sb}}}{V} \times 100\%$$

$\Rightarrow$

$$M = \left(V - \frac{UVC \times V}{100}\right) G_{sb}$$

$\Rightarrow$

$$M = \left(100 - \frac{50 \times 100}{100}\right) 2.590 = 129.5 \text{ g } (0.286 \text{ lb})$$

The MS Excel worksheet for this problem is shown in Figure 8.5.

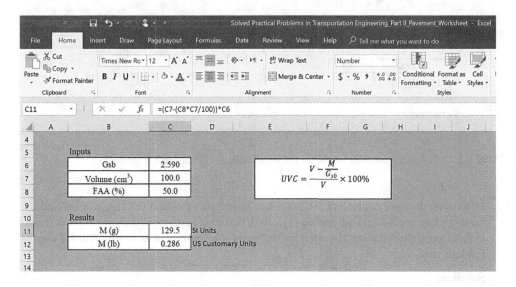

FIGURE 8.5 MS Excel worksheet image for the computations of Problem 8.5.

8.6 A 134-g sample of fine aggregate is tested for FAA to have uncompacted void content of 44%. Determine the specific gravity of the fine aggregate.

Solution:

Using the same formula

$$UVC = \frac{V - \dfrac{M}{G_{sb}}}{V} \times 100\%$$

$\Rightarrow$

$$G_{sb} = \frac{M}{\left(V - \dfrac{UVC \times V}{100}\right)}$$

$\Rightarrow$

$$G_{sb} = \frac{134}{\left(100 - \frac{44 \times 100}{100}\right)} = 2.393$$

The MS Excel worksheet used to solve this problem is shown in Figure 8.6.

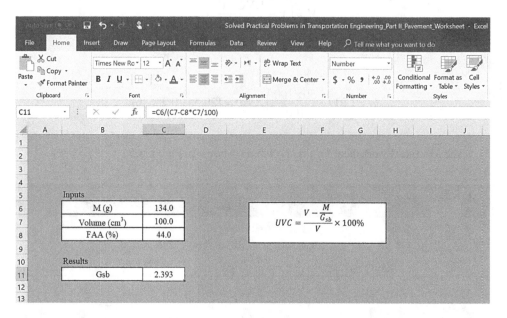

FIGURE 8.6 MS Excel worksheet image for the computations of Problem 8.6.

8.7 Determine the sand equivalent (SE) of a fine aggregate sample if the clay reading and the sand reading are 5.1 and 4.6 inches (11.7 and 13.0 cm), respectively. Based on this result, is this aggregate suitable to be used for the same road pavement as in Problem 8.3 above? Explain why.

Solution:

$$SE = \frac{\text{Sand Reading}}{\text{Clay Reading}} \times 100\% \tag{8.5}$$

Where:

SE = sand equivalent (%),

Sand reading = the reading of the sand in the SE graduated container (inches),

Clay reading = the reading of the clay in the SE graduated container (inches); this reading is composed of the sand reading plus the height of the clay in the container.

$\Rightarrow$

$$SE = \frac{4.6}{5.1} \times 100\% = 90.2\%$$

The Superpave SE criteria specify a minimum value of 45% for the SE at a traffic level of 3 – < 10 million ESALs as shown Table 8.6. Since the SE value obtained is 90.2%, which is higher than the Superpave minimum requirement (45%), the fine aggregate can be used for the specified surface layer.

TABLE 8.6

Superpave Sand Equivalent (SE) Criteria

Traffic (million ESALs)	Minimum SE Value (%)
< 3	40
3 – < 30	45
≥ 30	50

Reproduced with permission from *Superpave Mix Design SP-2*, 2001, by Asphalt Institute, Lexington, Kentucky, USA.

The MS Excel worksheet used to solve this problem is shown in Figure 8.7.

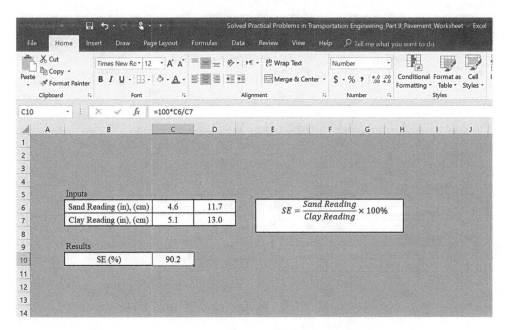

FIGURE 8.7 MS Excel worksheet image for the computations of Problem 8.7.

8.8 In a sand equivalent (SE) test for a fine aggregate sample, the SE is determined to be 60%. If the sand reading is 4.5 inches (11.4 cm), determine the clay height.

Solution:

$$SE = \frac{Sand\ Reading}{Clay\ Reading} \times 100\%$$

⇒

$$Clay\ Reading = \frac{Sand\ Reading \times 100}{SE}$$

$\Rightarrow$

$$\text{Clay Reading} = \frac{4.5 \times 100}{60} = 7.5 \text{ in } (19.1 \text{ cm})$$

But the clay reading is composed of the sand reading plus the clay height. Therefore:

$$\text{Clay Reading} = \text{Sand Reading} + \text{Clay Height} \qquad (8.6)$$

$\Rightarrow$

$$7.5 = 4.5 + \text{Clay Height}$$

$\Rightarrow$

$$\text{Clay Height} = 3.0 \text{ in } (7.6 \text{ cm})$$

The MS Excel worksheet used to solve this problem is shown in Figure 8.8.

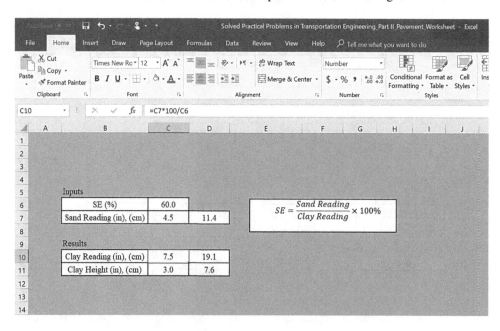

FIGURE 8.8 MS Excel worksheet image for the computations of Problem 8.8.

8.9 In Table 8.7, the percent passing for an aggregate blend is given. Determine the following:
   a. The nominal maximum aggregate size (NMAS)
   b. The maximum aggregate size of this aggregate
   c. The percentage of coarse aggregate
   d. The percentage of fine aggregate
   e. The percentage of filler material

TABLE 8.7

Gradation (Sieve Size and Percent Passing) for an Aggregate Blend

Sieve Size (mm)	% Passing
25.0	100
19.0	96
12.5	92
9.5	75
4.75	65
2.36	55
1.18	44
0.600	33
0.300	22
0.150	12
0.075	4

Solution:

a. The nominal maximum aggregate size (NMAS) is defined as the first sieve size above the sieve size that retains more than 10% (percent passing < 90%). The sieve size that retains more than 10% is 9.5 mm. The first sieve above this sieve is 12.5 mm. Hence, 12.5 mm is the NMAS.

b. The maximum aggregate size is the sieve size above the NMAS. Therefore, it is 19.0 mm.

c. The coarse aggregate is that portion of the material retaining on sieve No. 4 (4.75 mm). Therefore, the percentage of coarse aggregate in this case = 100−65 = 35%.

d. The fine aggregate is that part of the material passing sieve No. 4 (4.75 mm). Hence, the percentage of fine aggregate in this case = 65%.

e. The filler material is that portion passing sieve No. 200 (0.075 μm). In other words, it is equal to 4%.

9 Superpave Asphalt Binder Properties and Specifications

Chapter 9 is composed of questions and problems, which deal with the properties of asphalt binders including the Superpave properties and the specifications required for these properties. Despite that asphalt binder composes only about 5% of the asphalt mixture, but this material is considered an important part in the mixture due to the binding effect and the natural characteristics it has including durability, viscoelasticity, adhesion, and strength. Accordingly, the characterization of this material and the control of its properties are essential to optimize the properties of the asphalt mixture used for the construction of highway pavements. Although there have been lot of developments in the Superpave asphalt binder tests in the last decade including the Multiple Stress Creep Recovery (MSCR) test and the Linear Amplitude Sweep (LAS) test, but this chapter focuses on the main characterization tests required to evaluate the asphalt binder and its suitability for the use in asphalt paving mixtures based on the Superpave specifications and the well-known test methods. The Multiple Stress Creep Recovery (MSCR) test and the Linear Amplitude Sweep (LAS) test which are heavily used in research studies are conducted using the Dynamic Shear Rheometer (DSR) that is one of the devices used for the characterization methods in this chapter. The MSCR test evaluates the asphalt binder's potential for permanent deformation, while, the LAS test evaluates the ability of asphalt binder to resist fatigue damage.

9.1 In Table 9.1, there are different asphalt binders classified either according to penetration, viscosity, or performance grade (PG) values. Compare the set in the first column with that in the third column of the table in terms of the property value in the fourth column (use >, <, or =).

TABLE 9.1

Asphalt Binders with Penetration, Viscosity, and PG Classifications

Asphalt Binder	>, <, =	Asphalt Binder	Property
AC-40		AC-5	Penetration
AC-40		AC-5	Viscosity
AC-10		AC-5	High PG
60/70		85/100	Penetration
60/70		85/100	Viscosity
85/100		40/50	High PG
PG 64-10		PG 70-10	Penetration
PG 64-10		PG 70-10	Viscosity

Solution:

See Table 9.2.

TABLE 9.2

Ranking of Asphalt Binders Based on Penetration, Viscosity, and PG Classifications

Asphalt Binder	>, <, =	Asphalt Binder	Property
AC-40	<	AC-5	Penetration
AC-40	>	AC-5	Viscosity
AC-10	>	AC-5	High PG
60/70	<	85/100	Penetration
60/70	>	85/100	Viscosity
85/100	<	40/50	High PG
PG 64-10	>	PG 70-10	Penetration
PG 64-10	<	PG 70-10	Viscosity

9.2 Write down the standard test conditions for each of the asphalt binder tests in Table 9.3.

TABLE 9.3

Asphalt Binder Tests

Test	Test Conditions
Penetration	
Ductility	
Rotational Viscosity (RV)	
Rolling Thin-Film Oven (RTFO)	
Pressure Aging Vessel (PAV)	

Solution:

See Table 9.4.

TABLE 9.4

Test Conditions for Asphalt Binder Tests

Test	Test Conditions
Penetration	25°C; 100 g; 5 s
Ductility	25°C; 5 cm/min
Rotational Viscosity (RV)	135°C; 20 rpm
Rolling Thin-Film Oven (RTFO)	163°C; 75 min
Pressure Aging Vessel (PAV)	90, 100, or 110; 300 psi (2.1 MPa); 20 hr

9.3 Write down the specification for each of the Superpave asphalt binder test parameters shown in Table 9.5.

TABLE 9.5
Asphalt Binder Tests

Test	Specification
Flash Point	
Rotational Viscosity (RV)	
Rolling Thin-Film Oven (RTFO) Mass Loss	
Dynamic Shear Rheometer (DSR) G*/sin δ at high temperatures for original binder	
Dynamic Shear Rheometer (DSR) G*/sin δ at high temperatures for RTFO binder	
Dynamic Shear Rheometer (DSR) G*sin δ intermediate temperatures for PAV binder	
Bending Beam Rheometer (BBR) Stiffness (S) at 60 seconds	
Bending Beam Rheometer (BBR) m-Value at 60 seconds	
Direct Tension Test (DTT) Failure Strain (ε_f)	

Solution:

See Table 9.6.

TABLE 9.6
Specifications for Asphalt Binder Tests

Test	Specification*
Flash Point	≥ 230°C
Rotational Viscosity (RV)	≤ 3 Pa.s
Rolling Thin-Film Oven (RTFO) Mass Loss	≤ 1.00%
Dynamic Shear Rheometer (DSR) G*/sin δ at high temperatures for original binder	≥ 1.00 kPa
Dynamic Shear Rheometer (DSR) G*/sin δ at high temperatures for RTFO binder	≥ 2.20 kPa
Dynamic Shear Rheometer (DSR) G*sin δ intermediate temperatures for PAV binder	≤ 5000 kPa
Bending Beam Rheometer (BBR) Stiffness (S) at 60 seconds	≤ 300 MPa
Bending Beam Rheometer (BBR) m-Value at 60 seconds	≥ 0.300
Direct Tension Test (DTT) Failure Strain (ε_f)	≥ 1.00%

* Reference: *Superpave* Performance Graded *Asphalt Binder* Specifications and *Testing*, Asphalt Institute Superpave Series No. 1 (SP-1), 2003

9.4 Write down the correct performance parameter and specification value in the proper place for each of the following Superpave asphalt binder tests (see Table 9.7).

TABLE 9.7
Superpave Asphalt Binder Tests

Test	Rutting	Fatigue	Low-Temperature	Workability
RV				
DSR				
BBR				
DT				

Solution:

See Table 9.8.

TABLE 9.8
Performance Parameters and Specifications for Superpave Asphalt Binder Tests

Test	Rutting	Fatigue	Low-Temperature	Workability
RV				$RV \leq 3$ Pa.s
DSR	$G^*/\sin\delta \geq 1.0$ kPa $G^*/\sin\delta \geq 2.2$ kPa	$G^*\sin\delta \leq 5,000$ kPa		
BBR			$S \leq 300$ MPa m-Value ≥ 0.300	
DT			$\varepsilon_f \geq 1.0\%$	

9.5 Write down a check mark ($\sqrt{}$) in the appropriate space for each of the Superpave tests shown in Table 9.9 based on the temperature range at which each test is performed.

TABLE 9.9
Superpave Asphalt Binder Tests

Test	Mixing and Laydown	High	Intermediate	Low
RV				
DSR				
BBR				
DT				

Solution:

See Table 9.10.

TABLE 9.10

Temperature Range for Superpave Asphalt Binder Tests

Test	Mixing and Laydown	High	Intermediate	Low
		Temperature Range		
RV	√			
DSR		√	√	
BBR				√
DT				√

9.6 Write down a check mark (√) in the appropriate space for each of the Superpave tests shown in Table 9.11 based on the aging condition of the asphalt binder that each test uses.

TABLE 9.11

Superpave Asphalt Binder Tests

Test	Original	RTFO	PAV
		Aging	
RV			
DSR			
BBR			
DT			

Solution:

See Table 9.12.

TABLE 9.12

Proper Aging Condition for Superpave Asphalt Binder Tests

Test	Original	RTFO	PAV
		Aging	
RV	√		
DSR	√	√	√
BBR			√
DT			√

9.7   In Table 9.13, match the standard asphalt binder test in the left column with the correct property in the right column that suits the test.

TABLE 9.13
Asphalt Binder Tests and Properties

Test Name	Property	No.
Absolute Viscosity	Elastic and viscous properties at intermediate and high temperatures	1
Penetration	Aging due to hot mixing and construction	2
Flash Point	Workability at mixing and laydown temperatures	3
Ductility	Consistency at maximum surface temperature	4
Rolling Thin-Film Oven (RTFO) Test	The temperature at which the asphalt binder sparks	5
Pressure Aging Vessel (PAV) Test for Cold Climate	Low-temperature behavior for PAV-aged asphalt binder	6
Rotational Viscosity (RV)	Consistency at the average service temperature	7
Dynamic Shear Rheometer (DSR) Test	Tensile properties	8
Bending Beam Rheometer (BBR) Test	Long-term aging	9

Solution:

See Table 9.14.

TABLE 9.14
Matching of Asphalt Binder Tests with the Right Properties

Test Name	Property	No.
Absolute Viscosity	Elastic and viscous properties at intermediate and high temperatures	1
Penetration	Aging due to hot mixing and construction	2
Flash Point	Workability at mixing and laydown temperatures	3
Ductility	Consistency at maximum surface temperature	4
Rolling Thin-Film Oven (RTFO) Test	The temperature at which the asphalt binder sparks	5
Pressure Aging Vessel (PAV) Test for Cold Climate	Low-temperature behavior for PAV-aged asphalt binder	6
Rotational Viscosity (RV)	Consistency at the average service temperature	7
Dynamic Shear Rheometer (DSR) Test	Tensile properties	8
Bending Beam Rheometer (BBR) Test	Long-term aging	9

9.8 In a standard Superpave rotational viscosity (RV) test of an asphalt binder sample, if the effective length of the spindle is 33.02 mm, the radius of the spindle is 9.525 mm, and the radius of the container is 11.76 mm, determine the shear strain rate (sec^{-1}).

Solution:

The following formula is used to determine the shear strain rate (γ) of the asphalt binder:

$$\gamma = \frac{2\omega R_c^2 R_s^2}{x^2 \left(R_c^2 - R_s^2\right)} \tag{9.1}$$

Where:
γ = shear rate (sec^{-1})
ω = rotational speed (rad/s)
R_c = radius of the container (m)
R_s = radius of the spindle (m)
x = radial distance from axis of the container to the point where shear rate is being calculated (m).

In this problem, Rc = 11.76 mm = 0.01176 m, Rs = 9.525 mm = 0.009525 m.
The standard rotational speed in a standard Superpave RV test is equal to:

$$\omega = 20 \text{ rpm} = \frac{20 \times 2\pi}{60} = 2.094 \text{ rad/s}$$

A typical value for the shear strain rate is calculated at the surface of the spindle; and therefore, x is equal to the radius of the spindle in this case; i.e., x = 9.525 mm = 0.009525 m.

$\Rightarrow$

$$\gamma = \frac{2(2.094)(0.01176)^2(0.009525)^2}{(0.009525)^2\left[(0.01176)^2 - (0.009525)^2\right]} = 12.2 \text{ sec}^{-1}$$

The MS Excel worksheet used to solve this problem is shown in Figure 9.1.

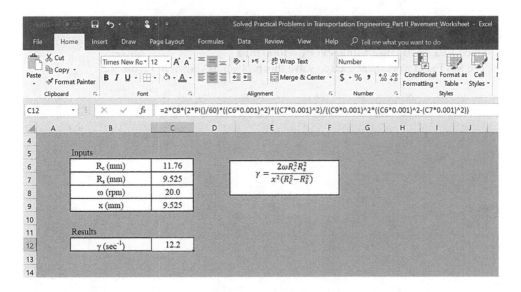

FIGURE 9.1   MS Excel worksheet image for the computations of Problem 9.8.

9.9 In a rotational viscosity test, if the radius of the spindle is 9.525 mm and the radius of the container is 11.76 mm, determine the rotational viscosity (Pa.s) at a shear stress of 5 Pa and a standard rotational speed (ω) of 20 rpm.

Solution:

The same formula for the shear strain rate (γ) shown below is used:

$$\gamma = \frac{2\omega R_c^2 R_s^2}{x^2 \left(R_c^2 - R_s^2\right)}$$

The standard rotational speed in a standard Superpave RV test is equal to:

$$\omega = 20 \text{ rpm} = \frac{20 \times 2\pi}{60} = 2.094 \text{ rad/s}$$

$\Rightarrow$

$$\gamma = \frac{2(2.094)(0.01176)^2 (0.009525)^2}{(0.009525)^2 \left[(0.01176)^2 - (0.009525)^2\right]} = 12.2 \text{ sec}^{-1}$$

The rotational viscosity is determined using the following formula:

$$\eta = \frac{\tau}{\gamma} \tag{9.2}$$

Where:
  η = rotational viscosity (Pa.s)
  τ = applied shear stress (Pa)
  γ = resulting shear strain rate (sec^{-1})

$\Rightarrow$

$$\eta = \frac{5}{12.2} = 0.411 \text{ Pa.s} = 411 \text{ cPoise} = 411 \text{ mPa.s}$$

Since 1 Pa.s = 10 Poise = 1000 cPoise = 1000 mPa.s
The MS Excel worksheet used to solve this problem is shown in Figure 9.2.

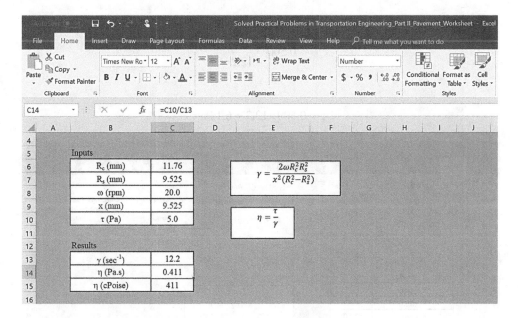

FIGURE 9.2 MS Excel worksheet image for the computations of Problem 9.9.

9.10 Determine the shear stress applied on an asphalt binder sample having a rotational viscosity of 500 cPoise in a standard Superpave rotational viscosity (RV) test, if the radius of the spindle is 9.525 mm and the radius of the container is 11.76 mm.

Solution:

The same formula for the shear strain rate (γ) shown below is used:

$$\gamma = \frac{2\omega R_c^2 R_s^2}{x^2\left(R_c^2 - R_s^2\right)}$$

The standard rotational speed in a standard Superpave RV test is equal to:

$$\omega = 20 \text{ rpm} = \frac{20 \times 2\pi}{60} = 2.094 \text{ rad/s}$$

$\Rightarrow$

$$\gamma = \frac{2(2.094)(0.01176)^2(0.009525)^2}{(0.009525)^2\left[(0.01176)^2 - (0.009525)^2\right]} = 12.2 \text{ sec}^{-1}$$

Using the formula shown below, the shear stress can be determined (see Figure 9.3):

$$\eta = \frac{\tau}{\gamma}$$

$$\eta = 500 \text{ cPoise} = 0.500 \text{ Pa}$$

$\Rightarrow$

$$0.500 = \frac{\tau}{12.2}$$

$\Rightarrow$

$$\tau = 0.500 \times 12.2 = 6.1 \text{ Pa}$$

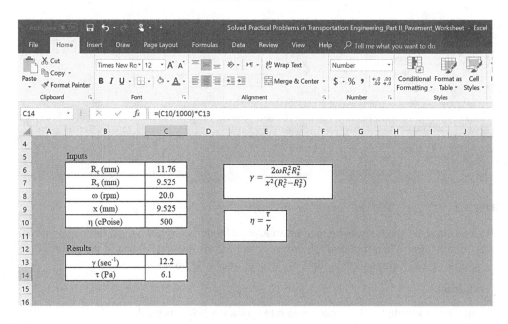

FIGURE 9.3 MS Excel worksheet image for the computations of Problem 9.10.

9.11 Determine the torque applied to an asphalt binder sample having a rotational viscosity of 800 cPoise in a standard Superpave rotational viscosity (RV) test if the effective length of the spindle is 33.02 mm, the radius of the spindle is 9.525 mm, and the radius of the container is 11.76 mm.

Solution:

The same formula for the shear strain rate (γ) shown below is used:

$$\gamma = \frac{2\omega R_c^2 R_s^2}{x^2\left(R_c^2 - R_s^2\right)}$$

The standard rotational speed in a standard Superpave RV test is equal to:

$$\omega = 20 \text{ rpm} = \frac{20 \times 2\pi}{60} = 2.094 \text{ rad/s}$$

$\Rightarrow$

$$\gamma = \frac{2(2.094)(0.01176)^2(0.009525)^2}{(0.009525)^2\left[(0.01176)^2 - (0.009525)^2\right]} = 12.2 \text{ sec}^{-1}$$

Using the formula shown below, the shear stress can be determined:

$$\eta = \frac{\tau}{\gamma}$$

$$\eta = 800 \text{ cPoise} = 0.800 \text{ Pa}$$

$\Rightarrow$

$$0.800 = \frac{\tau}{12.2}$$

$\Rightarrow$

$$\tau = 0.800 \times 12.2 = 9.7 \text{ Pa}$$

Using the following formula, the torque applied to the asphalt binder sample is calculated:

$$\tau = \frac{T}{2\pi R_s^2 L} \tag{9.3}$$

Where:
   τ = applied shear stress (Pa)
   T = torque applied to the asphalt binder sample (N.m)
   R_s = radius of the spindle (m)
   L = effective length of the spindle (m)

$\Rightarrow$

$$\tau = \frac{T}{2\pi R_s^2 L}$$

$$9.7 = \frac{T}{2\pi (0.009525)^2 (0.03302)}$$

$\Rightarrow$

$$T = 0.000183 \text{ N.m} = 0.183 \text{ N.mm}$$

The computations of the results of this problem are performed using the MS Excel worksheet shown in Figure 9.4.

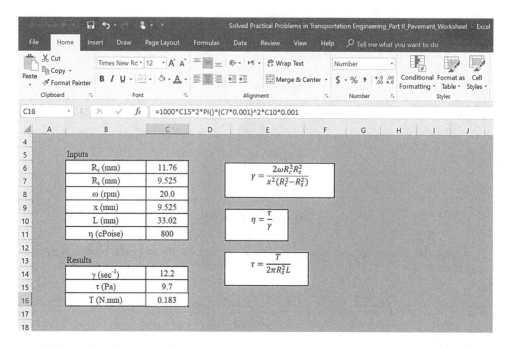

FIGURE 9.4   MS Excel worksheet image for the computations of Problem 9.11.

9.12 If the relationship between an asphalt binder's shear stress, τ (Pa) and shear strain rate, γ (sec^{-1}) at a specific temperature is given by: $\tau = 1.2\gamma$, determine the viscosity of the asphalt binder and describe the behavior of the asphalt binder based on this relationship.

Solution:

Since the relationship between the shear stress and the shear strain rate of the asphalt binder is linear and the viscosity is defined as the shear stress divided by the shear strain rate, the viscosity in this case is constant and equal to the slope of this relationship; i.e., the viscosity $= 1.2$ Pa.s. Consequently, the behavior of the asphalt binder is Newtonian behavior.

9.13 The relationship between a polymer-modified asphalt binder's rotational viscosity, RV (cP) and shear strain rate, γ (sec^{-1}) at a specific temperature is given by: $RV = 226.6 - 50.3 \ln \gamma$, describe the behavior of the modified asphalt binder in this case.

Solution:

From the relationship between the rotational viscosity and the shear strain rate; as the shear strain rate increases, the rotational viscosity decreases. This behavior is called a non-Newtonian behavior (shear-softening or thinning behavior).

By plotting this relationship over a rotational speed range of 1 to 100 rpm (shear strain rate range between 0.609 and 60.9 rad/s) using an arithmetic scale, the following figure is obtained (see Figure 9.5).

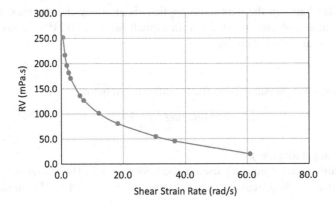

FIGURE 9.5   Relationship between shear strain rate and rotational viscosity for a polymer-modified asphalt binder.

Using a semi-log scale, the relationship is plotted as in Figure 9.6.

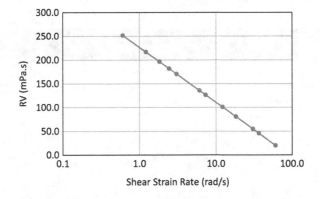

FIGURE 9.6   Shear strain rate versus rotational viscosity for a polymer-modified asphalt binder (on a semi-log scale).

The data used to plot the above relationship in both figures is summarized in Table 9.15.

TABLE 9.15
Superpave Rotational Viscosity Test Data at Different Shear Strain Rates for a Polymer-Modified Asphalt Binder

ω (rpm)	γ (rad/s)	RV (mPa.s)
1	0.609	251.5
2	1.218	216.7
3	1.827	196.3
4	2.435	181.8
5	3.044	170.6
10	6.089	135.7
12	7.306	126.6
20	12.18	100.9
30	18.27	80.5
50	30.44	54.8
60	36.53	45.6
100	60.89	19.9

9.14 If the time lag between the maximum applied shear stress and the maximum resulting shear strain in a standard DSR test for an asphalt binder is 0.135 seconds, compute the phase angle for this asphalt binder.

Solution:

The phase angle is computed using the formula shown below:

$$\delta = \text{time lag} \times f \times 360° \qquad\qquad (9.4)$$

Where:
δ = phase angle (degrees)
f = standard loading frequency used in the DSR test (1.59 Hz = 10 rad/s).
The standard loading frequency in a Superpave DSR test = 1.59 Hz (10 rad/s)

$\Rightarrow$

$$\delta = 0.135 \times 1.59 \times 360° = 77.3°$$

Figure 9.7 shows an image of the MS Excel worksheet computations for this problem.

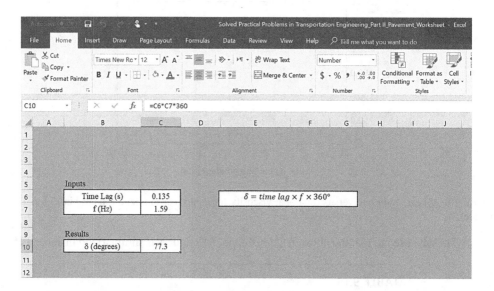

FIGURE 9.7   MS Excel worksheet image for the computations of Problem 9.14.

9.15 What is the Superpave performance grade (PG) of an asphalt binder that is passing the Superpave DSR test at a maximum temperature of 54°C and the BBR test at a minimum temperature of −14°C assuming that it passes the intermediate temperature DSR test?

Solution:

Since the asphalt binder passes the DSR test at a maximum temperature of 54°C, the Superpave high PG grade is 52. And since the asphalt binder passes the BBR test at a minimum temperature of −14°C, it will pass a minimum temperature of −14−10=−24°C in the field. There is a −10°C-shift between the lab and the field for low-temperature specifications. Therefore, the Superpave low PG grade is −22. In conclusion, the performance grade of this asphalt binder is PG 52-22.

9.16 What is the Superpave performance grade (PG) of an asphalt binder that is passing the Superpave DSR test at a maximum temperature of 62°C and the BBR test at a minimum temperature of −8°C assuming that it passes the intermediate temperature DSR test?

Solution:

Since the asphalt binder passes the DSR test at a maximum temperature of 62°C, the Superpave high PG grade is 58. And since the asphalt binder passes the BBR test at a minimum temperature of −12°C, it will pass a minimum temperature of −8−10=−18°C in the field. There is a −10°C-shift between the lab and the field for low-temperature specifications. Therefore, the Superpave low PG grade is −16. In conclusion, the performance grade of this asphalt binder is PG 58-16.

9.17 What would be the Superpave performance grade (PG) of an asphalt binder that is passing the Superpave asphalt binder grading tests in the laboratory at a maximum temperature of 66°C and at a minimum temperature of −7°C?

Solution:

Since the asphalt binder passes the Superpave asphalt binder grading tests at a maximum temperature of 66°C, the Superpave high PG grade is 64. And since the asphalt binder passes the Superpave asphalt binder grading tests at a minimum temperature of −7°C, it will pass a minimum temperature of −7−10=−17°C in the field. There is a −10°C-shift between the lab and the field for low-temperature specifications. Therefore, the Superpave low PG grade is −16. In conclusion, the performance grade of this asphalt binder is PG 64-16.

9.18 What is the high PG grade of an asphalt binder that passes the Superpave DSR test for both unaged and RTFO-aged conditions at a maximum test temperature of 67°C?

Solution:

Since the asphalt binder passes the DSR test at a maximum temperature of 67°C, the Superpave high PG grade is 64.

9.19 What is the low PG grade of an asphalt binder that passes the Superpave BBR test for PAV conditions at a minimum test temperature of −9°C?

Solution:

Since the asphalt binder passes the BBR test at a minimum temperature of −9°C, it will pass a minimum temperature of −9−10=−19°C in the field due to the −10°C-shift between the lab and the field specifications. Therefore, the Superpave low PG grade is −16.

9.20 Rank the following five Superpave asphalt binders: PG 76-10, PG 64-10, PG 58-16, and PG 70-16, PG 52-10 according to viscosity (or stiffness) from highest to lowest.

Solution:

The viscosity (or stiffness) of the asphalt binder is correlated to the high PG grade of the binder. As the high PG grade increases, the viscosity (or stiffness) increases as well, and vice versa. Therefore, the ranking of these five asphalt binders from highest to lowest would be as follows:

PG 76-10, PG 70-16, PG 64-10, PG 58-16, PG 52-10

9.21 If the BBR creep stiffness of an asphalt binder, S (MPa) with time, t (s) is given by the relationship $\log S(t) = 2.23 - 0.85\log(t) + 0.15(\log(t))^2$, determine the creep stiffness and the m-value at 60 seconds. Does this asphalt binder pass the Superpave specifications? And why?

Solution:

Substituting in the formula for S(t) using t=60 s, the stiffness is obtained as shown below:

$$\log S(t) = 2.23 - 0.85 \log(t) + 0.15(\log(t))^2$$

⇒

$$\log S(t) = 2.23 - 0.85 \log(60) + 0.15(\log(60))^2 = 1.193$$

⇒

$$S(t) = 10^{1.193} = 15.6 \text{ MPa}$$

The m-value is the slope of the relationship between log(S) and log(t). The slope is equal to the absolute value of the first derivative of the relationship with respect to log(t). Therefore:

$$\text{Slope} = \left| \frac{d}{d\log(t)}(\log S(t)) \right|$$

⇒

$$\text{Slope} = \left| \frac{d}{d\log(t)} \left[2.23 - 0.85 \log(t) + 0.15(\log(t))^2 \right] \right|$$

⇒

$$\text{m-Value} = \left| -0.85 + 2(0.15) \log(t) \right|$$

At t=60 seconds,

$$\text{m-Value} = \left| -0.85 + 2(0.15) \log(60) \right| = 0.32$$

The MS Excel worksheet used to solve this problem is shown in Figure 9.8.

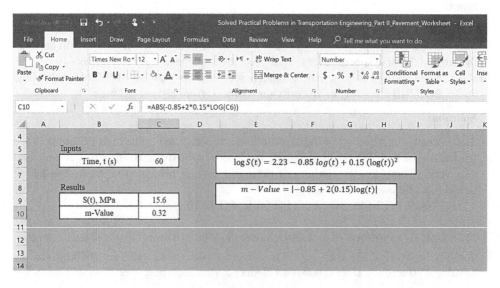

FIGURE 9.8   MS Excel worksheet image for the computations of Problem 9.21.

Yes, the asphalt binder passes the Superpave specifications because the stiffness S at 60 seconds = 15.6 MPa ≤ 300 MPa (the maximum requirement for S in the Superpave criteria), and the m-value = 0.32 ≥ 0.300 (the minimum requirement for the m-value in the Superpave criteria).

9.22 The results of a DSR test conducted on an asphalt binder under three conditions (fresh before RTFO aging, after RTFO aging, and after PAV aging) are summarized in Table 9.16.

TABLE 9.16

Superpave DSR Test Data for an Asphalt Binder Under Three Aging Conditions

Temperature (°C)	G* Value (kPa)	δ (degrees)
	Before RTFO	
70	2.77	71
	After RTFO	
70	5.87	66
	After PAV	
31	532.9	46

Where:

G* = complex shear modulus

δ = phase angle

Determine the fatigue parameter (|G*| sin δ) and the rutting parameter (|G*|/sin δ) for this asphalt binder.

Solution:

The fatigue parameter (|G*| sin δ) is determined for the asphalt binder after PAV aging. Therefore, the results for the PAV-aged asphalt binder are used as shown below:

$$\left|G^*\right|\sin\delta = 532.9 \times \sin(46°) = 383.3 \text{ kPa}$$

On the other hand, the rutting parameter (|G*|/sin δ) is determined for the fresh asphalt binder (before RTFO aging) and for the RTFO-aged asphalt binder as follows:

For the fresh asphalt binder:

$$\frac{\left|G^*\right|}{\sin\delta} = \frac{2.77}{\sin(71°)} = 2.93 \text{ kPa}$$

And for the RTFO-aged asphalt binder:

$$\frac{\left|G^*\right|}{\sin\delta} = \frac{5.87}{\sin(66°)} = 6.43 \text{ kPa}$$

The computations of the results in this problem are performed using the MS Excel worksheet shown in Figure 9.9.

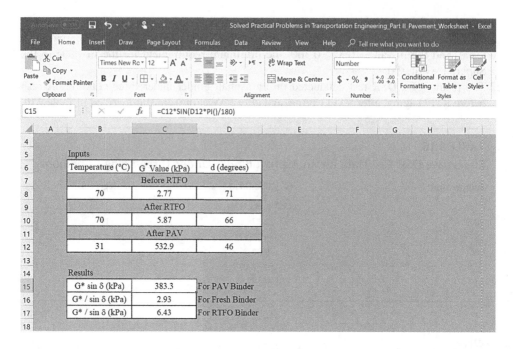

FIGURE 9.9 MS Excel worksheet image for the computations of Problem 9.22.

9.23 An asphalt binder having a performance grade of PG 70-16 is tested in the DSR test to verify the high PG grade. The test results are shown in Table 9.17.

TABLE 9.17
Superpave DSR Test Data for a PG70-16 Asphalt Binder

Temperature (°C)	G* Value (kPa)	δ (degrees)
	Before RTFO	
70	1.10	71
	After RTFO	
70	2.40	66
	After PAV	
31	480.0	46

Based on these results, verify that this asphalt binder complies with the Superpave specifications for the high PG 70.

Solution:

The rutting parameter (|G*|/sin δ) is determined for the fresh asphalt binder (before RTFO aging) and for the RTFO-aged asphalt binder as follows:
For the fresh asphalt binder:

$$\frac{\left|G^*\right|}{\sin\delta} = \frac{1.10}{\sin(71°)} = 1.16 \text{ kPa} \geq 1.0 \text{ kPa}$$

⇒ OK because the Superpave minimum requirement = 1.0 kPa for fresh asphalt binders.

For the RTFO-aged asphalt binder:

$$\frac{\left|G^*\right|}{\sin\delta} = \frac{2.40}{\sin(66°)} = 2.63 \text{ kPa} \geq 2.20 \text{ kPa}$$

$\Rightarrow$ OK because the Superpave minimum requirement = 2.20 kPa for RTFO-aged asphalt binders.

For PG 70-16, according to the Superpave, the intermediate temperature for fatigue cracking is equal to:

$$T_{\text{intermediate}} = \frac{\text{High PG} + \text{Low PG}}{2} + 4 \quad (9.5)$$

Where:

$T_{\text{intermediate}}$ = the intermediate temperature (°C) at which PAV-asphalt binder is tested in the DSR to check the fatigue cracking parameter ($|G^*| \sin \delta$).

High PG = high temperature in the PG grade

Low PG = low temperature in the PG grade

$\Rightarrow$

$$T_{\text{intermediate}} = \frac{70 - 16}{2} + 4 = 31°C$$

Therefore, the $|G^*| \sin \delta$ for the PAV-aged asphalt binder is calculated as below:

$$\left|G^*\right| \sin\delta = 480.0 \times \sin(46°) = 345.3 \text{ kPa} \leq 5000 \text{ kPa}$$

$\Rightarrow$ OK because the Superpave maximum requirement = 5000 kPa for PAV-aged asphalt binders.

In conclusion, the asphalt binder complies with the Superpave specifications for the high PG of the performance grade PG 70-16.

The computations of the results in this problem are performed using the MS Excel worksheet shown in Figure 9.10.

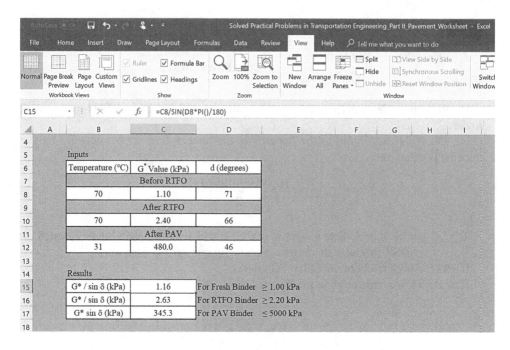

FIGURE 9.10   MS Excel worksheet image for the computations of Problem 9.23.

9.24 In a standard Superpave Bending Beam Rheometer (BBR) test for an asphalt binder, the following test results are obtained at time = 60 seconds (see Table 9.18).

TABLE 9.18

Superpave BBR Test Data for an Asphalt Binder

Test Temperature (°C)	Creep Stiffness, S (MPa)	m-Value
0	10.6	0.341
−6	15.8	0.322
−12	20.4	0.271

Based on these results, what is the lowest temperature at which the asphalt binder passes the Superpave criteria for low-temperature cracking?

Solution:

The Superpave criteria state that the BBR creep stiffness (S) ≤ 300 MPa and the m-value ≥ 0.300 at time = 60 seconds (see Table 9.19).

TABLE 9.19

Superpave BBR Test Criteria Compared to the BBR Test Data for the Given Asphalt Binder

Test Temperature (°C)	Creep Stiffness, S (MPa)	Superpave Criteria	Passes Criteria?	m-Value	Superpave Criteria	Passes Criteria?
0	10.6	≤ 300 MPa	OK	0.341	≥ 0.300	OK
−6	15.8		OK	0.322		OK
−12	20.4		OK	0.271		Not OK

Based on the above results, the BBR creep stiffness (S) for the three test temperatures are all lower than 300 MPa. However, the lowest temperature at which the m-value is higher than 0.300 is −6°C. Consequently, the lowest temperature in the lab at which this asphalt binder passes the Superpave criteria for low-temperature cracking is −6°C. In the field, the lowest temperature at which this asphalt binder will pass the Superpave criteria for low-temperature cracking is −16°C since there is a −10°C-shift between the lab and the field for low-temperature specifications.

9.25 The test results for an asphalt binder tested using the standard Superpave Bending Beam Rheometer (BBR) test at four low temperatures are summarized in Table 9.20.

TABLE 9.20

Superpave BBR Test Data for an Asphalt Binder at Four Low Temperatures

Test Temperature (°C)	Creep Stiffness, S (MPa)	m-Value
0	12.4	0.412
−6	16.5	0.360
−12	24.2	0.311
−18	36.8	0.263

Based on these results, determine the low performance grade (PG) for this asphalt binder assuming that the binder passes the Superpave specifications for intermediate temperatures.

Solution:

The Superpave criteria for low temperatures are: the BBR creep stiffness $(S) \leq 300$ MPa and the m-value ≥ 0.300 at time $= 60$ seconds (see Table 9.21).

TABLE 9.21

Superpave BBR Test Criteria Compared to the BBR Test Data for the Given Asphalt Binder at Four Low Temperatures

Test Temperature (°C)	Creep Stiffness, S (MPa)	Superpave Criteria	Passes Criteria?	m-Value	Superpave Criteria	Passes Criteria?
0	12.4	≤ 300 MPa	OK	0.412	≥ 0.300	OK
−6	16.5		OK	0.360		OK
−12	24.2		OK	0.311		OK
−18	36.8		OK	0.263		Not OK

Based on the above results, the asphalt binder passes the Superpave specifications for creep stiffness (S) at the four test temperatures. However, for m-value, the asphalt binder passes the Superpave specifications at a lowest temperature of −12°C. In the field, the lowest temperature at which this asphalt binder will pass the Superpave criteria for low-temperature cracking is −22°C since there is a −10°C-shift between the lab and the field for low-temperature specifications. Hence, the low PG grade for this asphalt binder would be PG-22.

9.26 Five asphalt binders have been tested in the BBR creep test at −6°C. The test results for creep stiffness (S) and m-value at 60 seconds are provided for the five asphalt binders in the table below, respectively (see Table 9.22).

TABLE 9.22

Superpave BBR Test Results (Stiffness and m-Value) for Five Asphalt Binders

Asphalt Binder #	[S (MPa), m-Value]
1	[280, 0.316]
2	[650, 0.330]
3	[340, 0.284]
4	[395, 0.314]
5	[650, 0.242]

In this case and according to the Superpave requirements, which asphalt binder should be tested using the Superpave Direct Tension Test (DTT) to ensure that the asphalt binder meets the Superpave specifications at −6°C?

Solution:

According to the Superpave requirements, the asphalt binder should be tested using the DTT as a complementary test after it has been tested in the BBR creep test for low temperature cracking under the following condition:

300 MPa < BBR creep stiffness, S (MPa) at 60 seconds ≤ 600 MPa, and m-value ≥ 0.300

In other words, the asphalt binder must pass the Superpave criteria for m-value (m-value must be ≥ 0.300); and when the creep stiffness, S fails the Superpave specifications (S > 300 MPa), it must not exceed 600 MPa so that the asphalt binder could be tested in the DTT to ensure that it meets the Superpave criteria for low temperature cracking. In case the asphalt binder's creep stiffness (S) exceeds 600 MPa, the asphalt binder is considered failing the Superpave specifications and there is no need to perform the DTT for the asphalt binder.

In this case, the following results are obtained (see Table 9.23). Therefore, asphalt binder #4 requires further testing in the Superpave DTT.

TABLE 9.23

Checking the Superpave BBR Test Criteria and Whether the DTT is Needed for Five Asphalt Binders

Asphalt Binder #	[S (MPa), m-Value]	m-Value	Stiffness, S	DTT Needed?
1	[280, 0.316]	Pass	Pass	No
2	[650, 0.330]	Pass	Not; S > 600 MPa	No
3	[340, 0.284]	Not	Not; 300 < S ≤ 600 MPa	No
4	[395, 0.314]	Pass	Not; 300 < S ≤ 600 MPa	Yes
5	[650, 0.242]	Not	Not; S > 600 MPa	No

9.27 In a Superpave DTT for an asphalt binder, the elongation at failure was obtained to be 450 microns. If the effective gage length is 33.8 mm, determine the tensile strain at failure for this asphalt binder. Does the asphalt binder pass the Superpave specifications?

Solution:

$$\varepsilon_f = \frac{\Delta L_f}{GL} \times 100\% \qquad (9.6)$$

Where:

ε_f = tensile strain of asphalt binder sample at failure (%)
ΔL_f = elongation of asphalt binder sample at failure (mm)
GL = effective gage length (mm)
ΔL_f = 450 microns = 0.450 mm
GL = 33.8 mm

⇒

$$\varepsilon_f = \frac{0.450}{33.8} \times 100\% = 1.33\%$$

The MS Excel worksheet used to perform the computations of this problem is shown in Figure 9.11.

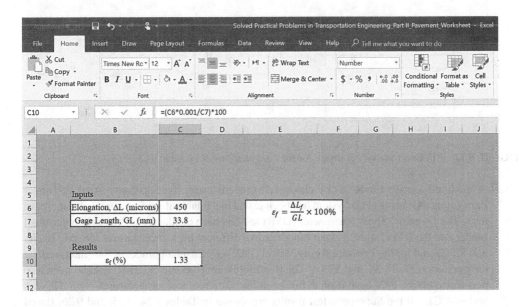

FIGURE 9.11   MS Excel worksheet image for the computations of Problem 9.27.

9.28 In a Superpave DTT for an asphalt binder, the tensile strain at failure for this asphalt binder is 0.80%, determine the elongation at failure for the asphalt binder if the effective gage length is 33.8 mm.

Solution:

$$\varepsilon_f = \frac{\Delta L_f}{GL} \times 100\%$$

ε_f = 0.80%
GL = 33.8 mm

⇒

$$0.80\% = \frac{\Delta L_f}{33.8} \times 100\%$$

$\Rightarrow$

$$\Delta L_f = \frac{0.80 \times 33.8}{100} = 0.270 \text{ mm} = 270 \text{ microns}$$

The MS Excel worksheet used to perform the computations of this problem is shown in Figure 9.12.

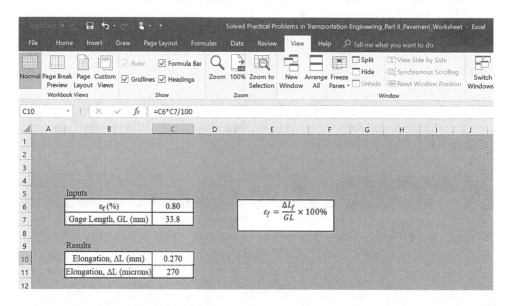

FIGURE 9.12   MS Excel worksheet image for the computations of Problem 9.28.

9.29 A modified asphalt binder to be classified (graded) using the Superpave Asphalt Binder Performance Grading (PG) System has been tested in the laboratory for a series of Superpave tests. The Dynamic Shear Rheometer (DSR) test has been conducted for the original asphalt binder and the RTFO-aged asphalt binder at five different high temperatures (58, 64, 70, 76, and 82°C), and for the Pressure Aging Vessel (PAV)-aged asphalt binder at four intermediate temperatures (28, 31, 34, and 37°C). The Bending Beam Rheometer (BBR) test has been also performed on the PAV-aged asphalt binder at four different low temperatures (−6, −12, −18, and −24°C). All the Superpave test results are shown in Tables 9.24, 9.25, and 9.26. Based on these results, required is to grade (classify) the asphalt binder using the Superpave asphalt binder PG System (see Tables 9.24, 9.25, and 9.26).

TABLE 9.24

Superpave DSR Test Results for a Modified Asphalt Binder (Original and RTFO-Aged Samples)

Temperature (°C)	Fresh Asphalt Binder		RTFO Asphalt Binder	
	G* (Pa)	δ (degrees)	G* (Pa)	δ (degrees)
58	6378	59.2	10534	58.1
64	3356	59.8	5859	59.0
70	1995	60.6	3361	59.6
76	1204	61.9	1959	60.5
82	1066	62.3	1108	61.2

TABLE 9.25

Superpave DSR Test Results for a Modified Asphalt Binder (PAV-Aged Sample)

Temperature (°C)	PAV Asphalt Binder	
	G* (kPa)	δ (degrees)
28	781.6	44.1
31	539.1	45.6
34	421.1	46.2
37	335.2	47.9

TABLE 9.26

Superpave BBR Test Results for a Modified Asphalt Binder (PAV-Aged Sample)

Temperature (°C)	Stiffness, S (MPa)	m-Value
−6	33.2	0.482
−12	45.6	0.364
−18	66.5	0.302
−24	80.5	0.256

Solution:

For rutting, the DSR |G*|/sin δ is calculated for the original asphalt binder and for the RTFO-aged asphalt binder at the high temperatures as shown in Tables 9.27 and 9.28.

TABLE 9.27

Superpave DSR |G*|/sin δ for a Modified Asphalt Binder (Unaged Sample)

Temperature (°C)	G*/sin δ (kPa)	Superpave Specification	Pass
58	7.4		Yes
64	3.9		Yes
70	2.3	≥ 1.0 kPa	Yes
76	1.4		Yes
82	1.2		Yes

TABLE 9.28

Superpave DSR |G*|/sin δ for a Modified Asphalt Binder (RTFO-Aged Sample)

Temperature (°C)	G*/sin δ (kPa)	Superpave Specification	Pass
58	12.4		Yes
64	6.8		Yes
70	3.9	≥ 2.20 kPa	Yes
76	2.3		Yes
82	1.3		No

Based on the above results, the highest temperature at which the asphalt binder passes the Superpave specifications for rutting is 76°C. In other words, the maximum temperature at which both the original asphalt binder's |G*|/sin δ must be higher than 1.00 kPa and the RTFO-aged asphalt binder's |G*|/sin δ must be higher than 2.20 kPa is 76°C. Therefore, the high-performance grade is PG 76.

For low-temperature cracking, the BBR creep stiffness (S) and m-value are summarized for the PAV-aged asphalt binder at the low temperatures as shown in Table 9.29.

TABLE 9.29

Superpave BBR Test Results for a Modified Asphalt Binder (PAV-Aged Sample)

Temperature (°C)	Stiffness, S (MPa)	Superpave Specification	m-Value	Superpave Specification	Pass
−6	33.2		0.482		Yes
−12	45.6	≤ 300 MPa	0.364	≥ 0.300	Yes
−18	66.5		0.302		Yes
−24	80.5		0.256		No

Based on these results, the minimum temperature at which the asphalt binder passes the Superpave specifications for low-temperature cracking is −18°C. Since there is a −10°C-shift between the lab and the field for low-temperature specifications, the low performance grade is PG-28.

Now, the criteria for fatigue cracking must be checked. The DSR |G*| sin δ is calculated for the PAV-aged asphalt binder at the intermediate temperatures as shown in Table 9.30.

TABLE 9.30

Checking the DSR Test Results against the Superpave
Criteria (PAV-Aged Sample)

Temperature (°C)	G*sin δ (kPa)	Superpave Specification
28	543.9	
31	385.2	≤ 5000 kPa
34	303.9	
37	248.7	

Since the performance grade is PG 76-28, the fatigue cracking parameter, |G*| sin δ is verified at an intermediate temperature equal to:

$$T_{intermediate} = \frac{76-28}{2} + 4 = 28°C$$

The |G*| sin δ at 28°C = 543.9 ≤ 5000 kPa; and therefore, the performance grade of this asphalt binder is PG 76-28 (see Table 9.31).

TABLE 9.31

Superpave High and Low
Performance Grades for
a Modified Asphalt
Binder

High PG	76
Low PG	−28
Performance Grade	PG 76-28

9.30 Select the appropriate performance grade (PG) asphalt binder among the following asphalt binders that will suit a Superpave highway pavement project located in an area with a high pavement temperature of 54°C and a low pavement temperature of −14°C.
PG 64-16; PG 64-10; PG 58-16; PG 58-10; PG 52-16; PG 52-10

Solution:

Since the high pavement temperature is 54°C, the selected asphalt binder must have a high-performance grade (PG) of 54 or higher to cover the high pavement temperature range in the area. In addition, since the low pavement temperature is −14°C, the selected asphalt binder must have a low performance grade (PG) of −14 or lower to cover the low pavement temperature range in the area. In Superpave, the high-performance grades are following the trend:

PG 52	PG 58	PG 64	PG 70	PG 76	PG 82

And the low performance grades also follow the trend:

PG –10	PG –16	PG –22	PG –28	PG –34

Therefore, the appropriate asphalt binder for this project in this area is PG 58-16.

9.31 A highway pavement is to be constructed in an area having high pavement temperature of 72°C and low pavement temperature of −12°C, which asphalt binder among the following Superpave asphalt binders—PG 82-22, PG 76-10, PG 76-16, PG 70-10, PG 70-16, PG 64-10, PG 64-16—would be most suitable for this highway pavement according to the Superpave asphalt binder grading system?

Solution:

Since the high pavement temperature is 72°C, the selected asphalt binder must have a high-performance grade (PG) of 72 or higher to cover the high pavement temperature range in the area. In addition, since the low pavement temperature is −12°C, the selected asphalt binder must have a low performance grade (PG) of −12 or lower to cover the low pavement temperature range in the area. Therefore, the appropriate asphalt binder among the given asphalt binders for this project in this area is PG 76-16.

10 Superpave Gyratory Compaction

Chapter 10 includes questions and problems dealing with the compaction properties for asphalt mixtures in the Superpave gyratory compactor (SGC), the compactive effort, the Superpave criteria for gyratory compaction, and the computations related to gyratory compaction of asphalt mixtures. The Superpave gyratory compactor simulates the compaction that takes place in field. Three levels of compaction are typically used in Superpave: $N_{initial}$, N_{design}, and $N_{maximum}$. $N_{initial}$ measures mixture compactability during construction and can easily verify if the mixture is tender during construction and unstable under traffic. N_{design} measures the density of the mixture that is equivalent to the density in the field after specific traffic loadings. On the other hand, $N_{maximum}$ measures the maximum mixture in field that should never be exceeded. Asphalt mixtures with very low air voids at $N_{maximum}$ indicate that these mixtures may experience excessive compaction under traffic resulting in low air voids, bleeding, and rutting. Therefore, during gyratory compaction, the behavior of asphalt mixture is monitored, and density and air voids are calculated for the asphalt mixture. The compaction properties provide a good idea about the tenderness and stiffness of the mixture and can predict the behavior of the mixture in pavement under traffic throughout the service life of the pavement.

10.1 In asphalt mixture with the following data is being compacted using the Superpave gyratory compactor (SGC). The compaction data (the mixture height versus the number of gyrations) is also given in Table 10.1.
- Diameter of gyratory specimen = 150 mm ($\cong$ 6 in)
- Maximum theoretical specific gravity (G_{mm}) for the loose mixture = 2.537
- Specific gravity for coarse aggregate used in the mixture = 2.605
- Specific gravity for fine aggregate used in the mixture = 2.507
- Percent passing sieve No. 4 (4.75 mm) for the aggregate gradation used in the mixture = 62%
- Date on the bulk specific gravity for the compacted mixture is given in Table 10.2:

TABLE 10.1

Superpave Gyratory Compaction Data of an Asphalt Mixture

Number of Gyrations, N	Height (mm)
1	135.2
2	132.8
3	131.1
4	130.0
5	128.8
6	128.0
7	127.2
8	126.6
9	125.9

(Continued)

TABLE 10.1   (CONTINUED)
Superpave Gyratory Compaction Data of an Asphalt Mixture

Number of Gyrations, N	Height (mm)
10	125.5
20	122.2
30	120.5
40	119.3
50	118.4
60	117.8
70	117.3
80	116.8
90	116.5
100	116.2
109	115.9
110	115.9
120	115.7
130	115.5
140	115.3
150	115.2
160	115.0
170	114.9
174	114.9

TABLE 10.2
Bulk Specific Gravity Data of a Compacted Asphalt Mixture

Weight of Specimen in Air (g)	4795.6
Weight of Saturated-Surface Dry (SSD) Specimen (g)	4802.2
Weight of Specimen in Water (g)	2790.0

Compute the following parameters during compaction:

a. The estimated bulk specific gravity ($G_{mb\text{-}Estimated}$)
b. The correction factor for bulk specific gravity
c. The corrected bulk specific gravity ($G_{mb\text{-}Corrected}$)
d. The percent maximum specific gravity ($\%G_{mm}$)
e. The percentage of air voids (V_a)

Solution:

In the beginning, the bulk specific gravity of the compacted specimen is computed as shown below:

$$G_{mb\text{-}Measured} = \frac{A}{B - C} \qquad (10.1)$$

Where:

$G_{mb\text{-}Measured}$ = bulk specific gravity of compacted specimen measured in the lab
A = weight of specimen in air (g)
B = weight of saturated-surface dry (SSD) specimen (g)
C = weight of specimen in water (g)

$\Rightarrow$

$$G_{\text{mb-Measured}} = \frac{4795.6}{4802.2 - 2790.0} = 2.383$$

The bulk specific gravity of the aggregate used in the mixture is calculated using the formula below:

$$G_{sb} = \frac{P_{CA} + P_{FA}}{\dfrac{P_{CA}}{G_{sb\text{-}CA}} + \dfrac{P_{FA}}{G_{sb\text{-}FA}}} = \frac{100}{\dfrac{P_{CA}}{G_{sb\text{-}CA}} + \dfrac{P_{FA}}{G_{sb\text{-}FA}}} \qquad (10.2)$$

Where:

P_{CA} = percentage of coarse aggregate in the aggregate gradation
P_{FA} = percentage of fine aggregate in the aggregate gradation
$G_{sb\text{-}CA}$ = bulk specific gravity of coarse aggregate
$G_{sb\text{-}FA}$ = bulk specific gravity of fine aggregate
P_{FA} = percentage of material passing sieve No. 4 = 62%

$$P_{CA} = 100 - P_{FA} \qquad (10.3)$$

$\Rightarrow$

$$P_{CA} = 100 - 62 = 38\%$$

$$G_{sb} = \frac{P_{CA} + P_{FA}}{\dfrac{P_{CA}}{G_{sb\text{-}CA}} + \dfrac{P_{FA}}{G_{sb\text{-}FA}}}$$

$\Rightarrow$

$$G_{sb} = \frac{38 + 62}{\dfrac{38}{2.605} + \dfrac{62}{2.507}} = 2.543$$

a. To determine the estimated bulk specific gravity ($G_{\text{mb-Estimated}}$) for the asphalt mixture during compaction, the volume of the mixture should be computed as shown below:

$$V = \pi \frac{D^2}{4} H \qquad (10.4)$$

Where:

V = volume of mixture
D = diameter of specimen (150 mm)
H = height of mixture inside mold during compaction

$$G_{\text{mb-Estimated}} = \frac{W}{V\gamma_w} \qquad (10.5)$$

Where:

W = weight of mixture or compacted specimen (g)
γ_w = density of water (1 g/cm³)

Sample Calculation:

At number of gyrations (N) = 5, the height of the mixture (H) = 128.8 mm.

$$V = \pi \frac{D^2}{4} H$$

$\Rightarrow$

$$V = \pi \frac{(150 \times 0.1)^2}{4} (128.8 \times 0.1) = 2276.1 \text{ cm}^3$$

W = weight of the mixture = weight of compacted specimen in air = 4795.6 g

$$G_{mb\text{-Estimated}} = \frac{W}{V\gamma_w}$$

$\Rightarrow$

$$G_{mb\text{-Estimated}} = \frac{4795.6}{2276.1 \times 1} = 2.107$$

The estimated bulk specific gravity at all other values for the number of gyrations (N) is computed in a similar manner as above.

b.   The correction factor for bulk specific gravity is determined according to the procedure shown below:

The estimated bulk specific gravity is determined at the end of compaction (at number of gyrations, N = 174). The measured bulk specific gravity for the compacted specimen is also determined using Equation (10.1). The correction factor for bulk specific gravity is then determined by dividing the measured bulk specific gravity by the estimated bulk specific gravity according to the following formula:

$$CF = \frac{G_{mb\text{-Measured}}}{G_{mb\text{-Estimated}}} \tag{10.6}$$

Where:

CF = correction factor for bulk specific gravity

$G_{mb\text{-Measured}}$ = measured bulk specific gravity of compacted specimen

$G_{mb\text{-Estimated}}$ = estimated bulk specific gravity of asphalt mixture at the end of compaction determined using Equation (10.5).

At the end of compaction when the number of gyrations (N) = 174, the height of the mixture (H) = 114.9 mm

$$V = \left(\pi \frac{D^2}{4} \right) H$$

$\Rightarrow$

$$V = \pi \frac{(150 \times 0.1)^2}{4} (114.9 \times 0.1) = 2030.5 \text{ cm}^3$$

W = weight of the mixture = weight of compacted specimen in air = 4795.6 g

$$G_{mb\text{-Estimated}} = \frac{W}{V\gamma_w}$$

$\Rightarrow$

$$G_{mb\text{-Estimated}} = \frac{4795.6}{2030.5 \times 1} = 2.362$$

$G_{\text{mb-Measured}} = 2.383$ determined above.

$$CF = \frac{G_{\text{mb-Measured}}}{G_{\text{mb-Estimated}}}$$

$\Rightarrow$

$$CF = \frac{2.383}{2.362} = 1.009$$

c. The corrected bulk specific gravity ($G_{\text{mb-Corrected}}$) at a specific number of gyrations is determined by multiplying the estimated bulk specific gravity at that number of gyrations by the correction factor:

Sample Calculation:

$$G_{\text{mb-Corrected}} = CF \times G_{\text{mb-Estimated}} \qquad (10.7)$$

At N=5 gyrations, the estimated bulk specific gravity is determined above as:

$$G_{\text{mb-Estimated}} = 2.107$$

$$G_{\text{mb-Corrected}} = CF \times G_{\text{mb-Estimated}}$$

$\Rightarrow$

$$G_{\text{mb-Corrected}} = 1.009 \times 2.107 = 2.126$$

The corrected bulk specific gravity is determined at all other values for the number of gyrations (N) following the same procedure.

d. The percent maximum specific gravity ($\%G_{\text{mm}}$) is defined as the corrected bulk specific gravity of the mixture during compaction divided by the maximum theoretical specific gravity (G_{mm}) of the mixture as a percentage. In other words, it represents the density of the mixture as a percentage by comparing the density of the mixture during compaction to the maximum density of the mixture when it is loose.

Therefore,

$$\%G_{\text{mm}} = \frac{G_{\text{mb-Corrected}}}{G_{\text{mm}}} \times 100\% \qquad (10.8)$$

Where:

$\%G_{\text{mm}}$ = percentage of maximum theoretical specific gravity of the mixture at a specific number of gyrations during compaction

G_{mm} = maximum theoretical specific gravity of the loose mixture

$G_{\text{mb-Corrected}}$ = corrected bulk specific gravity at a specific number of gyrations during compaction

Sample Calculation:
At N=5 gyrations,

$$\%G_{\text{mm}} = \frac{G_{\text{mb-Corrected}}}{G_{\text{mm}}} \times 100\%$$

$\Rightarrow$

$$\%G_{mm} = \frac{2.126}{2.537} \times 100\% = 83.8\%$$

The $\%G_{mm}$ is determined at all other values for the number of gyrations (N) following the same procedure.

e.   The percentage of air voids (V_a) is determined as:

$$V_a = 100 - \%G_{mm} \qquad\qquad (10.9)$$

Where:

V_a = percentage of air voids in the mixture at a specific number of gyrations during compaction

$\%G_{mm}$ = percentage of maximum theoretical specific gravity of the mixture at a specific number of gyrations during compaction

Sample Calculation:

At N = 5 gyrations,

$$V_a = 100 - \%G_{mm}$$

$\Rightarrow$

$$V_a = 100 - 83.8 = 16.2\%$$

The MS Excel worksheet used to perform the computations of this problem is shown in Figure 10.1.

The D23 cell formula bar shows: `=(PI()*((0.1*$C$6)^2)/4)*(C23*0.1)`

D (mm)	γ_w (g/cm³)	G_{mm}	G_{sb}
150.0	1.000	2.537	2.543

$G_{sb\text{-}CA}$	$G_{sb\text{-}FA}$	% CA	% FA
2.605	2.507	38.0	62.0

A (g)	4795.6
B (g)	4802.2
C (g)	2790.0
G_{mb} Measured	2.383

N	Height (mm)	Volume (cm³)	G_{mb} Estimated	CF	G_{mb} Corrected	% G_{mm}	V_a (%)
1	135.2	2389.2	2.007		2.025	79.8	20.2
2	132.8	2346.8	2.043		2.062	81.3	18.7
3	131.1	2316.7	2.070		2.089	82.3	17.7
4	130.0	2297.3	2.088		2.106	83.0	17.0
5	128.8	2276.1	2.107		2.126	83.8	16.2
6	128.0	2261.9	2.120		2.139	84.3	15.7
7	127.2	2247.8	2.133		2.153	84.9	15.1
8	126.6	2237.2	2.144		2.163	85.3	14.7
9	125.9	2224.8	2.155		2.175	85.7	14.3
10	125.5	2217.8	2.162		2.182	86.0	14.0
11	125.1	2210.7	2.169		2.189	86.3	13.7

FIGURE 10.1   MS Excel worksheet image for the computations of Problem 10.1.

10.2 The detailed computations for the compaction properties of the asphalt mixture in Problem 10.1 are shown in Table 10.3.

TABLE 10.3
Computations of the Compaction Properties of the Asphalt Mixture

N	Height (mm)	Volume (cm³)	G_{mb} Estimated	CF	G_{mb} Corrected	% G_{mm}	V_a (%)
1	135.2	2389.2	2.007		2.025	79.8	20.2
2	132.8	2346.8	2.043		2.062	81.3	18.7
3	131.1	2316.7	2.070		2.089	82.3	17.7
4	130.0	2297.3	2.088		2.106	83.0	17.0
5	128.8	2276.1	2.107		2.126	83.8	16.2
6	128.0	2261.9	2.120		2.139	84.3	15.7
7	127.2	2247.8	2.133		2.153	84.9	15.1
8	126.6	2237.2	2.144		2.163	85.3	14.7
9	125.9	2224.8	2.155		2.175	85.7	14.3
10	125.5	2217.8	2.162		2.182	86.0	14.0
11	125.1	2210.7	2.169		2.189	86.3	13.7
12	124.6	2201.9	2.178		2.198	86.6	13.4
13	124.3	2196.6	2.183		2.203	86.8	13.2
14	123.9	2189.5	2.190		2.210	87.1	12.9
15	123.6	2184.2	2.196		2.216	87.3	12.7
16	123.3	2178.9	2.201		2.221	87.5	12.5
17	123.0	2173.6	2.206		2.226	87.8	12.2
18	122.7	2168.3	2.212		2.232	88.0	12.0
19	122.5	2164.8	2.215		2.235	88.1	11.9
20	122.2	2159.5	2.221		2.241	88.3	11.7
21	122.0	2155.9	2.224		2.245	88.5	11.5
22	121.8	2152.4	2.228		2.248	88.6	11.4
23	121.6	2148.8	2.232		2.252	88.8	11.2
24	121.4	2145.3	2.235		2.256	88.9	11.1
25	121.3	2143.5	2.237		2.258	89.0	11.0
26	121.1	2140.0	2.241		2.261	89.1	10.9
27	120.9	2136.5	2.245		2.265	89.3	10.7
28	120.8	2134.7	2.246		2.267	89.4	10.6
29	120.6	2131.2	2.250		2.271	89.5	10.5
30	120.5	2129.4	2.252		2.273	89.6	10.4
31	120.3	2125.9	2.256		2.276	89.7	10.3
32	120.2	2124.1	2.258		2.278	89.8	10.2
33	120.1	2122.3	2.260		2.280	89.9	10.1
34	119.9	2118.8	2.263		2.284	90.0	10.0
35	119.8	2117.0	2.265		2.286	90.1	9.9
36	119.7	2115.3	2.267		2.288	90.2	9.8
37	119.6	2113.5	2.269		2.290	90.2	9.8
38	119.5	2111.7	2.271		2.292	90.3	9.7
39	119.4	2110.0	2.273		2.293	90.4	9.6
40	119.3	2108.2	2.275		2.295	90.5	9.5
41	119.2	2106.4	2.277		2.297	90.6	9.4
42	119.1	2104.7	2.279		2.299	90.6	9.4
43	119.0	2102.9	2.280		2.301	90.7	9.3
44	118.9	2101.1	2.282		2.303	90.8	9.2

(*Continued*)

TABLE 10.3   (CONTINUED)

Computations of the Compaction Properties of the Asphalt Mixture

N	Height (mm)	Volume (cm³)	G_{mb} Estimated	CF	G_{mb} Corrected	% G_{mm}	V_a (%)
45	118.8	2099.4	2.284		2.305	90.9	9.1
46	118.7	2097.6	2.286		2.307	90.9	9.1
47	118.7	2097.6	2.286		2.307	90.9	9.1
48	118.6	2095.8	2.288		2.309	91.0	9.0
49	118.5	2094.1	2.290		2.311	91.1	8.9
50	118.4	2092.3	2.292		2.313	91.2	8.8
51	118.4	2092.3	2.292		2.313	91.2	8.8
52	118.3	2090.5	2.294		2.315	91.2	8.8
53	118.2	2088.8	2.296		2.317	91.3	8.7
54	118.1	2087.0	2.298		2.319	91.4	8.6
55	118.1	2087.0	2.298		2.319	91.4	8.6
56	118.0	2085.2	2.300		2.321	91.5	8.5
57	118.0	2085.2	2.300		2.321	91.5	8.5
58	117.9	2083.5	2.302		2.323	91.5	8.5
59	117.8	2081.7	2.304		2.325	91.6	8.4
60	117.8	2081.7	2.304		2.325	91.6	8.4
61	117.7	2079.9	2.306		2.327	91.7	8.3
62	117.7	2079.9	2.306		2.327	91.7	8.3
63	117.6	2078.2	2.308		2.329	91.8	8.2
64	117.6	2078.2	2.308		2.329	91.8	8.2
65	117.5	2076.4	2.310		2.331	91.9	8.1
66	117.4	2074.6	2.312		2.333	91.9	8.1
67	117.4	2074.6	2.312		2.333	91.9	8.1
68	117.3	2072.9	2.314		2.334	92.0	8.0
69	117.3	2072.9	2.314		2.334	92.0	8.0
70	117.3	2072.9	2.314		2.334	92.0	8.0
71	117.2	2071.1	2.315		2.336	92.1	7.9
72	117.2	2071.1	2.315		2.336	92.1	7.9
73	117.1	2069.3	2.317		2.338	92.2	7.8
74	117.1	2069.3	2.317		2.338	92.2	7.8
75	117.0	2067.6	2.319		2.340	92.3	7.7
76	117.0	2067.6	2.319		2.340	92.3	7.7
77	116.9	2065.8	2.321		2.342	92.3	7.7
78	116.9	2065.8	2.321		2.342	92.3	7.7
79	116.9	2065.8	2.321		2.342	92.3	7.7
80	116.8	2064.0	2.323		2.344	92.4	7.6
81	116.8	2064.0	2.323		2.344	92.4	7.6
82	116.8	2064.0	2.323		2.344	92.4	7.6
83	116.7	2062.3	2.325		2.347	92.5	7.5
84	116.7	2062.3	2.325		2.347	92.5	7.5
85	116.6	2060.5	2.327	1.00907	2.349	92.6	7.4
86	116.6	2060.5	2.327		2.349	92.6	7.4
87	116.6	2060.5	2.327		2.349	92.6	7.4
88	116.5	2058.7	2.329		2.351	92.7	7.3
89	116.5	2058.7	2.329		2.351	92.7	7.3

(Continued)

TABLE 10.3   (CONTINUED)
Computations of the Compaction Properties of the Asphalt Mixture

N	Height (mm)	Volume (cm³)	G_{mb} Estimated	CF	G_{mb} Corrected	% G_{mm}	V_a (%)
90	116.5	2058.7	2.329		2.351	92.7	7.3
91	116.4	2057.0	2.331		2.353	92.7	7.3
92	116.4	2057.0	2.331		2.353	92.7	7.3
93	116.4	2057.0	2.331		2.353	92.7	7.3
94	116.3	2055.2	2.333		2.355	92.8	7.2
95	116.3	2055.2	2.333		2.355	92.8	7.2
96	116.3	2055.2	2.333		2.355	92.8	7.2
97	116.3	2055.2	2.333		2.355	92.8	7.2
98	116.2	2053.4	2.335		2.357	92.9	7.1
99	116.2	2053.4	2.335		2.357	92.9	7.1
100	116.2	2053.4	2.335		2.357	92.9	7.1
101	116.1	2051.7	2.337		2.359	93.0	7.0
102	116.1	2051.7	2.337		2.359	93.0	7.0
103	116.1	2051.7	2.337		2.359	93.0	7.0
104	116.1	2051.7	2.337		2.359	93.0	7.0
105	116.0	2049.9	2.339		2.361	93.0	7.0
106	116.0	2049.9	2.339		2.361	93.0	7.0
107	116.0	2049.9	2.339		2.361	93.0	7.0
108	116.0	2049.9	2.339		2.361	93.0	7.0
109	115.9	2048.1	2.341		2.363	93.1	6.9
110	115.9	2048.1	2.341		2.363	93.1	6.9
111	115.9	2048.1	2.341		2.363	93.1	6.9
112	115.9	2048.1	2.341		2.363	93.1	6.9
113	115.8	2046.4	2.343		2.365	93.2	6.8
114	115.8	2046.4	2.343		2.365	93.2	6.8
115	115.8	2046.4	2.343		2.365	93.2	6.8
116	115.8	2046.4	2.343		2.365	93.2	6.8
117	115.7	2044.6	2.346		2.367	93.3	6.7
118	115.7	2044.6	2.346		2.367	93.3	6.7
119	115.7	2044.6	2.346		2.367	93.3	6.7
120	115.7	2044.6	2.346		2.367	93.3	6.7
121	115.7	2044.6	2.346		2.367	93.3	6.7
122	115.6	2042.8	2.348		2.369	93.4	6.6
123	115.6	2042.8	2.348		2.369	93.4	6.6
124	115.6	2042.8	2.348		2.369	93.4	6.6
125	115.6	2042.8	2.348		2.369	93.4	6.6
126	115.6	2042.8	2.348		2.369	93.4	6.6
127	115.5	2041.1	2.350		2.371	93.5	6.5
128	115.5	2041.1	2.350		2.371	93.5	6.5
129	115.5	2041.1	2.350		2.371	93.5	6.5
130	115.5	2041.1	2.350		2.371	93.5	6.5
131	115.5	2041.1	2.350		2.371	93.5	6.5
132	115.4	2039.3	2.352		2.373	93.5	6.5
133	115.4	2039.3	2.352		2.373	93.5	6.5
134	115.4	2039.3	2.352		2.373	93.5	6.5

(Continued)

TABLE 10.3   (CONTINUED)

Computations of the Compaction Properties of the Asphalt Mixture

N	Height (mm)	Volume (cm³)	G_{mb} Estimated	CF	G_{mb} Corrected	% G_{mm}	V_a (%)
135	115.4	2039.3	2.352		2.373	93.5	6.5
136	115.4	2039.3	2.352		2.373	93.5	6.5
137	115.4	2039.3	2.352		2.373	93.5	6.5
138	115.3	2037.5	2.354		2.375	93.6	6.4
139	115.3	2037.5	2.354		2.375	93.6	6.4
140	115.3	2037.5	2.354		2.375	93.6	6.4
141	115.3	2037.5	2.354		2.375	93.6	6.4
142	115.3	2037.5	2.354		2.375	93.6	6.4
143	115.3	2037.5	2.354		2.375	93.6	6.4
144	115.2	2035.8	2.356		2.377	93.7	6.3
145	115.2	2035.8	2.356		2.377	93.7	6.3
146	115.2	2035.8	2.356		2.377	93.7	6.3
147	115.2	2035.8	2.356		2.377	93.7	6.3
148	115.2	2035.8	2.356		2.377	93.7	6.3
149	115.2	2035.8	2.356		2.377	93.7	6.3
150	115.2	2035.8	2.356		2.377	93.7	6.3
151	115.1	2034.0	2.358		2.379	93.8	6.2
152	115.1	2034.0	2.358		2.379	93.8	6.2
153	115.1	2034.0	2.358		2.379	93.8	6.2
154	115.1	2034.0	2.358		2.379	93.8	6.2
155	115.1	2034.0	2.358		2.379	93.8	6.2
156	115.1	2034.0	2.358		2.379	93.8	6.2
157	115.1	2034.0	2.358		2.379	93.8	6.2
158	115.0	2032.2	2.360		2.381	93.9	6.1
159	115.0	2032.2	2.360		2.381	93.9	6.1
160	115.0	2032.2	2.360		2.381	93.9	6.1
161	115.0	2032.2	2.360		2.381	93.9	6.1
162	115.0	2032.2	2.360		2.381	93.9	6.1
163	115.0	2032.2	2.360		2.381	93.9	6.1
164	115.0	2032.2	2.360		2.381	93.9	6.1
165	115.0	2032.2	2.360		2.381	93.9	6.1
166	114.9	2030.5	2.362		2.383	93.9	6.1
167	114.9	2030.5	2.362		2.383	93.9	6.1
168	114.9	2030.5	2.362		2.383	93.9	6.1
169	114.9	2030.5	2.362		2.383	93.9	6.1
170	114.9	2030.5	2.362		2.383	93.9	6.1
171	114.9	2030.5	2.362		2.383	93.9	6.1
172	114.9	2030.5	2.362		2.383	93.9	6.1
173	114.9	2030.5	2.362		2.383	93.9	6.1
174	114.9	203 0.5	2.362		2.383	93.9	6.1

Plot the relationship between the number of gyrations (N) and the $\%G_{mm}$ during compaction for the asphalt mixture in Problem 5.1. Comment on the behavior of this mixture during compaction and compare the properties $\%G_{mm}$ @$N_{initial}$, $\%G_{mm}$ at N_{design}, and $\%G_{mm}$ at $N_{maximum}$ with the Superpave criteria knowing that $N_{initial}=8$ gyrations, $N_{design}=109$ gyrations, and $N_{maximum}=174$ gyrations, and the traffic level that this mixture is designed for is 12 million ESALs.

Solution:

The number of gyrations, N is plotted against the $\%G_{mm}$ during compaction for the asphalt mixture as shown in Figure 10.2. A semi-log scale is used in this figure.

According to the Superpave criteria, the $\%G_{mm}$ @$N_{initial} \leq 89\%$, $\%G_{mm}$ at $N_{design}=96\%$, and $\%G_{mm}$ at $N_{maximum} \leq 98\%$. Therefore, the two borders at $N_{initial}=8$ gyrations and $N_{maximum}=109$ gyrations are plotted as shown in Figure 10.1.

The behavior of this asphalt mixture during gyrator compaction does well and passes the Superpave specifications for the $\%G_{mm}$ @$N_{initial}=85.3\% \leq 89\%$, and for the $\%G_{mm}$ at $N_{maximum}=93.9\% \leq 98\%$. However, the mixture does not pass the Superpave specifications for the $\%G_{mm}$ at $N_{design}=93.1\% \neq 96\%$. In other words, the air-void content is equal to 6.9% at N_{design}; it should be 4%. In conclusion, the mixture does not meet the Superpave criteria.

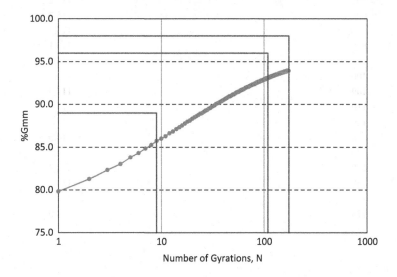

FIGURE 10.2 Number of gyrations versus $\%G_{mm}$ for the asphalt mixture.

10.3 Perform the necessary computations for the asphalt mixture with the data given below to determine the air-void content during gyratory compaction:
- Diameter of gyratory specimen = 150 mm ($\cong$ 6 in)
- Maximum theoretical specific gravity (G_{mm}) for the loose mixture = 2.497
- Bulk specific gravity of aggregate used in the mixture = 2.543
- Bulk specific gravity of compacted specimen = 2.407
- Weight of asphalt mixture = 4776.9 g
- Gyratory compaction data (see Table 10.4).

TABLE 10.4

Superpave Gyratory Compaction Data for an Asphalt Mixture

N	Height (mm)
1	132.8
2	130.4
3	128.6
4	127.3
5	126.2
6	125.3
7	124.5
8	**123.8**
9	123.2
10	122.7
20	119.5
30	117.9
40	116.9
50	116.2
60	115.7
70	115.3
80	115.1
90	114.8
100	114.7
109	**114.5**
110	114.5
120	114.4
130	114.3
140	114.2
150	114.2
160	114.1
170	114.1
174	**114.0**

Solution:

• The estimated bulk specific gravity ($G_{mb\text{-}Estimated}$) for the asphalt mixture during compaction is determined.

   Sample Calculation:

   At number of gyrations (N)=8, the height of the mixture (H)=123.8 mm.

$$V = \pi \frac{D^2}{4} H$$

⇒

$$V = \pi \frac{(150 \times 0.1)^2}{4} (123.8 \times 0.1) = 2187.7 \text{ cm}^3$$

W=weight of the mixture=weight of compacted specimen in air=4776.9 g

$$G_{mb\text{-}Estimated} = \frac{W}{V\gamma_w}$$

⇒

$$G_{\text{mb-Estimated}} = \frac{4776.9}{2187.7 \times 1} = 2.183$$

The estimated bulk specific gravity at all other values for the number of gyrations (N) is computed in a similar manner as above.

- The correction factor for bulk specific gravity is determined as described in the procedure below:

The correction factor for bulk specific gravity is defined as the measured bulk specific gravity divided by the estimated bulk specific gravity at N_{maximum} (at the end of gyrations or gyratory compaction):

At the end of compaction when the number of gyrations (N) = 174, the height of the mixture (H) = 114.0 mm.

$$V = \left(\pi \frac{D^2}{4} \right) H$$

⇒

$$V = \pi \frac{(150 \times 0.1)^2}{4} (114.0 \times 0.1) = 2014.5 \text{ cm}^3$$

W = weight of the mixture = weight of compacted specimen in air = 4795.6 g

$$G_{\text{mb-Estimated}} = \frac{W}{V \gamma_w}$$

⇒

$$G_{\text{mb-Estimated}} = \frac{4776.9}{2014.5 \times 1} = 2.371$$

$G_{\text{mb-Measured}} = 2.407$ (given in the problem).

$$CF = \frac{G_{\text{mb-Measured}}}{G_{\text{mb-Estimated}}}$$

⇒

$$CF = \frac{2.407}{2.371} = 1.015$$

- The corrected bulk specific gravity ($G_{\text{mb-Corrected}}$) at a specific number of gyrations is determined by multiplying the estimated bulk specific gravity at that number of gyrations by the correction factor:

Sample Calculation:

$$G_{\text{mb-Corrected}} = CF \times G_{\text{mb-Estimated}}$$

At N = 8 gyrations, the estimated bulk specific gravity is determined above as:

$$G_{\text{mb-Estimated}} = 2.183$$

$$G_{mb\text{-Corrected}} = CF \times G_{mb\text{-Estimated}}$$

$\Rightarrow$

$$G_{mb\text{-Corrected}} = 1.015 \times 2.183 = 2.216$$

The corrected bulk specific gravity is determined at all other values for the number of gyrations (N) following the same procedure.

• The percent maximum specific gravity ($\%G_{mm}$) is determined using the formula below:

$$\%G_{mm} = \frac{G_{mb\text{-Corrected}}}{G_{mm}} \times 100\%$$

Sample Calculation:
At N=8 gyrations,

$$\%G_{mm} = \frac{G_{mb\text{-Corrected}}}{G_{mm}} \times 100\%$$

$\Rightarrow$

$$\%G_{mm} = \frac{2.216}{2.497} \times 100\% = 88.8\%$$

The $\%G_{mm}$ is determined at all other values for the number of gyrations (N) following the same procedure.

• The percentage of air voids (V_a) is determined using the formula shown below:

$$V_a = 100 - \%G_{mm}$$

Sample Calculation:
At N=8 gyrations,

$$V_a = 100 - \%G_{mm}$$

$\Rightarrow$

$$V_a = 100 - 88.8 = 11.2\%$$

The MS Excel worksheet used to perform the computations of this problem is shown in Figure 10.3.

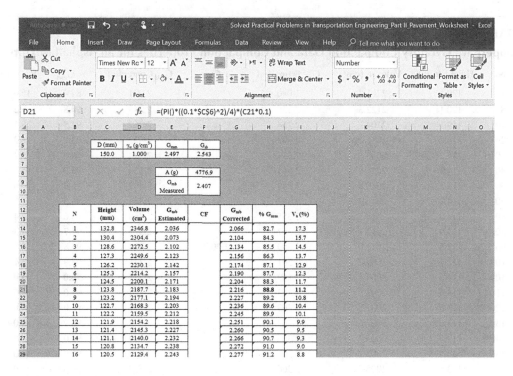

FIGURE 10.3 MS Excel worksheet image for the computations of Problem 10.3.

10.4 The Superpave gyratory compaction data and analysis for an asphalt mixture are given in Table 10.5. Determine the volumetric properties at N_{design}: air voids (V_a), voids in mineral aggregate (VMA), voids filled with asphalt (VFA), and dust proportion (DP) if the following data are also provided:

- Diameter of gyratory specimen = 150 mm ($\cong$ 6 in)
- Maximum theoretical specific gravity (G_{mm}) for the loose mixture = 2.477
- Bulk specific gravity of aggregate used in the mixture = 2.543
- Bulk specific gravity of compacted specimen = 2.381
- Specific gravity of asphalt binder = 1.00
- Weight of asphalt mixture = 4744.6 g
- % Passing No. 200 (75 μm) sieve in aggregate gradation used in the mixture = 4%
- N_{design} = 109 gyrations
- Asphalt binder content = 5.5%

TABLE 10.5

Superpave Gyratory Compaction Data and Compaction Properties for an Asphalt Mixture

N	Height (mm)	Volume (cm³)	G_{mb} Estimated	CF	G_{mb} Corrected	% G_{mm}	V_a (%)
1	129.1	2281.4	2.080		2.091	84.4	15.6
8	120.1	2122.3	2.236		2.248	90.8	9.2
10	119.1	2104.7	2.254		2.267	91.5	8.5
20	116.3	2055.2	2.309		2.322	93.7	6.3
30	115.2	2035.8	2.331		2.344	94.6	5.4
40	114.7	2026.9	2.341		2.354	95.0	5.0
50	114.3	2019.8	2.349		2.362	95.4	4.6
60	114.1	2016.3	2.353		2.366	95.5	4.5
70	114.0	2014.5	2.355		2.368	95.6	4.4
80	113.9	2012.8	2.357		2.371	95.7	4.3
90	113.8	2011.0	2.359	1.00565	2.373	95.8	4.2
100	113.7	2009.2	2.361		2.375	95.9	4.1
109	113.6	2007.5	2.363		2.377	96.0	4.0
110	113.6	2007.5	2.363		2.377	96.0	4.0
120	113.6	2007.5	2.363		2.377	96.0	4.0
130	113.6	2007.5	2.363		2.377	96.0	4.0
140	113.5	2005.7	2.366		2.379	96.0	4.0
150	113.5	2005.7	2.366		2.379	96.0	4.0
160	113.5	2005.7	2.366		2.379	96.0	4.0
170	113.4	2003.9	2.368		2.381	96.1	3.9
174	113.4	2003.9	2.368		2.381	96.1	3.9

Solution:

- The percentage of air voids (V_a) in the asphalt mixture at N_{design} is determined using the following formula:

$$V_a = 100 - \%G_{mm\text{-}N_{design}}$$

$\%G_{mm\text{-}N_{design}} = 96.0\%$ from the compaction data analysis table (see Table 10.5).

⇒

$$V_a = 100 - 96.0 = 4.0\%$$

- The voids in mineral aggregate (VMA) in the asphalt mixture at N_{design} is calculated using the formula below:

$$\text{VMA} = 100 - \left(\frac{\%G_{mm\text{-}N_{design}} \times G_{mm} \times \dfrac{P_s}{100}}{G_{sb}} \right) \tag{10.10}$$

Where:

VMA = voids in mineral aggregate at N_{design}, percent by total volume of asphalt mixture

$\%G_{mm\text{-}Ndesign}$ = percentage of G_{mm} at N_{design}

G_{mm} = theoretical maximum specific gravity

P_s = percentage of aggregate in asphalt mixture
G_{sb} = bulk specific gravity of aggregate
At N_{design} = 109 gyrations,
$\%G_{mm\text{-}Ndesign}$ = 96.0% from the compaction data analysis table given above.

$$P_s = 100 - P_b \tag{10.11}$$

Where:
P_s = percentage of aggregate in the asphalt mixture
P_b = percentage of asphalt binder in the asphalt mixture

$\Rightarrow$

$$P_s = 100 - 5.5 = 94.5\%$$

$$G_{mm} = 2.477$$

$$G_{sb} = 2.543$$

$\Rightarrow$

$$VMA = 100 - \left(\frac{96.0 \times 2.477 \times \dfrac{94.5}{100}}{2.543} \right) = 11.7\%$$

- The voids filled with asphalt (VFA) in the asphalt mixture at N_{design} is calculated using the formula below:

$$VFA = 100 \left(\frac{VMA - V_a}{VMA} \right) \tag{10.12}$$

Where:
VFA = voids filled with asphalt at N_{design}, percent by volume of total voids in asphalt mixture
VMA = voids in mineral aggregate at N_{design}, percent by total volume of asphalt mixture
V_a = air voids at N_{design}, percent by total volume of asphalt mixture

$\Rightarrow$

$$VFA = 100 \left(\frac{11.7 - 4.0}{11.7} \right) = 65.4\%$$

- The dust proportion (DP) in the asphalt mixture is computed using the formula below:

$$DP = \frac{P_{0.075}}{P_{be}} \tag{10.13}$$

Where:
$P_{0.075}$ = percentage of material passing sieve No. 200 (0.075 mm)
P_{be} = percentage of effective asphalt binder content in the asphalt mixture (%)

$$P_{be} = P_b - \left(\frac{P_{ba}}{100}\right)P_s \qquad (10.14)$$

Where:

P_{be} = effective asphalt binder content, percent by total weight of asphalt mixture

P_b = percentage of asphalt binder content, by total weight of asphalt mixture

P_{ba} = absorbed asphalt content, percent by total weight of aggregate

P_s = percentage of aggregate, by total weight of asphalt mixture

And:

$$P_{ba} = 100\left(\frac{G_{se} - G_{sb}}{G_{se}G_{sb}}\right)G_b \qquad (10.15)$$

Where:

P_{ba} = absorbed asphalt content, percent by weight of aggregate

G_{se} = effective specific gravity of aggregate

G_{sb} = bulk specific gravity of aggregate

G_b = specific gravity of asphalt binder

The effective specific gravity of the aggregate (G_{se}) can be determined from the formula below:

$$G_{mm} = \frac{P_{mm}}{\dfrac{P_s}{G_{se}} + \dfrac{P_b}{G_b}} \qquad (10.16)$$

Where:

P_{mm} = percentage of loose mixture by total weight of mixture (100%)

P_s = percentage of aggregate in the asphalt mixture (by total weight of mixture)

G_{se} = effective specific gravity of the aggregate

P_b = percentage of asphalt binder in the asphalt mixture (by total weight of mixture)

G_b = specific gravity of asphalt binder

$\Rightarrow$

$$2.477 = \frac{100}{\dfrac{94.5}{G_{se}} + \dfrac{5.5}{1.00}}$$

$\Rightarrow$

$$\frac{94.5}{G_{se}} + \frac{5.5}{1.00} = \frac{100}{2.477}$$

$\Rightarrow$

$$G_{se} = 2.710$$

$$P_{ba} = 100\left(\frac{G_{se} - G_{sb}}{G_{se}G_{sb}}\right)G_b$$

$\Rightarrow$

$$P_{ba} = 100 \left(\frac{2.710 - 2.543}{2.710 \times 2.543} \right) 1.00 = 2.42\%$$

$$P_{be} = P_b - \left(\frac{P_{ba}}{100} \right) P_s$$

$\Rightarrow$

$$P_{be} = 5.5 - \left(\frac{2.42}{100} \right) 94.5 = 3.21\%$$

$$DP = \frac{P_{0.075}}{P_{be}}$$

$\Rightarrow$

$$DP = \frac{4}{3.21} = 1.2$$

The MS Excel worksheet used to perform the computations of this problem is shown in Figure 10.4.

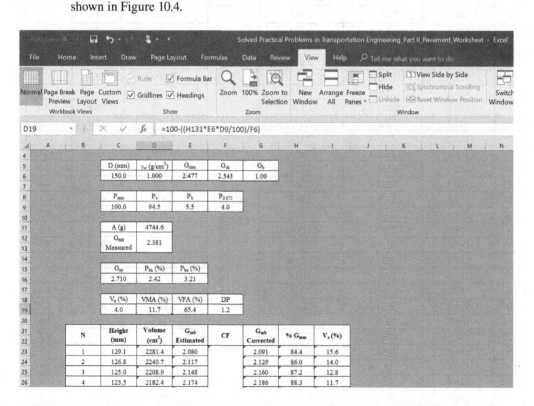

FIGURE 10.4 A screen image of the MS Excel worksheet used for the computations of Problem 10.4.

10.5 Plot the relationship between the number of gyrations (N) and the $\%G_{mm}$ during compaction for the three asphalt mixtures with the compaction data shown in Tables 10.6 through 10.11. The $\%G_{mm}$ is calculated from the given mixture data and the compaction data following the same procedure in Problem 10.1. Comment on the behavior of the three asphalt mixtures during compaction and compare the properties $\%G_{mm}$ @$N_{initial}$, $\%G_{mm}$ at N_{design}, and $\%G_{mm}$ at $N_{maximum}$ with the Superpave criteria knowing that $N_{initial} = 8$ gyrations, $N_{design} = 109$ gyrations, and $N_{maximum} = 174$ gyrations, and the traffic level these mixtures are designed for is 15 million ESALs.

TABLE 10.6

Data for Mixture #1

D (mm)	γ_w (g/cm³)	G_{mm}	G_{sb}	Weight of Mixture (g)	G_{mb} Measured
150.0	1.000	2.467	2.575	4810.0	2.361

TABLE 10.7

Compaction Properties for Asphalt Mixture #1

N	Height (mm)	Volume (cm³)	G_{mb} Estimated	CF	G_{mb} Corrected	% G_{mm}	V_a (%)
1	137.0	2421.0	1.987		2.001	81.1	18.9
2	134.5	2376.8	2.024		2.038	82.6	17.4
3	132.6	2343.2	2.053		2.067	83.8	16.2
4	131.1	2316.7	2.076		2.091	84.8	15.2
5	129.9	2295.5	2.095		2.110	85.5	14.5
6	128.8	2276.1	2.113		2.128	86.3	13.7
7	128.0	2261.9	2.126		2.142	86.8	13.2
8	127.1	2246.0	2.142		2.157	87.4	12.6
9	126.5	2235.4	2.152		2.167	87.8	12.2
10	125.8	2223.1	2.164		2.179	88.3	11.7
20	122.1	2157.7	2.229		2.245	91.0	9.0
30	120.3	2125.9	2.263		2.279	92.4	7.6
40	119.2	2106.4	2.283		2.300	93.2	6.8
50	118.5	2094.1	2.297	1.00720	2.313	93.8	6.2
60	118.0	2085.2	2.307		2.323	94.2	5.8
70	117.6	2078.2	2.315		2.331	94.5	5.5
80	117.4	2074.6	2.318		2.335	94.7	5.3
90	117.1	2069.3	2.324		2.341	94.9	5.1
100	116.9	2065.8	2.328		2.345	95.1	4.9
109	116.8	2064.0	2.330		2.347	95.1	4.9
110	116.8	2064.0	2.330		2.347	95.1	4.9
120	116.6	2060.5	2.334		2.351	95.3	4.7
130	116.5	2058.7	2.336		2.353	95.4	4.6
140	116.4	2057.0	2.338		2.355	95.5	4.5
150	116.3	2055.2	2.340		2.357	95.6	4.4
160	116.2	2053.4	2.342		2.359	95.6	4.4
170	116.2	2053.4	2.342		2.359	95.6	4.4
174	116.1	2051.7	2.344		2.361	95.7	4.3

TABLE 10.8

Data for Asphalt Mixture #2

D (mm)	γ_w (g/cm³)	G_{mm}	G_{sb}	Weight of Mixture (g)	G_{mb} Measured
150.0	1.000	2.449	2.575	4841.0	2.355

TABLE 10.9

Compaction Properties for Asphalt Mixture #2

N	Height (mm)	Volume (cm³)	G_{mb} Estimated	CF	G_{mb} Corrected	% G_{mm}	V_a (%)
1	134.5	2376.8	2.037		2.046	83.6	16.4
2	132.0	2332.6	2.075		2.085	85.1	14.9
3	130.1	2299.1	2.106		2.116	86.4	13.6
4	128.5	2270.8	2.132		2.142	87.5	12.5
5	127.2	2247.8	2.154		2.164	88.4	11.6
6	126.3	2231.9	2.169		2.179	89.0	11.0
7	125.5	2217.8	2.183		2.193	89.6	10.4
8	124.6	2201.9	2.199		2.209	90.2	9.8
9	124.0	2191.3	2.209		2.220	90.6	9.4
10	123.4	2180.7	2.220		2.231	91.1	8.9
20	120.3	2125.9	2.277		2.288	93.4	6.6
30	119.0	2102.9	2.302		2.313	94.4	5.6
40	118.3	2090.5	2.316		2.327	95.0	5.0
50	117.9	2083.5	2.324	1.00476	2.335	95.3	4.7
60	117.7	2079.9	2.327		2.339	95.5	4.5
70	117.5	2076.4	2.331		2.343	95.7	4.3
80	117.4	2074.6	2.333		2.345	95.7	4.3
90	117.3	2072.9	2.335		2.347	95.8	4.2
100	117.2	2071.1	2.337		2.349	95.9	4.1
109	117.2	2071.1	2.337		2.349	95.9	4.1
110	117.2	2071.1	2.337		2.349	95.9	4.1
120	117.1	2069.3	2.339		2.351	96.0	4.0
130	117.1	2069.3	2.339		2.351	96.0	4.0
140	117.0	2067.6	2.341		2.353	96.1	3.9
150	117.0	2067.6	2.341		2.353	96.1	3.9
160	116.9	2065.8	2.343		2.355	96.1	3.9
170	116.9	2065.8	2.343		2.355	96.1	3.9
174	116.9	2065.8	2.343		2.355	96.1	3.9

TABLE 10.10

Data for Asphalt Mixture #3

D (mm)	γ_w (g/cm³)	G_{mm}	G_{sb}	Weight of Mixture (g)	G_{mb} Measured
150.0	1.000	2.439	2.575	4860.0	2.341

TABLE 10.11

Compaction Properties for Asphalt Mixture #3

N	Height (mm)	Volume (cm³)	G_{mb} Estimated	CF	G_{mb} Corrected	% G_{mm}	V_a (%)
1	129.8	2293.8	2.119		2.126	87.2	12.8
2	127.9	2260.2	2.150		2.158	88.5	11.5
3	126.2	2230.1	2.179		2.187	89.7	10.3
4	124.9	2207.2	2.202		2.210	90.6	9.4
5	123.8	2187.7	2.221		2.229	91.4	8.6
6	123.1	2175.4	2.234		2.242	91.9	8.1
7	122.3	2161.2	2.249		2.257	92.5	7.5
8	121.9	2154.2	2.256		2.264	92.8	7.2
9	121.5	2147.1	2.264		2.272	93.1	6.9
10	121.2	2141.8	2.269		2.277	93.4	6.6
20	119.4	2110.0	2.303		2.312	94.8	5.2
30	118.8	2099.4	2.315		2.323	95.3	4.7
40	118.5	2094.1	2.321		2.329	95.5	4.5
50	118.3	2090.5	2.325	1.00360	2.333	95.7	4.3
60	118.2	2088.8	2.327		2.335	95.7	4.3
70	118.2	2088.8	2.327		2.335	95.7	4.3
80	118.1	2087.0	2.329		2.337	95.8	4.2
90	118.1	2087.0	2.329		2.337	95.8	4.2
100	118.0	2085.2	2.331		2.339	95.9	4.1
109	118.0	2085.2	2.331		2.339	95.9	4.1
110	118.0	2085.2	2.331		2.339	95.9	4.1
120	118.0	2085.2	2.331		2.339	95.9	4.1
130	117.9	2083.5	2.333		2.341	96.0	4.0
140	117.9	2083.5	2.333		2.341	96.0	4.0
150	117.9	2083.5	2.333		2.341	96.0	4.0
160	117.9	2083.5	2.333		2.341	96.0	4.0
170	117.9	2083.5	2.333		2.341	96.0	4.0
174	117.9	2083.5	2.333		2.341	96.0	4.0

Solution:

The number of gyrations, N is plotted against the %G_{mm} during compaction for the three asphalt mixtures as shown in Figure 10.5. The compaction properties of the three asphalt mixtures are shown in Table 10.12.

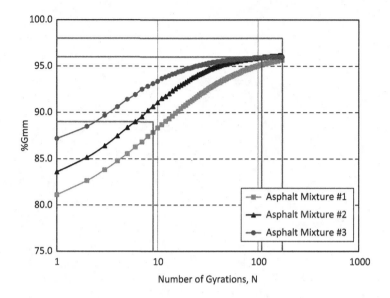

FIGURE 10.5   Number of gyrations versus %G$_{mm}$ for the three asphalt mixtures.

TABLE 10.12

Compaction Properties for the Three Mixtures Versus Superpave Criteria

Asphalt Mixture #	%G$_{mm}$ at N$_{initial}$	%G$_{mm}$ at N$_{design}$	%G$_{mm}$ at N$_{maximum}$
1	87.4	95.1	95.7
2	90.2	95.9	96.1
3	92.8	95.9	96.0
Superpave Criteria*	≤ 89.0	$= 96.0$	≤ 98.0
Pass/Fail	#1 Passes; #2 and #3 Fail	All Fail	All Pass
Conclusion	None of the three mixtures passes the Superpave criteria.		

* Source for Superpave Criteria: Superpave Mix Design, Asphalt Institute Superpave Series No.2 (SP-2), 2001.

11 Superpave Mixture Design Method

Chapter 11 discusses the Superpave mixture design method with all the steps required in the design process. The Superpave gyratory compaction discussed in the previous chapter is an important step in the Superpave mixture design. The selection of aggregate, asphalt binder, and aggregate gradation is done according to the Superpave criteria prior to mixture design. Throughout the mixture design method, typically three trial aggregate blends are used to check whether any of the blends meets the Superpave criteria. Volumetric properties including air voids (V_a), voids in mineral aggregate (VMA), voids filled with asphalt (VFA), and dust proportion (DP), and compaction properties including $\%G_{mm}$ at $N_{initial}$, $\%G_{mm}$ at N_{design}, $\%G_{mm}$ at $N_{maximum}$, are determined after the compaction process. If the air-void content is 4%, the values of the volumetric and compaction properties for each of the trial mixtures are checked against the Superpave criteria. If the criteria are met, the blend is acceptable; otherwise, a redesign of the aggregate blend is required. If the air-void content is altered from 4%, the asphalt binder content is reestimated to correct for the air-void content using a special formula. Afterwards, the other volumetric and compaction properties are recalculated accordingly and compared with the Superpave criteria to check whether they meet the criteria. If the criteria are met, the blend is acceptable; otherwise, a redesign of the aggregate blend is needed. The next step after the selection of the design aggregate blend, asphalt binder contents of $\pm0.5\%$ and $\pm1.0\%$ from the estimated asphalt binder content are used to compact at least two specimens at each binder content using the Superpave gyratory compactor (SGC). The volumetric and compaction properties are calculated again after compaction for each asphalt binder content and plotted versus the asphalt binder content. The asphalt binder content at 4% air voids is selected as the design asphalt binder content. The volumetric and compaction properties at the design asphalt binder content are then determined for the mixture and compared with the Superpave criteria. The moisture sensitivity test is also conducted on gyratory-compacted specimens of the mixture following a standard procedure to determine the indirect tensile strength ratio (TSR) of the specimens that should be equal or higher than 0.80.

This chapter includes questions and problems dealing with the compaction properties for asphalt mixtures in the Superpave gyratory compactor (SGC), the compactive effort, the Superpave criteria for gyratory compaction, and the computations related to gyratory compaction of asphalt mixtures. The Superpave gyratory compactor simulates the compaction that takes place in field. Three levels of compaction are typically used in Superpave: $N_{initial}$, N_{design}, and $N_{maximum}$. $N_{initial}$ measures mixture compactability during construction and can easily verify if the mixture is tender during construction and unstable under traffic. N_{design} measures the density of the mixture that is equivalent to the density in the field after specific traffic loadings. On the other hand, $N_{maximum}$ measures the maximum mixture in field that should never be exceeded. Asphalt mixtures with very low air voids at $N_{maximum}$ indicate that these mixtures may experience excessive compaction under traffic resulting in low air voids, bleeding, and rutting. Therefore, during gyratory compaction, the behavior of asphalt mixture is monitored, and density and air voids are calculated for the asphalt mixture. The compaction properties provide a good idea about the tenderness and stiffness of the mixture and can predict the behavior of the mixture in pavement under traffic throughout the service life of the pavement. This part covers questions and problems that deal with the Superpave mixture design including the computation of the volumetric and compaction properties of asphalt mixtures, plotting the different curves required in the mixture design, determining the design asphalt binder content, and comparing the mixture properties against the Superpave criteria.

11.1　In a Superpave mixture design method, an initial asphalt binder content of 5.0 percent is used to prepare two gyratory specimens using the Superpave gyratory compactor (SGC) in an attempt to determine the estimated asphalt binder content to be used in the mixture design. The following volumetric properties at N_{design} and compaction properties are obtained at this initial asphalt binder content as shown in Table 11.1. Perform the necessary computations to readjust the asphalt binder content to determine an estimated value for the asphalt binder content. Based on that, recalculate the volumetric and compaction properties of the mixture.

TABLE 11.1

Volumetric Properties at N_{design} and Compaction Properties for an Asphalt Mixture

V_a (%)	VMA (%)	VFA (%)	DP	% G_{mm} at $N_{initial}$	% G_{mm} at N_{design}	% G_{mm} at $N_{maximum}$
5.2	12.3	57.8	1.0	87.2	94.8	95.2

- Maximum theoretical specific gravity of asphalt mixture = 2.485
- Bulk specific gravity of aggregate used in the mixture = 2.553
- Effective specific gravity of aggregate = 2.683
- Specific gravity of asphalt binder = 1.035
- Percent passing sieve No. 200 (75 µm) for the aggregate gradation used in the mixture = 3.0%

Solution:

Since the air-void content (V_a) at N_{design} is not equal to 4%, an adjustment for the initial asphalt binder content is performed to determine an estimated value for the binder content using the following formula:

$$P_{b\text{-estimated}} = P_{b\text{-initial}} - \left[0.4(4 - V_a)\right] \qquad (11.1)$$

Where:

$P_{b\text{-estimated}}$ = estimated asphalt binder content based on adjustment for the initial binder content (percent by total weight of mixture)

$P_{b\text{-initial}}$ = initial asphalt binder content used (percent by total weight of mixture)

0.4 = adjustment factor

Therefore,

$$P_{b\text{-estimated}} = 5.0 - \left[0.4(4 - 5.2)\right] = 5.48\%$$

$$VMA_{estimated} = VMA_{initial} + \left[C(4 - V_a)\right] \qquad (11.2)$$

Where:

$VMA_{estimated}$ = estimated voids in mineral aggregate based on the estimated asphalt binder content (percent by total volume of mixture)

$VMA_{initial}$ = initial voids in mineral aggregate at the initial asphalt binder content (percent by total volume of mixture)

C = adjustment factor; 0.1 if V_a < 4%, and 0.2 if V_a > 4%.

Therefore,

C = 0.2 since V_a = 5.2% > 4%

$$\mathrm{VMA}_{\mathrm{estimated}} = 12.3 + \left[0.2(4 - 5.2) \right] = 12.06\%$$

$$\mathrm{VFA}_{\mathrm{estimated}} = 100 \left(\frac{\mathrm{VMA}_{\mathrm{estimated}} - 4.0}{\mathrm{VMA}_{\mathrm{estimated}}} \right) \qquad (11.3)$$

Where:

$\mathrm{VFA}_{\mathrm{estimated}}$ = estimated voids filled with asphalt based on the estimated asphalt binder content (percent by total volume of voids in mineral aggregate)

$\mathrm{VMA}_{\mathrm{estimated}}$ = estimated voids in mineral aggregate based on the estimated asphalt binder content (percent by total volume of mixture)

$\Rightarrow$

$$\mathrm{VFA}_{\mathrm{estimated}} = 100 \left(\frac{12.06 - 4.0}{12.06} \right) = 66.83\%$$

$$\%G_{\mathrm{mm\text{-}estimated}} @ N_{\mathrm{initial}} = \%G_{\mathrm{mm\text{-}initial}} @ N_{\mathrm{initial}} - (4 - V_a) \qquad (11.4)$$

Where:

$\%G_{\mathrm{mm\text{-}estimated}} @ N_{\mathrm{initial}}$ = percentage of maximum theoretical specific gravity at N_{initial} using the estimated asphalt binder content

$\%G_{\mathrm{mm\text{-}initial}} @ N_{\mathrm{initial}}$ = percentage of maximum theoretical specific gravity at N_{initial} using the initial asphalt binder content

Therefore,

$$\%G_{\mathrm{mm\text{-}estimated}} @ N_{\mathrm{initial}} = 87.2 - (4 - 5.2) = 88.4\%$$

$$\%G_{\mathrm{mm\text{-}estimated}} @ N_{\mathrm{maximum}} = \%G_{\mathrm{mm\text{-}initial}} @ N_{\mathrm{maximum}} - (4 - V_a) \qquad (11.5)$$

Where:

$\%G_{\mathrm{mm\text{-}estimated}} @ N_{\mathrm{maximum}}$ = percentage of maximum theoretical specific gravity at N_{maximum} using the estimated asphalt binder content

$\%G_{\mathrm{mm\text{-}initial}} @ N_{\mathrm{maximum}}$ = percentage of maximum theoretical specific gravity at N_{maximum} using the initial asphalt binder content

Therefore,

$$\%G_{\mathrm{mm\text{-}estimated}} @ N_{\mathrm{maximum}} = 95.2 - (4 - 5.2) = 96.4\%$$

$$P_{\mathrm{be}} = P_{\mathrm{b\text{-}estimated}} - P_s \left(\frac{G_{\mathrm{se}} - G_{\mathrm{sb}}}{G_{\mathrm{se}} G_{\mathrm{sb}}} \right) G_b \qquad (11.6)$$

Where:

P_{be} = effective asphalt binder content based on the estimated binder content (percent by total weight of mixture)

$P_{\mathrm{b\text{-}estimated}}$ = estimated asphalt binder content (percent by total weight of mixture)

P_s = percentage of aggregate in asphalt mixture (percent by total weight of mixture)

G_{se} = effective specific gravity of aggregate
G_{sb} = bulk specific gravity of aggregate
G_b = specific gravity of asphalt binder

$\Rightarrow$

$P_s = 100 - 5.0 = 95\%$

$$P_{be} = 5.48 - 95 \left(\frac{2.683 - 2.553}{2.683 \times 2.553} \right) 1.035 = 3.61\%$$

If the effective specific gravity of aggregate is not given, it can be calculated using the following formula:

$$G_{mm} = \frac{P_{mm}}{\dfrac{P_s}{G_{se}} + \dfrac{P_b}{G_b}}$$

(11.7)

Where:
P_{mm} = percentage of loose mixture by total weight of mixture (100%)
P_s = percentage of aggregate in the asphalt mixture (by total weight of mixture)
G_{se} = effective specific gravity of the aggregate
P_b = percentage of asphalt binder in the asphalt mixture (by total weight of mixture)
G_b = specific gravity of asphalt binder

$$DP = \frac{P_{0.075}}{P_{be}}$$

(11.8)

Where:
DP = dust proportion (represented as a ratio)
$P_{0.075}$ = percentage of material passing sieve No. 200 (0.075 mm)
P_{be} = effective asphalt binder content based on the estimated binder content (percent by total weight of mixture)

Therefore,

$$DP = \frac{3}{3.61} = 0.83$$

These volumetric and compaction properties are compared with the Superpave criteria to decide whether the aggregate blend of this mixture is acceptable or not.

The MS Excel worksheet used to solve this problem is shown in Figure 11.1.

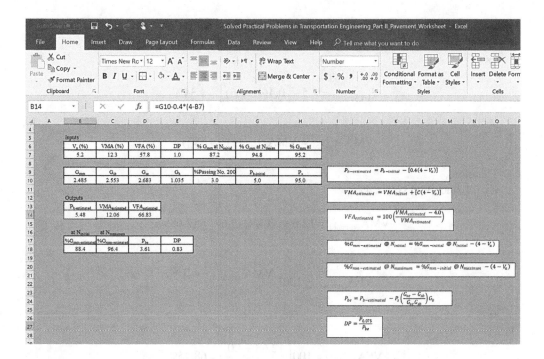

FIGURE 11.1 MS Excel worksheet image for the computations of Problem 11.1.

11.2 An initial asphalt binder content of 4.5% is used in a Superpave mixture design method
to prepare Superpave gyratory-compacted specimens in an attempt to determine the estimated asphalt binder content that will be used in the mixture design. The following volumetric properties at N_{design} and compaction properties are obtained at this initial asphalt
binder content as shown in Table 11.2. Adjust the initial asphalt binder content and the
volumetric and compaction properties that will provide an air-void content of 4% instead of
3.6% at N_{design}. Compare the adjusted properties of the mixture with the Superpave criteria
if the mixture is designed for a traffic level of 16 million ESALs and check whether the
aggregate blend used is acceptable or not!

TABLE 11.2
Volumetric Properties at N_{design} and Compaction Properties for an Asphalt Mixture

V_a (%)	VMA (%)	VFA (%)	DP	% G_{mm} at $N_{initial}$	% G_{mm} at N_{design}	% G_{mm} at $N_{maximum}$
3.6	12.6	71.7	1.5	86.9	96.4	97.2

- Maximum theoretical specific gravity of asphalt mixture = 2.438
- Bulk specific gravity of aggregate used in the mixture = 2.568
- Effective specific gravity of aggregate = 2.604
- Specific gravity of asphalt binder = 1.035
- Nominal maximum aggregate size (NMAS) for the aggregate gradation = 19.0 mm
- Percent passing sieve No. 200 (75 µm) for the aggregate gradation used in the mixture = 5.8%

Solution:

Since the air-void content (V_a) at N_{design} is not equal to 4%, an adjustment for the initial
asphalt binder content is performed to determine an estimated value for the binder content
using the following formula:

$$P_{\text{b-estimated}} = P_{\text{b-initial}} - \left[0.4(4 - V_a)\right]$$

$\Rightarrow$

$$P_{\text{b-estimated}} = 4.5 - \left[0.4(4 - 3.6)\right] = 4.34\%$$

$$\text{VMA}_{\text{estimated}} = \text{VMA}_{\text{initial}} + \left[C(4 - V_a)\right]$$

Therefore,
$C = 0.1$ since $V_a = 3.6\% < 4\%$

$$\text{VMA}_{\text{estimated}} = 12.6 + \left[0.1(4 - 3.6)\right] = 12.68\%$$

$$\text{VFA}_{\text{estimated}} = 100\left(\frac{\text{VMA}_{\text{estimated}} - 4.0}{\text{VMA}_{\text{estimated}}}\right)$$

$\Rightarrow$

$$\text{VFA}_{\text{estimated}} = 100\left(\frac{12.68 - 4.0}{12.68}\right) = 68.45\%$$

$$\%G_{\text{mm-estimated}} @ N_{\text{initial}} = \%G_{\text{mm-initial}} @ N_{\text{initial}} - (4 - V_a)$$

Therefore,

$$\%G_{\text{mm-estimated}} @ N_{\text{initial}} = 86.9 - (4 - 3.6) = 86.5\%$$

$$\%G_{\text{mm-estimated}} @ N_{\text{maximum}} = \%G_{\text{mm-initial}} @ N_{\text{maximum}} - (4 - V_a)$$

Therefore,

$$\%G_{\text{mm-estimated}} @ N_{\text{maximum}} = 97.2 - (4 - 3.6) = 96.8\%$$

$$P_{\text{be}} = P_{\text{b-estimated}} - P_s\left(\frac{G_{\text{se}} - G_{\text{sb}}}{G_{\text{se}} G_{\text{sb}}}\right)G_b$$

$\Rightarrow$

$$P_s = 100 - 4.5 = 95.5\%$$

$$P_{\text{be}} = 4.34 - 95.5\left(\frac{2.604 - 2.568}{2.604 \times 2.568}\right)1.035 = 3.81\%$$

$$DP = \frac{P_{0.075}}{P_{\text{be}}}$$

Therefore,

$$DP = \frac{5.8}{3.81} = 1.52$$

These volumetric and compaction properties are compared with the Superpave criteria to decide whether the aggregate blend of this mixture is acceptable or not. The Superpave SGC criteria is shown in Table 11.3.

TABLE 11.3
Superpave Criteria for Mixture Design Properties

Traffic Level (million ESALs)	%G_{mm}			VMA (%)	VFA (%)	DP
	@ $N_{initial}$	@ N_{design}	@ $N_{maximum}$			
< 0.3	≤ 91.5				70–80	
0.3 – < 3	≤ 90.5	96.0	≤ 98.0	NA	65–78	0.6–1.2
≥ 3	≤ 89.0				65–75	

* Reproduced with permission from *Superpave Mix Design SP-2*, 2001, by Asphalt Institute, Lexington, Kentucky, USA.

The Superpave VMA criteria is based on the nominal maximum aggregate size (NMAS) as shown in Table 11.4.

TABLE 11.4
Superpave VMA Criteria

NMAS (mm)	Minimum VMA (%)
9.5	15
12.5	14
19.0	13
25.0	12
37.5	11

* Reproduced with permission from *Superpave Mix Design SP-2*, 2001, by Asphalt Institute, Lexington, Kentucky, USA.

The adjusted volumetric and compaction properties are summarized in Table 11.5.

TABLE 11.5
Adjusted Volumetric and Compaction Properties for the Asphalt Mixture

V_a (%)	VMA (%)	VFA (%)	DP	% G_{mm} at $N_{initial}$	% G_{mm} at N_{design}	% G_{mm} at $N_{maximum}$
4.0	12.68	68.45	1.52	86.5	96.0	96.8
4.0	≥ 13	65–75	0.6–1.2	≤ 89.0	96.0	≤ 98.0
⇒	Not OK	OK	Not OK	OK	OK	OK

Since the traffic level is 16 million ESALs and the NMAS for the aggregate gradation is 19.0 mm, the Superpave criteria would be as shown in the third row of Table 11.5.

From this table and after comparing the volumetric and compaction properties with the Superpave SGC criteria, it is revealed that all the properties are ok except for the VMA and DP, which do not meet the criteria. In conclusion, this aggregate blend would be not acceptable to proceed with in the Superpave mixture design. Another aggregate blend should be attempted.

The MS Excel worksheet used to solve this problem is shown in Figure 11.2.

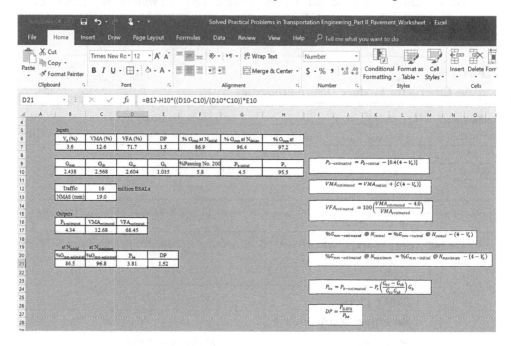

FIGURE 11.2 MS Excel worksheet image for the computations of Problem 11.2.

11.3 An acceptable 19.0-mm (3/4 in) NMAS-aggregate blend is used in a Superpave mixture design method. Asphalt binder contents of ±0.5% and +1.0% from the estimated asphalt binder content (5.0%) are used in the mixture design procedure. These asphalt binder contents are: 4.5, 5.0, 5.5, and 6.0%. The volumetric and compaction properties are obtained at these binder contents as shown in Table 11.6. Plot the necessary curves needed in mixture design, determine the design asphalt binder content and the volumetric and compaction properties at the design asphalt content, and compare these properties with the Superpave criteria if the asphalt mixture is designed for a highway pavement to carry traffic loadings of 8 million ESALs.

TABLE 11.6

Volumetric and Compaction Properties for a 19.0-mm Asphalt Mixture in a Superpave Mixture Design Experiment

P_b (%)	V_a (%)	VMA (%)	VFA (%)	DP	% G_{mm} at $N_{initial}$	% G_{mm} at N_{design}	% G_{mm} at $N_{maximum}$
4.5	5.6	12.9	56.4	1.4	82.1	94.4	94.6
5.0	4.8	12.4	61.2	1.2	83.4	95.2	95.5
5.5	4.4	13.2	66.7	1.0	85.3	95.6	96.2
6.0	3.9	14.0	72.2	0.9	87.2	96.1	96.8

Solution:

• The different curves are plotted between each of the compaction and volumetric properties and the asphalt binder content as shown in Figure 11.3.

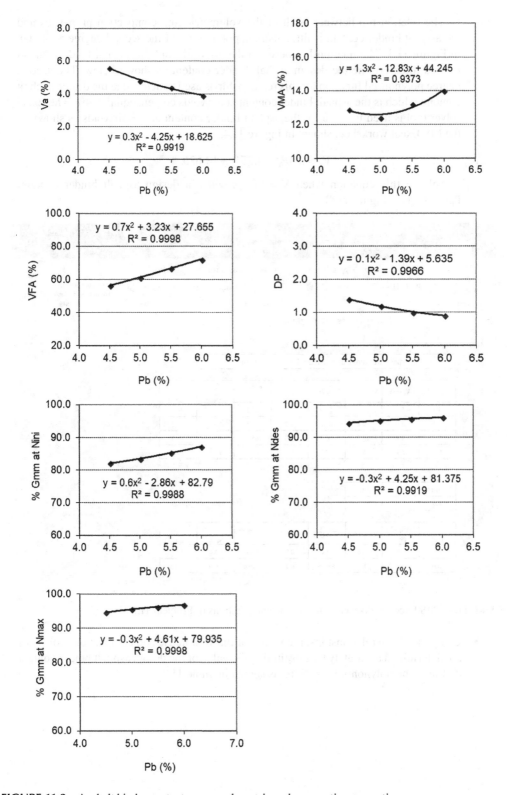

FIGURE 11.3   Asphalt binder content versus volumetric and compaction properties.

The relationship between each of the volumetric and compaction properties and the asphalt binder content is fitted using a polynomial of the second degree as shown in Figure 11.3. The relationship between the air voids and the asphalt binder content is used to determine the design asphalt binder content as shown below. The second degree polynomial function for this relationship is used to determine the design asphalt content, which is the asphalt binder content at air voids content equal to 4%. The Excel solver tool is used to solve for the asphalt binder content at 4% air voids as shown in the MS Excel worksheet shown in Figure 11.4.

$$V_a = 0.3P_b^2 - 4.25P_b + 18.625$$

Solving this equation when $V_a = 4\%$, provides a design asphalt binder content; $P_b = 5.9\%$ (see Figure 11.4).

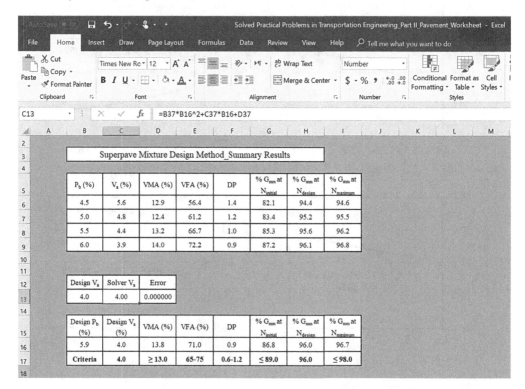

FIGURE 11.4 MS Excel worksheet image for the computations of Problem 11.3.

- Using the above relationships, the values of the volumetric and compaction properties are determined accurately by substituting the value of the design asphalt binder content (5.9%) in the polynomial functions as shown in Table 11.7.

TABLE 11.7
Volumetric and Compaction Properties at the Design Asphalt Binder Content

Design P_b (%)	Design V_a (%)	VMA (%)	VFA (%)	DP	% G_{mm} at $N_{initial}$	% G_{mm} at N_{design}	% G_{mm} at $N_{maximum}$
5.9	4.0	13.8	71.0	0.9	86.8	96.0	96.7
Criteria	4.0	≥ 13.0	65–75	0.6–1.2	≤ 89.0	96.0	≤ 98.0

- By comparing the volumetric and compaction properties obtained at $P_b=5.9\%$, it is noticed that all the values are within the Superpave criteria for a traffic level of 8 million ESALs. Therefore, the asphalt mixture in this problem passes successfully the Superpave criteria and can be used for the highway pavement with a traffic loading of 8 million ESALs.

11.4 The volumetric and compaction properties are obtained for an asphalt mixture with a nominal maximum aggregate size (NMAS) of 19.0 mm (3/4 in) in a Superpave mixture design method at four asphalt binder contents as shown in Table 11.8. Using this data, determine the design asphalt binder content and the volumetric and compaction properties at the design asphalt content. Does this asphalt mixture fulfill the Superpave criteria at a traffic level of 2 million ESALs?

TABLE 11.8
Volumetric and Compaction Properties for a 19.0-mm Asphalt Mixture in a Superpave Mixture Design Experiment

P_b (%)	V_a (%)	VMA (%)	VFA (%)	DP	% G_{mm} at $N_{initial}$	% G_{mm} at N_{design}	% G_{mm} at $N_{maximum}$
4.5	3.4	12.4	72.6	1.5	86.8	96.6	97.4
5.0	2.2	12.7	82.8	1.3	89.1	97.8	98.2
5.5	0.2	12.8	98.5	1.1	93.2	99.8	99.9
6.0	0.4	13.9	97.3	1.0	95.3	99.6	99.8

Solution:

- The relationships between each of the compaction and volumetric properties and the asphalt binder content are plotted as shown in Figure 11.5.

  The relationship between each of the volumetric and compaction properties and the asphalt binder content is fitted using a polynomial of the second degree as shown in Figure 11.5. The relationship between the air voids and the asphalt binder content is used to determine the design asphalt binder content as shown below. The sceond degree polynomial function for this relationship is used to determine the design asphalt content, which is the asphalt binder content at air voids content equal to 4%. The Excel solver tool is used to solve for the asphalt binder content at 4% air voids as shown in the MS Excel worksheet shown in Figure 11.6.

$$V_a = 1.3882P_b^2 - 16.794P_b + 51.01$$

  Solving this equation when $V_a=4\%$, provides a design asphalt binder content; $P_b=4.4\%$

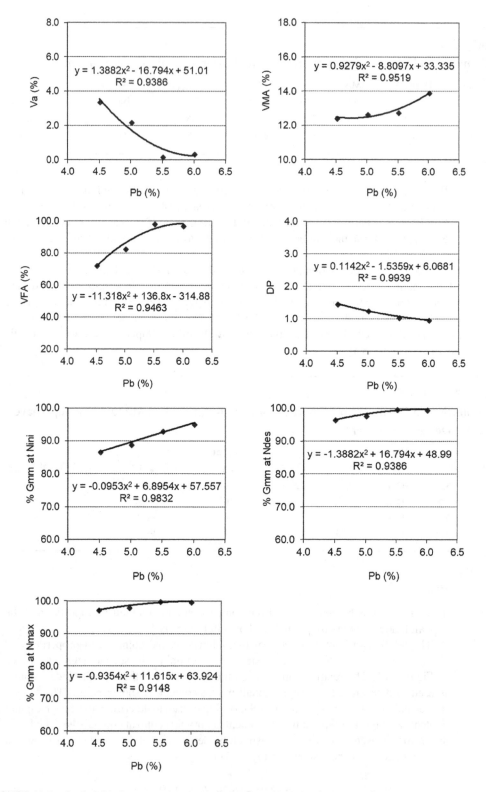

FIGURE 11.5   Asphalt binder content versus volumetric and compaction properties.

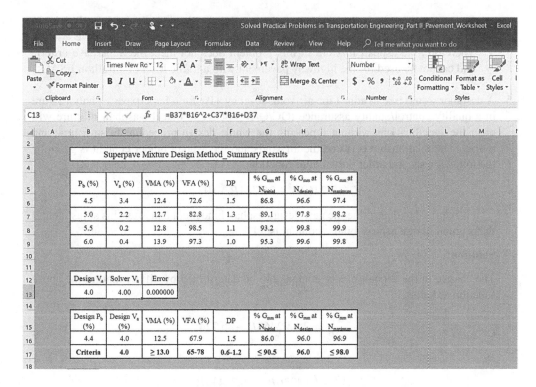

FIGURE 11.6 MS Excel worksheet image for the computations of Problem 11.4.

- Using the above relationships, the values of the volumetric and compaction properties are determined accurately by substituting the value of the design asphalt binder content (4.4%) in the polynomial functions as shown in Table 11.9.

TABLE 11.9
Volumetric and Compaction Properties at the Design Asphalt Binder Content

Design P_b (%)	Design V_a (%)	VMA (%)	VFA (%)	DP	% G_{mm} at $N_{initial}$	% G_{mm} at N_{design}	% G_{mm} at $N_{maximum}$
4.4	4.0	12.5	67.9	1.5	86.0	96.0	96.9
Criteria	4.0	≥ 13.0	65–78	0.6–1.2	≤ 90.5	96.0	≤ 98.0

- By comparing the volumetric and compaction properties obtained at $P_b = 4.4\%$, it is noticed that the VMA and the DP values are not within the Superpave criteria. Therefore, the asphalt mixture in this problem does not meet the Superpave criteria and cannot be used for the highway pavement with a traffic level of 2 million ESALs.

11.5 A designer would like to adjust the air voids level by +1% for a Superpave asphalt mixture in a quality control/quality assurance (QC/QA) process so that the asphalt mixture will meet the specifications without repeating the whole design procedure to save time and money. The relationship between the air voids and the asphalt binder content is given by the following second order polynomial function:

$$V_a = 3.2463P_b^2 - 36.724P_b + 107.58$$

It is required to determine the new design asphalt binder content after this adjustment if the original design asphalt binder content is 5.1%.

Solution:

The change in the asphalt binder content can be determined by differentiating the above function as below:

$$\frac{dV_a}{dP_b} = 2 \times 3.2463P_b - 36.724$$

$\Rightarrow$

$$dV_a = (6.4926P_b - 36.724)dP_b$$

$\Rightarrow$

$$dP_b = \frac{1\%}{6.4926(5.1) - 36.724} = -0.28\%$$

Therefore,

The new design asphalt binder content $= 5.1 - 0.28 = 4.8\%$

In other words, by reducing the design asphalt content by 0.28%, the air voids level will increase by 1%.

The MS Excel worksheet used to solve this problem is shown in Figure 11.7:

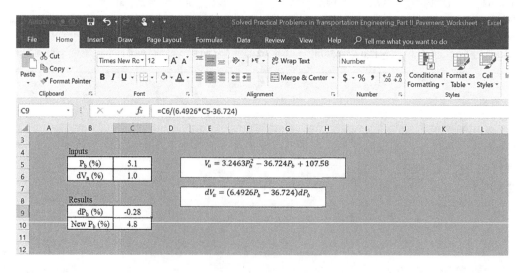

FIGURE 11.7 MS Excel worksheet image for the computations of Problem 11.5.

11.6 In a Superpave mixture design method for an asphalt mixture with 19.0 mm (3/4 in) nominal maximum aggregate size (NMAS), the following volumetric and compaction properties are obtained as shown in Table 11.10. The mixture is designed for a traffic level of 6 million ESALs. The relationships between each of the air voids and VMA and the asphalt binder content are also given in Figures 11.8 and 11.9, respectively. As shown in Table 11.10, the VMA does not meet the Superpave criteria for a NMAS of 19.0 mm ($\geq$ 13.0%). It is required to do a little adjustment in the design asphalt binder content so that the VMA will fulfill the criteria. What is the effect of this change in the binder content on the design air voids?

TABLE 11.10

Design Volumetric and Compaction Properties for an Asphalt Mixture in a Superpave Mixture Design Experiment

Design P_b (%)	Design V_a (%)	VMA (%)	VFA (%)	DP	% G_{mm} at $N_{initial}$	% G_{mm} at N_{design}	% G_{mm} at $N_{maximum}$
5.3	4.0	12.8	68.8	0.8	85.4	96.0	97.1
Criteria	4.0	$\geq$ 13.0	65–75	0.6–1.2	$\leq$ 89.0	96.0	$\leq$ 98.0

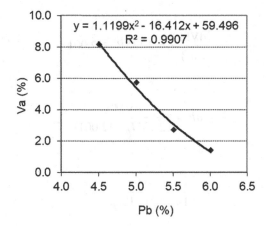

FIGURE 11.8 Asphalt binder content versus air voids.

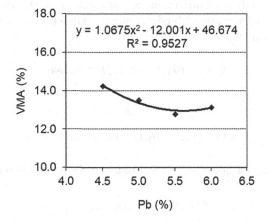

FIGURE 11.9 Asphalt binder content versus VMA.

Solution

The second order polynomial functions that fit the relationships between each of the VMA and air voids and the asphalt binder content are shown below:

$$V_a = 1.1199P_b^2 - 16.412P_b + 59.496$$

$$\text{VMA} = 1.0675P_b^2 - 12.001P_b + 46.674$$

To make the VMA value meets the Superpave criteria, the value must be at least 13.0%. Therefore,

The change in the VMA value is equal to:

$$\text{dVMA} = 13.0 - 12.8 = 0.2\%$$

But:

$$\frac{\text{dVMA}}{dP_b} = 2 \times 1.0675P_b - 12.001$$

Or:

$$\frac{\text{dVMA}}{dP_b} = 2.135P_b - 12.001$$

$\Rightarrow$

$$dP_b = \frac{\text{dVMA}}{\left(2.135P_b - 12.001\right)}$$

$\Rightarrow$

$$dP_b = \frac{0.2\%}{2.135(5.3) - 12.001} = -0.29\%$$

Therefore,

The new design asphalt binder content $= 5.3 - 0.29 = 5.0\%$

The effect of this change on the air voids level can be determined either from the polynomial function directly or by differentiating the function and determining the change.

By substituting the new asphalt binder content (P_b) in the polynomial function directly, the following is obtained:

$$V_a = 1.1199P_b^2 - 16.412P_b + 59.496$$

$\Rightarrow$

$$V_a = 1.1199(5.0)^2 - 16.412(5.0) + 59.496 = 5.4\%$$

By differentiating the above polynomial function, the following is obtained:

$$\frac{dV_a}{dP_b} = 2 \times 1.1199P_b - 16.412$$

$$\Rightarrow$$

$$dV_a = \left(2.2398P_b - 16.412\right)dP_b$$

$$\Rightarrow$$

$$dV_a = (2.2398 \times 5.3 - 16.412)(-0.29) = 1.32\%$$

Therefore,

The new air voids level is equal to:

$$V_a = 4.0 + 1.32 = 5.3\%$$

If this change is acceptable in the air voids level, the little adjustment in the asphalt binder content, to change the VMA value (an important volumetric property in the mixture design) so that it will meet the Superpave minimum requirement, would be acceptable instead of redesigning the asphalt mixture, changing the aggregate blend, and repeating the whole process.

The MS Excel worksheet used to solve this problem is shown in Figure 11.10.

Solved Practical Problems in Transportation Engineering_Part II_Pavement_Worksheet - Excel

C15 f_x =C12/(2.135*C11-12.001)

	Design P_b (%)	Design V_a (%)	VMA (%)	VFA (%)	DP	% G_{mm} at $N_{initial}$	% G_{mm} at N_{design}	% G_{mm} at $N_{maximum}$
	5.3	4.0	12.8	68.8	0.8	85.4	96	97.1
Criteria	4.0	≥ 13.0	65-75	0.6-1.2	≤ 89.0	96	≤ 98.0

Inputs

P_b (%)	5.3
dVMA (%)	0.20

$$V_a = 1.1199P_b^2 - 16.412P_b + 59.496$$

$$VMA = 1.0675P_b^2 - 12.001P_b + 46.674$$

Results

dP_b (%)	-0.29
New P_b (%)	5.0
New V_a (%)	5.4

$$\frac{dVMA}{dP_b} = 2.135P_b - 12.001$$

FIGURE 11.10 MS Excel worksheet image for the computations of Problem 11.6.

12 Volumetric Analysis of Asphalt Mixtures (Marshall and Superpave)

Chapter 12 focuses mainly on the calculation of the volumetric properties as well as compaction properties related to both Marshall mixture design procedure and Superpave mixture design method. This process is called the volumetric analysis of asphalt mixtures. In this analysis, the relationships between the volumetric properties are to be understood. The effect of one property on other properties will be assessed. The intercorrelation between all volumetric properties will be evaluated. In the volumetric analysis, the outputs of the asphalt mixture (all the volumetric properties) should be computed from raw data including weights, specific gravities, and percentages of asphalt binder and aggregate in addition to bulk and theoretical maximum specific gravities of asphalt mixture. Additionally, given some of the volumetric properties of the mixture, the rest of the volumetric properties should be easy to determine. The correlation between all the phases (by volume and by weight) of the composite material (the asphalt mixture) in the phasing diagram should be linked to the volumetric analysis so that the designer would precisely understand these phases and their relationship to mixture design. Therefore, the questions and problems in this part will deal with the volumetric analysis of asphalt mixtures in Marshall and Superpave mixture design methods, the interrelationships between the volumetric properties, and the determination of all the volumetric properties of asphalt mixtures in the design process.

12.1 In a Marshall test, the percentage of asphalt binder by total weight of aggregate is 5.0%. The bulk specific gravity of aggregate $(G_{sb}) = 2.624$, the specific gravity of asphalt binder $(G_b) = 1.000$, and the density of water $(\gamma_w) = 1.000$ g/cm^3. If the absorbed asphalt (P_{ba}) is 2.00% by total weight of aggregate, and the voids in total mixture (VTM) is 4.0%, determine the following properties:
   a. The effective asphalt content (P_{be})
   b. The bulk specific gravity of the compacted mixture (G_{mb})
   c. The voids in mineral aggregate (VMA)
   d. The theoretical maximum specific gravity of the loose mixture (G_{mm})
   e. The voids filled with asphalt (VFA)
   f. The effective specific gravity of aggregate solids (G_{se}) (see Figure 12.1).

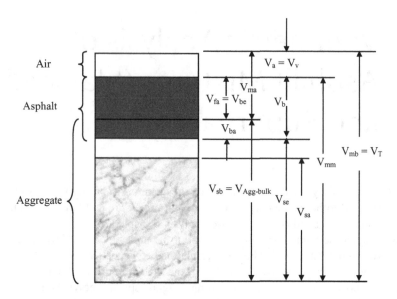

FIGURE 12.1   Phasing diagram of a compacted asphalt mixture.

Solution:

a.   The effective asphalt content (P_{be}) is determined using the following formula:

$$P_{be} = P_b - \left(\frac{P_{ba}}{100} \right) P_s \qquad (12.1)$$

Where:
P_{be} = effective asphalt binder content, percent by total weight of asphalt mixture
P_b = percentage of asphalt binder content, by total weight of asphalt mixture
P_{ba} = absorbed asphalt content, percent by total weight of aggregate
P_s = percentage of aggregate, by total weight of asphalt mixture

But before substituting in the formula, it has to be made sure that all inputs are ready to be used in the formula. The asphalt binder content that is given in the problem is the binder content by weight of aggregate and not by weight of mixture. Therefore, the value should be converted as shown in the procedure below:

$$P_s = \frac{100}{\left(1 + \dfrac{P_{b\text{-aggregate}}}{100} \right)} \qquad (12.2)$$

Or:

$$P_b = \frac{P_{b\text{-aggregate}}}{\left(1 + \dfrac{P_{b\text{-aggregate}}}{100} \right)} \qquad (12.3)$$

Where:
P_s = percentage of aggregate, by total weight of asphalt mixture
$P_{b\text{-aggregate}}$ = percentage of asphalt binder content, by total weight of aggregate

$\Rightarrow$

$$P_s = \frac{100}{\left(1+\dfrac{5.0}{100}\right)} = 95.24\%$$

But,

$$P_s = 100 - P_b \qquad\qquad (12.4)$$

Where:

P_s = percentage of aggregate, by total weight of asphalt mixture
P_b = percentage of asphalt binder content, by total weight of asphalt mixture

$\Rightarrow$

$$P_b = 100 - P_s = 100 - 95.24 = 4.76\%$$

Now, all the inputs can be substituted in the formula for effective asphalt binder content shown below:

$$P_{be} = P_b - \left(\frac{P_{ba}}{100}\right)P_s$$

$\Rightarrow$

$$P_{be} = 4.76 - \left(\frac{2.0}{100}\right)95.24 = 2.86\%$$

b. The bulk specific gravity of the compacted mixture (G_{mb}) is determined by doing some manipulation of the formulas and the definition of the VMA as explained in the procedure below:

Notice that the air voids content (VTM) is given, but the VMA and the VFA are both unknowns. Also, the G_{mb} for the asphalt mixture is unknown. Therefore, none of the available formulas will be ready directly to use to determine any of these unknowns. For this reason, the original definition of the VMA will be used. The definition of voids in mineral aggregate (VMA) is the percentage of total voids volume in the asphalt mixture by the total volume of the mixture. In other words, by looking at the phasing diagram of the asphalt mixture above, the VMA can be written as:

$$\text{VMA} = 100\left(\frac{V_a + V_{be}}{V_{mb}}\right) \qquad\qquad (12.5)$$

Where:

VMA = voids in mineral aggregate, percent by total volume of asphalt mixture
$V_a = V_v$ = volume of air voids in asphalt mixture
V_{be} = volume of effective asphalt binder in asphalt mixture
V_{mb} = total volume of asphalt mixture

Again, the voids in total mixture (VTM) is defined as in the following expression:

$$\text{VTM} = 100\left(\frac{V_a}{V_{mb}}\right) \qquad\qquad (12.6)$$

Where:

VTM = voids in total mixture, percent by total volume of asphalt mixture
$V_a = V_v$ = volume of air voids in asphalt mixture
V_{mb} = total volume of asphalt mixture

And the effective asphalt binder content is defined as in the following expression:

$$P_{be} = 100 \left(\frac{W_{be}}{W_{mb}} \right) \tag{12.7}$$

Where:

P_{be} = percentage of asphalt binder content, by total weight of asphalt mixture
W_{be} = weight of effective asphalt binder
W_{mb} = bulk or total weight of asphalt mixture

Therefore, the VMA can be re-written as:

$$VMA = 100 \left(\frac{V_a + V_{be}}{V_{mb}} \right)$$

$\Rightarrow$

$$VMA = 100 \left(\frac{V_a}{V_{mb}} \right) + 100 \left(\frac{V_{be}}{V_{mb}} \right) \tag{12.8}$$

Since the specific gravity of the asphalt binder and the bulk specific gravity of the asphalt mixture are defined as:

$$G_b = \frac{W_b}{V_b \gamma_w} \tag{12.9}$$

Or:

$$G_{be} = \frac{W_{be}}{V_{be} \gamma_w} \tag{12.10}$$

Where:

$G_b = G_{be}$ = specific gravity of asphalt binder
W_b = weight of asphalt binder
V_b = volume of asphalt binder
W_{be} = weight of effective asphalt binder
V_{be} = volume of effective asphalt binder
γ_w = density of water (1.000 g/cm³)

$$G_{mb} = \frac{W_{mb}}{V_{mb} \gamma_w} \tag{12.11}$$

Where:

G_{mb} = bulk specific gravity of asphalt mixture
W_{bm} = bulk or total weight of asphalt mixture
V_{mb} = bulk or total volume of asphalt mixture
γ_w = density of water (1.000 g/cm³)

Therefore,

$$VMA = 100\left(\frac{V_a}{V_{mb}}\right) + 100\left(\frac{\left(\dfrac{W_{be}}{\gamma_w G_b}\right)}{\left(\dfrac{W_{mb}}{\gamma_w G_{mb}}\right)}\right) \tag{12.12}$$

Equation 12.8 can be rewritten as:

$$VMA = 100\left(\frac{V_a}{V_{mb}}\right) + 100\left(\frac{W_{be}}{W_{mb}}\right)\left(\frac{G_{mb}}{G_b}\right) \tag{12.13}$$

But:

$$VTM = 100\left(\frac{V_a}{V_{mb}}\right)$$

And:

$$P_{be} = 100\left(\frac{W_{be}}{W_{mb}}\right)$$

Therefore,

$$VMA = VTM + P_{be}\left(\frac{G_{mb}}{G_b}\right) \tag{12.14}$$

This formula along with the original formula for VMA shown below can be used together to determine G_{mb}:

$$VMA = 100\left(1 - \frac{G_{mb} \times \dfrac{P_s}{100}}{G_{sb}}\right) \tag{12.15}$$

Where:
VMA = voids in mineral aggregate, percent by total volume of asphalt mixture
G_{mb} = bulk specific gravity of asphalt mixture
P_s = percentage of aggregate in asphalt mixture, by total weight of asphalt mixture
G_{sb} = bulk specific gravity of aggregate

From Equation 12.14 and Equation 12.15, the following formula can be derived:

$$VTM + P_{be}\left(\frac{G_{mb}}{G_b}\right) = 100\left(1 - \frac{G_{mb} \times \dfrac{P_s}{100}}{G_{sb}}\right) \tag{12.16}$$

This formula can be used now to determine the G_{mb} of the asphalt mixture (the only unknown):

$$4.0 + 2.86 \left(\frac{G_{mb}}{1.000} \right) = 100 \left(1 - \frac{G_{mb} \times \frac{95.24}{100}}{2.624} \right)$$

Solving for G_{mb} provides:

$$G_{mb} = 2.452$$

c. The voids in mineral aggregate (VMA) can be computed now easily from one of the two formulas above:

$$VMA = 100 \left(1 - \frac{G_{mb} \times \frac{P_s}{100}}{G_{sb}} \right)$$

$\Rightarrow$

$$VMA = 100 \left(1 - \frac{2.452 \times \frac{95.24}{100}}{2.624} \right) = 11.0\%$$

d. The theoretical maximum specific gravity of the loose mixture (G_{mm}) is determined using the formula for the air voids:

$$VTM = 100 \left(1 - \frac{G_{mb}}{G_{mm}} \right) \tag{12.17}$$

Where:
VTM = voids in total mixture (or air voids), percent by total volume of asphalt mixture
G_{mb} = bulk specific gravity of asphalt mixture (compacted mixture)
G_{mm} – theoretical maximum specific gravity of asphalt mixture (loose mixture)

$\Rightarrow$

$$4.0 = 100 \left(1 - \frac{2.452}{G_{mm}} \right)$$

$\Rightarrow$

$$G_{mm} = \frac{100 \times 2.452}{100 - 4.0} = 2.554$$

e. The voids filled with asphalt (VFA) is determined using the formula for VFA:

$$VFA = 100 \left(\frac{VMA - VTM}{VMA} \right) \tag{12.18}$$

Where:
VFA = voids filled with asphalt, percent by volume of total voids in asphalt mixture
VMA = voids in mineral aggregate, percent by total volume of asphalt mixture
VTM = $\%V_a$ = air voids content, percent by total volume of asphalt mixture

⇒

$$\text{VFA} = 100\left(\frac{11.0-4.0}{11.0}\right) = 63.6\%$$

f. The effective specific gravity of aggregate solids (G_{se}) is calculated using the formula for absorbed asphalt content (P_{ba}):

$$P_{ba} = 100\left(\frac{G_{se}-G_{sb}}{G_{se}G_{sb}}\right)G_b \qquad (12.19)$$

Where:

P_{ba} = absorbed asphalt content, percent by weight of aggregate
G_{se} = effective specific gravity of aggregate
G_{sb} = bulk specific gravity of aggregate
G_b = specific gravity of asphalt binder

⇒

$$2.0 = 100\left(\frac{G_{se}-2.624}{2.624G_{se}}\right)1.000$$

Solving for G_{se} provides the following value:
⇒

$$G_{se} = 2.769$$

The MS Excel worksheet used to solve this problem for a rapid and efficient solution is shown in Figure 12.2.

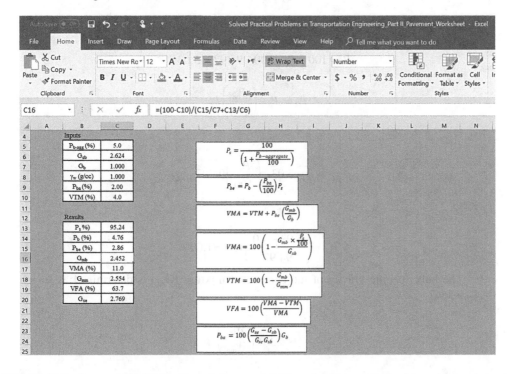

FIGURE 12.2 Image of the MS Excel worksheet used for the volumetric analysis of the asphalt mixture in Problem 12.1.

12.2 In a Marshall test, the percentage of asphalt binder by total weight of aggregate is 5.3%. The bulk specific gravity of aggregate $(G_{sb})=2.582$, the specific gravity of asphalt binder $(G_b)=1.000$, and the density of water $(\gamma_w)=1.000$ g/cm^3. If the voids filled with asphalt (VFA) is 70.0% and the effective asphalt binder (P_{be}) is 4.0% by total weight of mixture, determine the following properties:

a. The absorbed asphalt binder (P_{ba})
b. The bulk specific gravity of the compacted mixture (G_{mb})
c. The voids in mineral aggregate (VMA)
d. The voids in total mixture (VTM)
e. The theoretical maximum specific gravity of the loose mixture (G_{mm})
f. The effective specific gravity of the aggregate (G_{se})

Solution:

a. The absorbed asphalt binder (P_{ba}) is calculated as shown below:

$$P_s = \frac{100}{\left(1 + \dfrac{P_{b\text{-aggregate}}}{100}\right)}$$

$\Rightarrow$

$$P_s = \frac{100}{\left(1 + \dfrac{5.3}{100}\right)} = 94.97\%$$

But,

$$P_s = 100 - P_b$$

$\Rightarrow$

$$P_b = 100 - P_s = 100 - 94.97 = 5.03\%$$

$$P_{be} = P_b - \left(\frac{P_{ba}}{100}\right)P_s$$

$\Rightarrow$

$$4.0 = 5.03 - \left(\frac{P_{ba}}{100}\right)94.97$$

$\Rightarrow$

$$P_{ba} = \left(\frac{5.03 - 4.0}{94.97}\right)100 = 1.09\%$$

b. The bulk specific gravity of the compacted mixture (G_{mb}) is determined as shown below:

$$P_{be} = 100\left(\frac{W_{be}}{W_{mb}}\right)$$

But:

$$G_b = G_{be} = \frac{W_{be}}{V_{be}\gamma_w}$$

And:

$$G_{mb} = \frac{W_{mb}}{V_{mb}\gamma_w}$$

$\Rightarrow$

$$P_{be} = 100\left(\frac{G_{be}V_{be}\gamma_w}{G_{mb}V_{mb}\gamma_w}\right)$$

Or:

$$P_{be} = 100\left(\frac{G_{be}V_{be}}{G_{mb}V_{mb}}\right) \qquad (12.20)$$

Multiplying the above equation by V_{ma}/V_{ma} provides:

$$P_{be} = 100\left(\frac{G_{be}V_{be}V_{ma}}{G_{mb}V_{mb}V_{ma}}\right)$$

But:

$$VMA = 100\left(\frac{V_{ma}}{V_{mb}}\right)$$

And:

$$VFA = 100\left(\frac{V_{fa}}{V_{ma}}\right) = 100\left(\frac{V_{be}}{V_{ma}}\right)$$

$V_{fa} = V_{be}$ as a volume phase in the phasing diagram.
Therefore Equation 12.20 becomes:

$$P_{be} = \frac{(VFA)(VMA)}{100}\frac{G_b}{G_{mb}} \qquad (12.21)$$

$\Rightarrow$

$$VMA = 100\left(\frac{P_{be}}{VFA}\right)\left(\frac{G_{mb}}{G_b}\right) \qquad (12.22)$$

And the VMA is also given by:

$$VMA = 100\left(1 - \frac{G_{mb} \times \dfrac{P_s}{100}}{G_{sb}}\right)$$

Therefore, a new formula can be written as in the following expression:

$$100\left(\frac{P_{be}}{VFA}\right)\left(\frac{G_{mb}}{G_b}\right) = 100\left(1 - \frac{G_{mb} \times \dfrac{P_s}{100}}{G_{sb}}\right) \tag{12.23}$$

Using this formula, the G_{mb} can be calculated:

$$100\left(\frac{4.0}{70.0}\right)\left(\frac{G_{mb}}{1.000}\right) = 100\left(1 - \frac{G_{mb} \times \dfrac{94.97}{100}}{2.582}\right)$$

$\Rightarrow$

$$G_{mb} = \frac{100}{\left(\dfrac{4.0}{70.0}\right)\left(\dfrac{100}{1.000}\right) + \dfrac{94.97}{2.582}} = 2.353$$

c. The voids in mineral aggregate (VMA) can be determined easily using any of the formulas above:

$$VMA = 100\left(1 - \frac{G_{mb} \times \dfrac{P_s}{100}}{G_{sb}}\right)$$

$\Rightarrow$

$$VMA = 100\left(1 - \frac{2.353 \times \dfrac{94.97}{100}}{2.582}\right) = 13.4\%$$

d. The voids in total mixture (VTM) is also calculated using the formula below:

$$VFA = 100\left(\frac{VMA - VTM}{VMA}\right)$$

$\Rightarrow$

$$VTM = VMA\left(1 - \frac{VFA}{100}\right)$$

$\Rightarrow$

$$VTM = 13.4\left(1 - \frac{70.0}{100}\right) = 4.0\%$$

e. The theoretical maximum specific gravity of the loose mixture (G_{mm}) is computed using the following formula:

$$VTM = 100\left(1 - \frac{G_{mb}}{G_{mm}}\right)$$

$$\Rightarrow$$

$$4.0 = 100\left(1 - \frac{2.353}{G_{mm}}\right)$$

$$\Rightarrow$$

$$G_{mm} = 2.452$$

f. The effective specific gravity of the aggregate (G_{se}) is determined using the following formula:

$$P_{ba} = 100\left(\frac{G_{se} - G_{sb}}{G_{se}G_{sb}}\right)G_b$$

$$\Rightarrow$$

$$1.09 = 100\left(\frac{G_{se} - 2.582}{2.582 G_{se}}\right)1.000$$

$$\Rightarrow$$

$$G_{se} = \frac{1}{\left(\frac{1}{2.582} - \frac{1.05}{100 \times 1.000}\right)} = 2.657$$

The MS Excel worksheet used to solve this problem is shown in Figure 12.3. The worksheet provides a rapid and efficient solution.

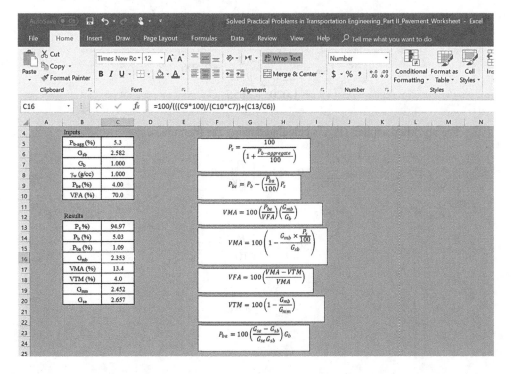

FIGURE 12.3 Image of the MS Excel worksheet used for the volumetric analysis of the asphalt mixture in Problem 12.2.

12.3 In a Marshall test, the percentage of asphalt binder by total weight of asphalt mixture is 4.8%. The bulk specific gravity of aggregate (G_{sb})=2.655, the specific gravity of asphalt binder (G_b)=1.020, and the density of water (γ_w)=1.000 g/cm^3. If the effective asphalt binder content (P_{be}) is 2.50% by total weight of mixture, and the voids in total mixture (VTM) is 4.0% by total volume of mixture, determine the following properties:
   a. The bulk specific gravity of the compacted mixture (G_{mb})
   b. The absorbed asphalt (P_{ba})
   c. The voids in mineral aggregate (VMA)
   d. The voids filled with asphalt (VFA)

Solution:

   a. The bulk specific gravity of the compacted mixture (G_{mb}) is determined as shown below:
   Equation 12.16 derived in Problem 12.1 is used to calculate G_{mb}.

$$VTM + P_{be}\left(\frac{G_{mb}}{G_b}\right) = 100\left(1 - \frac{G_{mb} \times \dfrac{P_s}{100}}{G_{sb}}\right)$$

$\Rightarrow$

$$4.0 + 2.50\left(\frac{G_{mb}}{1.020}\right) = 100\left(1 - \frac{G_{mb} \times \dfrac{95.2}{100}}{2.655}\right)$$

$\Rightarrow$

$$G_{mb} = \frac{100 - 4.0}{\dfrac{2.50}{1.020} + \dfrac{95.2}{2.655}} = 2.506$$

   b. The absorbed asphalt (P_{ba}) is calculated using the formula below:

$$P_{be} = P_b - \left(\frac{P_{ba}}{100}\right)P_s$$

$\Rightarrow$

$$2.50 = 4.8 - \left(\frac{P_{ba}}{100}\right)95.2$$

$\Rightarrow$

$$P_{ba} = 2.42\%$$

c. The voids in mineral aggregate (VMA) is determined using the following VMA formula:

$$\text{VMA} = 100\left(1 - \frac{G_{mb} \times \dfrac{P_s}{100}}{G_{sb}}\right)$$

⇒

$$\text{VMA} = 100\left(1 - \frac{2.506 \times \dfrac{95.2}{100}}{2.655}\right) = 10.14\%$$

d. The voids filled with asphalt (VFA) is calculated using the formula below:

$$\text{VFA} = 100\left(\frac{\text{VMA} - \text{VTM}}{\text{VMA}}\right)$$

⇒

$$\text{VFA} = 100\left(\frac{10.14 - 4.0}{10.14}\right) = 60.6\%$$

The MS Excel worksheet used to solve this problem is shown in Figure 12.4.

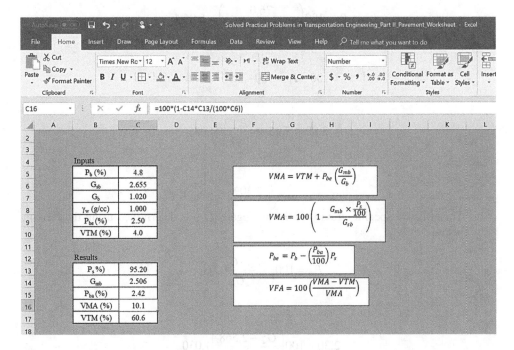

FIGURE 12.4 Image of the MS Excel worksheet used for the volumetric analysis of the asphalt mixture in Problem 12.3.

12.4 In a Marshall design procedure for an asphalt mixture, the weight of aggregate (W_s) is 1200.0 g, and the weight of asphalt binder (W_b) is 76.0 g. The voids filled with asphalt (VFA) value is determined to be 73.0%, and the effective asphalt content by total weight of mixture (P_{be}) is found to be 3.80%. If the bulk specific gravity of aggregate (G_{sb}) is 2.568, the specific gravity of asphalt binder (G_b) = 1.030, and the density of water (γ_w) = 1.000 g/cm³, determine the following properties of the asphalt mixture:

a. The absorbed asphalt binder content (P_{ba}), percent by weight of aggregate
b. The effective specific gravity of the aggregate (G_{se})
c. The bulk specific gravity of the compacted asphalt mixture (G_{mb})
d. The voids in mineral aggregate (VMA), percent by total volume of mixture
e. The voids in total mixture (VTM), percent by total volume of mixture
f. The theoretical maximum specific gravity of the loose mixture (G_{mm})

Solution:

a. The absorbed asphalt binder content (P_{ba}) is determined using the formula:
The asphalt binder content as a percentage of the asphalt mixture is defined as:

$$P_b = 100 \frac{W_b}{W_T} \tag{12.24}$$

Where:
P_b = percentage of asphalt binder by total weight of asphalt mixture
W_b = weight of asphalt binder
W_T = total weight of asphalt mixture

$\Rightarrow$

$$P_b = 100 \left(\frac{76.0}{1200.0 + 76.0} \right) = 5.96\%$$

$\Rightarrow$

$$P_s = 100 - P_b = 100 - 5.96 = 94.04\%$$

$$3.80 = 5.96 - \left(\frac{P_{ba}}{100} \right) 94.04$$

$\Rightarrow$

$$P_{ba} = 2.30\%$$

b. The effective specific gravity of the aggregate (G_{se}) is determined using the formula below:

$$P_{ba} = 100 \left(\frac{G_{se} - G_{sb}}{G_{se} G_{sb}} \right) G_b$$

$\Rightarrow$

$$2.30 = 100 \left(\frac{G_{se} - 2.568}{2.568 G_{se}} \right) 1.030$$

$\Rightarrow$

$$G_{se} = 2.724$$

c. The bulk specific gravity of the compacted asphalt mixture (G_{mb}) is determined using Equation 12.23 derived in Problem 12.2:

$$100\left(\frac{P_{be}}{VFA}\right)\left(\frac{G_{mb}}{G_b}\right) = 100\left(1 - \frac{G_{mb} \times \dfrac{P_s}{100}}{G_{sb}}\right)$$

$\Rightarrow$

$$100\left(\frac{3.80}{73.0}\right)\left(\frac{G_{mb}}{1.030}\right) = 100\left(1 - \frac{G_{mb} \times \dfrac{94.04}{100}}{2.568}\right)$$

$\Rightarrow$

$$G_{mb} = \frac{100}{\left(\dfrac{3.80}{73.0}\right)\left(\dfrac{100}{1.030}\right) + \dfrac{94.04}{2.568}} = 2.400$$

d. The voids in mineral aggregate (VMA) is calculated using the typical VMA formula shown below:

$$VMA = 100\left(1 - \frac{G_{mb} \times \dfrac{P_s}{100}}{G_{sb}}\right)$$

$\Rightarrow$

$$VMA = 100\left(1 - \frac{2.400 \times \dfrac{94.04}{100}}{2.568}\right) = 12.1\%$$

e. The voids in total mixture (VTM) is determined as below:

$$VFA = 100\left(\frac{VMA - VTM}{VMA}\right)$$

$\Rightarrow$

$$VTM = VMA\left(1 - \frac{VFA}{100}\right)$$

$\Rightarrow$

$$VTM = 12.1\left(1 - \frac{73.0}{100}\right) = 3.27\%$$

f. The theoretical maximum specific gravity of the loose mixture (G_{mm}) is computed as follows:

$$\text{VTM} = 100\left(1 - \frac{G_{mb}}{G_{mm}}\right)$$

$\Rightarrow$

$$G_{mm} = \frac{G_{mb}}{\left(1 - \dfrac{\text{VTM}}{100}\right)}$$

$\Rightarrow$

$$G_{mm} = \frac{2.400}{\left(1 - \dfrac{3.27}{100}\right)} = 2.481$$

The MS Excel worksheet shown in Figure 12.5 is used to solve the problem for quick and efficient solution.

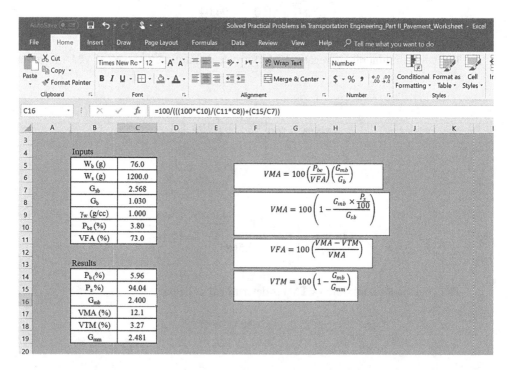

FIGURE 12.5 Image of the MS Excel worksheet used for the volumetric analysis of the asphalt mixture in Problem 12.4.

12.5 In a Superpave mixture design method for an asphalt mixture, the percentage of asphalt binder by total weight of asphalt mixture is 5.0%. The bulk specific gravity of aggregate (G_{sb}) = 2.462, the specific gravity of asphalt binder (G_b) = 1.030, and the density of water (γ_w) = 1.000 g/cm³. If the absorbed asphalt binder is 1.20% by total weight of asphalt mixture, the volume of the air voids in the mixture is 76.0 cm³, and the bulk volume of the aggregate is 1805.0 cm³, determine the following properties:

a. The effective asphalt binder content (P_{be})

b. The voids in mineral aggregate (VMA)
c. The voids filled with asphalt (VFA)
d. The air voids in the asphalt mixture (V_A or VTM)
e. The bulk specific gravity of the asphalt mixture (G_{mb})
f. The theoretical maximum specific gravity of the asphalt mixture (G_{mm})

Solution:

The phasing diagram of the asphalt mixture shown below will help in solving this problem.

Since volume values are given in the problem, to deal with volumes and weights will be easier in the volumetric analysis of this mixture.

The weight of aggregate is determined from the volume of aggregate given in the problem as shown below (see Figure 12.6):

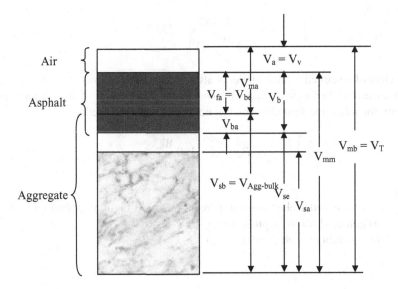

FIGURE 12.6 Phasing diagram of a compacted asphalt mixture for Problem 12.5.

$$G_{sb} = \frac{W_s}{\gamma_w V_{sb}} \tag{12.25}$$

Where:
G_{sb} = bulk specific gravity of aggregate
W_s = weight of aggregate in asphalt mixture (g)
V_{sb} = bulk or total volume of aggregate in asphalt mixture (cm³)
γ_w = density of water (1.000 g/cm³)

$\Rightarrow$

$$2.462 = \frac{W_s}{1.000\,(1805.0)}$$

$\Rightarrow$

$$W_s = 4443.9 \text{ g}$$

The weight of asphalt binder in the mixture is now determined as shown in the procedure below:

$$P_b = 100 \frac{W_b}{W_T}$$

$\Rightarrow$

$$5.0 = 100 \frac{W_b}{(4443.9) + W_b}$$

Since the total weight of asphalt mixture, $W_T = W_b + W_s$
$\Rightarrow$

$$W_b = \frac{4443.9}{\left(\dfrac{100}{5.0} - 1\right)} = 233.9 \text{ g}$$

The weight of absorbed asphalt binder is also calculated as follows:

Since the absorbed asphalt binder is given by the total weight of asphalt mixture in this problem, the following formula (which is not typical) will be used:

$$P_{\text{ba-mixture}} = 100 \frac{W_{ba}}{W_T} \tag{12.26}$$

Where:

$P_{\text{ba-mixture}}$ = absorbed asphalt content, percent by total weight of asphalt mixture
W_{ba} = weight of absorbed asphalt binder (g)
W_T = total weight of asphalt mixture (g)

$\Rightarrow$

$$1.20 = 100 \frac{W_{ba}}{(4443.9 + 233.9)}$$

$\Rightarrow$

$$W_{ba} = 56.1 \text{ g}$$

Now the following volumes are calculated:

$$G_b = \frac{W_b}{V_b \gamma_w}$$

$\Rightarrow$

$$V_b = \frac{W_b}{G_b \gamma_w} = \frac{233.9}{1.030(1.000)} = 227.1 \text{ cm}^3$$

Also, the definition of the specific gravity of the asphalt binder applies to the absorbed asphalt binder:

$$G_b = G_{ba} = \frac{W_{ba}}{V_{ba} \gamma_w} \tag{12.27}$$

⇒

$$V_{ba} = \frac{W_{ba}}{G_b \gamma_w} = \frac{56.1}{1.030(1.000)} = 54.5 \text{ cm}^3$$

Based on the phasing diagram of the three different phases of the asphalt mixture, the total (bulk) volume of asphalt mixture is determined as below:

$$V_{mb} = V_T = V_{sb} - V_{ba} + V_b + V_a \quad (12.28)$$

Where:

$V_{mb} = V_T =$ bulk or total volume of asphalt mixture
$V_{sb} =$ bulk or total volume of aggregate
$V_{ba} =$ volume of absorbed asphalt binder
$V_b =$ volume of asphalt binder
$V_a = V_v =$ volume of air voids in asphalt mixture

Therefore,

$$V_{mb} = V_T = 1805.0 - 54.5 + 227.1 + 76.0 = 2053.6 \text{ cm}^3$$

According to the phasing diagram, also the volume of effective asphalt binder is equal to:

$$V_{be} = V_b - V_{ba} \quad (12.29)$$

Where:

$V_{be} =$ volume of effective asphalt binder
$V_b =$ volume of asphalt binder
$V_{ba} =$ volume of absorbed asphalt binder

⇒

$$V_{be} = 227.1 - 54.5 = 172.6 \text{ cm}^3$$

$$G_b = G_{be} = \frac{W_{be}}{V_{be} \gamma_w} \quad (12.30)$$

⇒

$$W_{be} = G_b V_{be} \gamma_w = 1.030 \times 172.6 \times 1.000 = 177.8 \text{ g}$$

According to the phasing diagram, the volume of voids in mineral aggregate is equal to:

$$V_{ma} = V_{be} + V_a \quad (12.31)$$

Where:

$V_{ma} =$ volume of voids in mineral aggregate in asphalt mixture
$V_{be} = V_{fa} =$ volume of effective asphalt binder = volume of voids filled with asphalt
$V_a = V_v =$ volume of air voids in asphalt mixture

$\Rightarrow$

$$V_{ma} = 172.6 + 76.0 = 248.6 \text{ cm}^3$$

Now as all the needed volumes and weights in the phasing system of the asphalt mixture are determined, the required volumetric properties can be calculated easily based on the definitions of these properties as below:

a. The effective asphalt binder content (P_{be}) is determined based on its definition or from the asphalt binder content and the absorbed asphalt binder content since both are provided in the problem:

$$P_{be} = 100 \left(\frac{W_{be}}{W_{mb}} \right)$$

$\Rightarrow$

$$P_{be} = 100 \left(\frac{177.8}{(4443.9 + 233.9)} \right) = 3.80\%$$

Or:

$$P_{be} = P_b - \left(\frac{P_{ba}}{100} \right) P_s$$

And notice that:

$$P_{ba\text{-mixture}} = \left(\frac{P_{ba}}{100} \right) P_s \tag{12.32}$$

Where:
P_{ba} = absorbed asphalt binder content, percent by total weight of aggregate
P_s = percentage of aggregate by total weight of asphalt mixture
$P_{ba\text{-mixture}}$ = absorbed asphalt binder content, percent by total weight of asphalt mixture

In other words, the effective asphalt binder content formula can be re-written as:

$$P_{be} = P_b - P_{ba\text{-mixture}} \tag{12.33}$$

Therefore, the effective asphalt binder content can be determined directly from the formula since the absorbed asphalt binder content is given by total weight of asphalt mixture and not by total weight of aggregate as below:

$$P_{be} = 5.0 - 1.20 = 3.80\%$$

b. The voids in mineral aggregate (VMA) is determined based on the definition of these voids in asphalt mixture as follows:

$$VMA = 100 \left(\frac{V_{ma}}{V_{mb}} \right) = 100 \left(\frac{V_{ma}}{V_T} \right) \tag{12.34}$$

Where:

VMA = voids in mineral aggregate, percent by total volume of asphalt mixture
V_{ma} = volume of voids in mineral aggregate
V_{mb} = V_T = bulk or total volume of asphalt mixture

$\Rightarrow$

$$VMA = 100\left(\frac{248.6}{2053.6}\right) = 12.1\%$$

c. The voids filled with asphalt (VFA) is also determined in a similar manner using the definition of these voids in the phasing system of the asphalt mixture:

$$VFA = 100\left(\frac{V_{fa}}{V_{ma}}\right) = 100\left(\frac{V_{be}}{V_{ma}}\right) \qquad (12.35)$$

Where:
VFA = voids filled with asphalt, percent by total volume of voids in asphalt mixture
V_{fa} = V_{be} = volume of in voids filled with asphalt = volume of effective asphalt binder
V_{ma} = volume of voids in mineral aggregate

$\Rightarrow$

$$VFA = 100\left(\frac{172.6}{248.6}\right) = 69.4\%$$

d. The air voids in the asphalt mixture (V_A or VTM) is calculated using the definition as well:

$$V_A = VTM = 100\left(\frac{V_a}{V_{mb}}\right) = 100\left(\frac{V_a}{V_T}\right) \qquad (12.36)$$

Where:
V_A = VTM = air voids in asphalt mixture, percent by total volume of asphalt mixture
V_a = V_v = volume of air voids in asphalt mixture
V_{mb} = V_T = bulk or total volume of asphalt mixture

$\Rightarrow$

$$V_a = 100\left(\frac{76.0}{2053.6}\right) = 3.7\%$$

e. The bulk specific gravity of the asphalt mixture (G_{mb}) is determined from the formula of the VMA as follows:

$$VMA = 100\left(1 - \frac{G_{mb} \times \dfrac{P_s}{100}}{G_{sb}}\right)$$

$\Rightarrow$

$$G_{mb} = \frac{\left(1 - \dfrac{VMA}{100}\right)}{\left(\dfrac{P_s}{100 G_{sb}}\right)} = \frac{\left(1 - \dfrac{12.1}{100}\right)}{\left(\dfrac{95.0}{100 \times 2.462}\right)} = 2.278$$

Or it can be also determined using the definition of the bulk specific gravity (Equation 12.11) as follows:

$$G_{mb} = \frac{W_{mb}}{V_{mb}\gamma_w}$$

⇒

$$G_{mb} = \frac{(4443.9 + 233.9)}{2053.6(1.000)} = 2.278$$

f. The theoretical maximum specific gravity of the asphalt mixture (G_{mm}) is finally determined using the following formula:

$$VTM = V_a = 100\left(1 - \frac{G_{mb}}{G_{mm}}\right)$$

⇒

$$3.7 = 100\left(1 - \frac{2.278}{G_{mm}}\right)$$

⇒

$$G_{mm} = 2.365$$

Or, it can be also determined using the definition of the theoretical maximum specific gravity as follows:

$$G_{mm} = \frac{W_{mm}}{V_{mm}\gamma_w} \qquad (12.37)$$

Where:
G_{mm} = theoretical maximum specific gravity of asphalt mixture (loose mixture)
$W_{mm} = W_{mb}$ = total or bulk weight of asphalt mixture
V_{mm} = void-less volume of asphalt mixture = total volume of asphalt mixture less air voids volume

⇒

$$G_{mm} = \frac{(4443.9 + 233.9)}{2053.6 - 76.0} = 2.365$$

The MS Excel worksheet used to solve this problem in an efficient and easy way is shown in Figure 12.7.

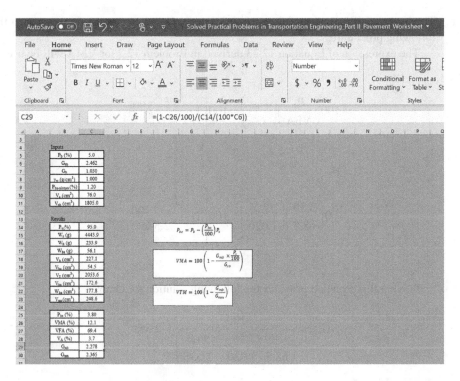

FIGURE 12.7 Image of the MS Excel worksheet used for the volumetric analysis of the asphalt mixture in Problem 12.5.

12.6 In a Superpave design for an asphalt mixture, the percentage of asphalt binder by total weight of asphalt mixture is 5.2%. The bulk specific gravity of aggregate $(G_{sb})=2.536$, the specific gravity of asphalt binder $(G_b)=1.020$, and the density of water $(\gamma_w)=1.000$ g/cm³. If the absorbed asphalt binder (P_{ba}) is 1.70% by total weight of aggregate, and the voids filled with asphalt (VFA) is 68.3%, determine the following properties of the asphalt mixture:

a. The effective asphalt binder content (P_{be})
b. The bulk specific gravity of the asphalt mixture (G_{mb})
c. The voids in mineral aggregate (VMA)
d. The air voids in the asphalt mixture $(V_a$ or VTM)
e. The theoretical maximum specific gravity of the asphalt mixture (G_{mm})
f. The effective specific gravity of the aggregate (G_{se})
g. The dust proportion (DP) if the percent passing No. 200 (75 μm) sieve is 4.5%.

Solution:

a. The effective asphalt binder content (P_{be}) is determined using the formula shown below:

$$P_{be} = P_b - \left(\frac{P_{ba}}{100}\right) P_s$$

$$P_s = 100 - P_b = 100 - 5.2 = 94.8\%$$

$\Rightarrow$

$$P_{be} = 5.2 - \left(\frac{1.70}{100}\right) 94.8 = 3.59\%$$

b. The bulk specific gravity of the asphalt mixture (G_{mb}) is determined as shown in the following procedure:

The expression of Equation 12.23 derived in Problem 12.2 for VMA will be used:

$$100\left(\frac{P_{be}}{VFA}\right)\left(\frac{G_{mb}}{G_b}\right) = 100\left(1 - \frac{G_{mb} \times \dfrac{P_s}{100}}{G_{sb}}\right)$$

$\Rightarrow$

$$100\left(\frac{3.59}{68.3}\right)\left(\frac{G_{mb}}{1.020}\right) = 100\left(1 - \frac{G_{mb} \times \dfrac{94.8}{100}}{2.536}\right)$$

Solving this equation provides the following solution for Gmb:

$$G_{mb} = \frac{100}{\left(\dfrac{100 \times 3.59}{68.3 \times 1.020} + \dfrac{94.8}{2.536}\right)} = 2.351$$

c. The voids in mineral aggregate (VMA) is determined using the typical VMA formula:

$$VMA = 100\left(1 - \frac{G_{mb} \times \dfrac{P_s}{100}}{G_{sb}}\right)$$

$\Rightarrow$

$$VMA = 100\left(1 - \frac{2.351 \times \dfrac{94.8}{100}}{2.536}\right) = 12.1\%$$

Or using the formula derived earlier in this chapter:

$$VMA = 100\left(\frac{P_{be}}{VFA}\right)\left(\frac{G_{mb}}{G_b}\right)$$

$\Rightarrow$

$$VMA = 100\left(\frac{3.59}{68.3}\right)\left(\frac{2.351}{1.020}\right) = 12.1\%$$

d. The air voids in the asphalt mixture (V_A or VTM) is determined as shown below:

$$VFA = 100\left(\frac{VMA - VTM}{VMA}\right)$$

$\Rightarrow$

$$68.3 = 100\left(\frac{12.1 - VTM}{12.1}\right)$$

Solving for VTM provides:

$\Rightarrow$

$$VTM \text{ or } V_A = 12.1\left(1 - \frac{68.3}{100}\right) = 3.84\%$$

e. The theoretical maximum specific gravity of the asphalt mixture (G_{mm}) is determined using the VTM (V_A) formula:

$$VTM = V_A = 100\left(1 - \frac{G_{mb}}{G_{mm}}\right)$$

$\Rightarrow$

$$3.84 = 100\left(1 - \frac{2.351}{G_{mm}}\right)$$

$\Rightarrow$

$$G_{mm} = \frac{2.351}{\left(1 - \frac{3.84}{100}\right)} = 2.445$$

f. The effective specific gravity of the aggregate (G_{se}) is calculated using the P_{ba} formula shown below:

$$P_{ba} = 100\left(\frac{G_{se} - G_{sb}}{G_{se}G_{sb}}\right)G_b$$

$\Rightarrow$

$$1.70 = 100\left(\frac{G_{se} - 2.536}{2.536G_{se}}\right)1.020$$

Solving for G_{se} provides:

$$G_{se} = 2.648$$

g. The dust proportion (DP) is determined using the typical DP formula:

$$DP = \frac{P_{0.075}}{P_{be}} \tag{12.38}$$

Where:
$P_{0.075}$ = percent aggregate portion passing sieve No. 200 (0.075 mm = 75 micron), %
P_{be} = effective asphalt binder content, percent by total weight of asphalt mixture

$\Rightarrow$

$$DP = \frac{4.5}{3.59} = 1.25$$

The MS Excel worksheet used to solve this problem in an efficient and easy way is shown in Figure 12.8.

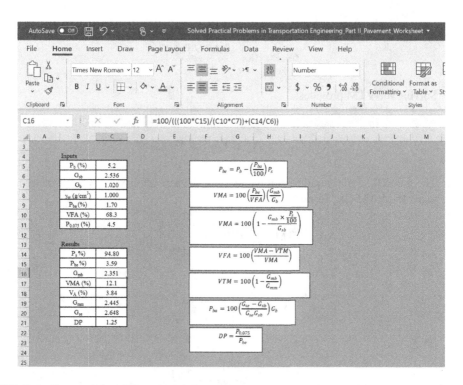

FIGURE 12.8 Image of the MS Excel worksheet used for the volumetric analysis of the asphalt mixture in Problem 12.6.

12.7 If the apparent volume of the aggregate in a Marshall specimen is 405.0 cm³, the volume of the impermeable voids inside the aggregate particles is 50.0 cm³, and the asphalt binder content by total weight of asphalt mixture is 5.1%, determine the effective specific gravity (G_{se}) of the aggregate, knowing that the weight of the asphalt binder is 65.0 g.

Solution:

According to the phasing system of the asphalt mixture, the effective volume of aggregate is equal to the apparent volume of aggregate plus the volume of impermeable voids in the aggregate. Therefore:

$$V_{se} = V_{sa} + V_{impermeable\ voids}$$ (12.39)

Where:

V_{se} = effective volume of aggregate in asphalt mixture
V_{sa} = apparent volume of aggregate in asphalt mixture
$V_{impermeable\ voids}$ = volume of impermeable voids of aggregate in asphalt mixture

⇒

$$V_{se} = 405.0 + 50.0 = 455.0 \text{ cm}^3$$

$$P_b = 100 \frac{W_b}{W_T}$$

$\Rightarrow$

$$5.1 = 100 \frac{65.0}{(65.0 + W_s)}$$

$\Rightarrow$

$$W_s = 1209.5 \text{ g}$$

The definition of G_{se} is given as:

$$G_{se} = \frac{W_s}{V_{se}\gamma_w} \tag{12.40}$$

Where:

G_{se} = effective specific gravity of aggregate in asphalt mixture
W_s = weight of aggregate
V_{se} = effective volume of aggregate in asphalt mixture
γ_w = density of water (1.000 g/cm³)

$\Rightarrow$

$$G_{se} = \frac{1209.5}{455\,(1.000)} = 2.658$$

The MS Excel worksheet used to solve this problem is shown in Figure 12.9.

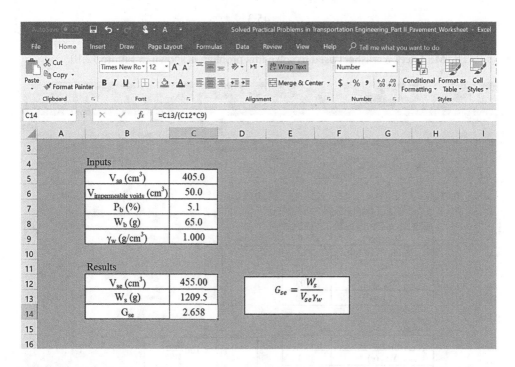

FIGURE 12.9 Image of the MS Excel worksheet used for the computations of Problem 12.7.

12.8 If the volume of the voids in mineral aggregate of a Marshall specimen (having a weight of 1270.0 g) is 55.0 cm³, and 20% of these voids are air voids, determine the effective asphalt binder content (P_{be}) knowing that the specific gravity of the asphalt binder is 1.030.

Solution:

Since 20% of the volume of voids in mineral aggregate is air voids, the remaining 80% of the volume of voids in mineral aggregate is the volume of the effective asphalt binder (voids filled with asphalt); in other words, it is equal to:

$$V_{be} = V_{fa} = \left(1 - \frac{20}{100}\right)V_{ma}$$

⇒

$$V_{be} = V_{fa} = 55.0\left(1 - \frac{20}{100}\right) = 44.0 \text{ cm}^3$$

$$G_b = G_{be} = \frac{W_{be}}{V_{be}\gamma_w}$$

⇒

$$W_{be} = G_b V_{be}\gamma_w = 1.030(44.0)(1.000) = 45.3 \text{ g}$$

$$P_{be} = 100\frac{W_{be}}{W_T}$$

⇒

$$P_{be} = 100\frac{45.3}{1270.0} = 3.57\%$$

The MS Excel worksheet used to solve this problem is shown in Figure 12.10.

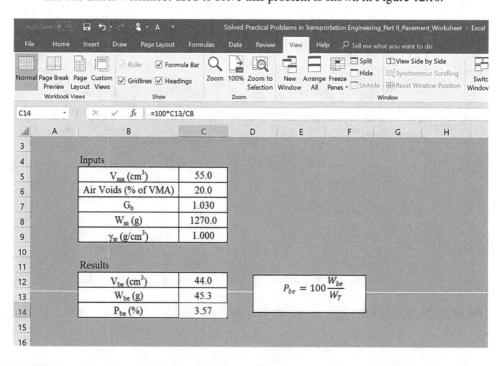

FIGURE 12.10 Image of the MS Excel worksheet used for the computations of Problem 12.8.

12.9 If the voids filled with asphalt of a Marshall specimen by total volume of the specimen is 9.0%, the bulk specific gravity of the specimen is 2.327, and the specific gravity of the asphalt binder is 1.025, determine the effective asphalt binder content (P_{be}).

Solution:

A formula for VMA derived earlier in Equation 12.22 in Problem 12.2 will be used:

$$VMA = 100 \left(\frac{P_{be}}{VFA} \right) \left(\frac{G_{mb}}{G_b} \right)$$

The voids filled with asphalt in this problem is given as a percent by total volume of the specimen. To understand the physical meaning of this term, the following derivations must be performed and remarked:

$$\text{Voids filled with asphalt, percent by total volume of mixture} = \frac{(VFA)(VMA)}{100} \quad (12.41)$$

Where:
VFA = voids filled with asphalt, percent by total volume of voids in asphalt mixture
VMA = voids in mineral aggregate, percent by total volume of asphalt mixture

The value (VFA)(VMA)/100 is the same value as the voids filled with asphalt as a percent by total volume of asphalt mixture. Using the definitions of VFA and VMA, this value can be written as:

$$\frac{(VFA)(VMA)}{100} = \frac{\left(100 \frac{V_{fa}}{V_{ma}} \right) \left(100 \frac{V_{ma}}{V_T} \right)}{100}$$

⇒

$$\frac{(VFA)(VMA)}{100} = 100 \frac{V_{fa}}{V_T} = \text{Voids filled with asphalt, \% by total volume of asphalt mixture}$$

$$(12.42)$$

Now from the formula:

$$VMA = 100 \left(\frac{P_{be}}{VFA} \right) \left(\frac{G_{mb}}{G_b} \right)$$

⇒

$$P_{be} = \left(\frac{(VMA)(VFA)}{100} \right) \left(\frac{G_b}{G_{mb}} \right) \quad (12.43)$$

And:

$$\left(\frac{(VMA)(VFA)}{100} \right) = 9.0\%$$

⇒

$$P_{be} = 9.0 \left(\frac{1.025}{2.327} \right) = 3.96\%$$

Again, the MS Excel worksheet used to solve this problem is shown in Figure 12.11.

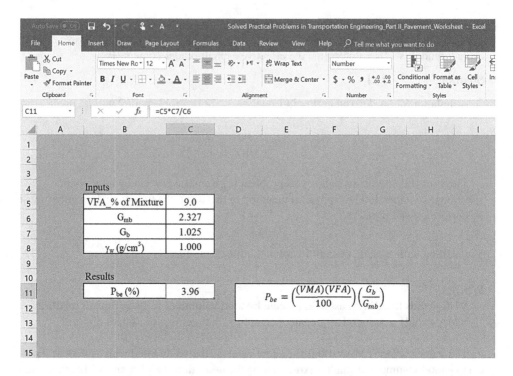

FIGURE 12.11   Image of the MS Excel worksheet used for the computations of Problem 12.9.

12.10 In a Superpave asphalt mixture design test, if the voids in mineral aggregate by total volume of asphalt mixture (VMA) is 13.0%, the bulk specific gravity of the asphalt mixture is 2.442, the specific gravity of the asphalt binder is 1.020, the absorbed asphalt binder by weight of aggregate is 1.30%, and the asphalt binder is mixed with the aggregate at a proportion of 1:20 by weight, determine the air voids content in the asphalt mixture (V_a or VTM), and the voids filled with asphalt (VFA).

Solution:

The percentage of asphalt binder by total weight of asphalt mixture is determined as below:

Since the proportion of asphalt binder to aggregate by weight = 1/20, this means that:

$$\frac{W_b}{W_s} = 0.05$$

⇒

$$W_b = 0.05\,W_s$$

From Equation 12.24 (in Problem 12.4), since:

$$P_b = 100\,\frac{W_b}{W_T}$$

Or:

$$P_b = 100\,\frac{W_b}{W_b + W_s}$$

⇒

$$P_b = 100 \frac{0.05W_s}{0.05W_s + W_s} = 4.76\%$$

$$P_s = 100 - P_b = 100 - 4.76 = 95.24\%$$

$$P_{be} = P_b - \left(\frac{P_{ba}}{100}\right) P_s$$

⇒

$$P_{be} = 4.76 - \left(\frac{1.30}{100}\right) 95.24 = 3.52\%$$

The formula derived in Equation 12.22 (in Problem 12.2) and used above in Problem 12.9 will be used to determine the VFA:

$$VMA = 100 \left(\frac{P_{be}}{VFA}\right) \left(\frac{G_{mb}}{G_b}\right)$$

⇒

$$VFA = 100 \left(\frac{P_{be}}{VMA}\right) \left(\frac{G_{mb}}{G_b}\right)$$

⇒

$$VFA = 100 \left(\frac{3.52}{13.0}\right) \left(\frac{2.442}{1.020}\right) = 64.8\%$$

And the formula for VFA will be used to determine the VTM or V_A:

$$VFA = 100 \left(\frac{VMA - VTM}{VMA}\right)$$

⇒

$$VTM \text{ or } V_A = VMA \left(1 - \frac{VFA}{100}\right)$$

⇒

$$VTM \text{ or } V_a = 13.0 \left(1 - \frac{64.8}{100}\right) = 4.6\%$$

The MS Excel worksheet used to solve this problem is shown in Figure 12.12.

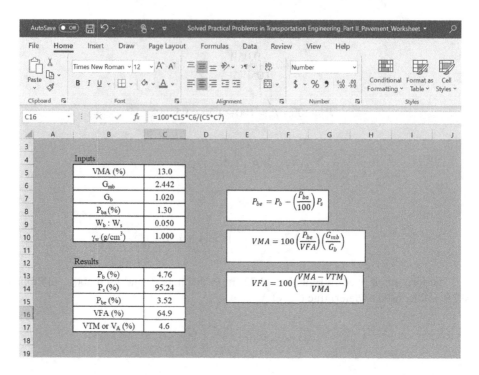

FIGURE 12.12 Image of the MS Excel worksheet used for the computations of Problem 12.10.

12.11 If the voids filled with asphalt (VFA) of a Superpave gyratory specimen is 68.0%, the absorbed asphalt binder by weight of aggregate is 1.80%, the asphalt binder content by weight of aggregate is 5.5%, the effective specific gravity of the aggregate is 2.715, and the specific gravity of the asphalt binder is 1.030, determine the following asphalt mixture properties:

   a.   The effective asphalt binder content (P_{be})
   b.   The bulk specific gravity of the specimen (G_{mb})
   c.   The voids in mineral aggregate (VMA)
   d.   The air voids or voids in total mixture (V_a or VTM)
   e.   The theoretical maximum specific gravity of the mixture (G_{mm})

Solution:

The bulk specific gravity of the aggregate is calculated using the formula shown below:

$$P_{ba} = 100 \left(\frac{G_{se} - G_{sb}}{G_{se} G_{sb}} \right) G_b$$

$\Rightarrow$

$$G_{sb} = \frac{1}{\left(\dfrac{P_{ba}}{100 G_b} + \dfrac{1}{G_{se}} \right)}$$

$\Rightarrow$

$$G_{sb} = \frac{1}{\left(\dfrac{1.80}{100(1.030)} + \dfrac{1}{2.715} \right)} = 2.592$$

The asphalt binder is given as a percent by total weight of aggregate; therefore, the asphalt binder must be determined as a percent by total weight of asphalt mixture as shown below:

$$P_s = \frac{100}{\left(1 + \dfrac{P_{b\text{-aggregate}}}{100}\right)}$$

Or:

$$P_b = \frac{P_{b\text{-aggregate}}}{\left(1 + \dfrac{P_{b\text{-aggregate}}}{100}\right)}$$

$\Rightarrow$

$$P_b = \frac{5.5}{\left(1 + \dfrac{5.5}{100}\right)} = 5.21\%$$

But,

$$P_s = 100 - P_b$$

$\Rightarrow$

$$P_s = 100 - P_b = 100 - 5.21 = 94.79\%$$

a. The effective asphalt binder content (P_{be}) is determined using the formula shown below:

$$P_{be} = P_b - \left(\frac{P_{ba}}{100}\right) P_s$$

$\Rightarrow$

$$P_{be} = 5.21 - \left(\frac{1.80}{100}\right) 94.79 = 3.51\%$$

b. The bulk specific gravity of the specimen (G_{mb}) is determined as shown in the procedure below:
The expression of Equation 12.23 derived in Problem 12.2 will be used to determine the G_{mb}:

$$100\left(\frac{P_{be}}{VFA}\right)\left(\frac{G_{mb}}{G_b}\right) = 100\left(1 - \frac{G_{mb} \times \dfrac{P_s}{100}}{G_{sb}}\right)$$

$\Rightarrow$

$$G_{mb} = \frac{100}{\left(\dfrac{P_{be}}{VFA}\right)\left(\dfrac{100}{G_b}\right) + \dfrac{P_s}{G_{sb}}} \qquad (12.44)$$

$\Rightarrow$

$$G_{mb} = \frac{100}{\left(\dfrac{3.51}{68.0}\right)\left(\dfrac{100}{1.030}\right) + \dfrac{94.79}{2.592}} = 2.405$$

c. The voids in mineral aggregate (VMA) is calculated using the VMA formula as follows:

$$VMA = 100\left(1 - \frac{G_{mb} \times \dfrac{P_s}{100}}{G_{sb}}\right)$$

$\Rightarrow$

$$VMA = 100\left(1 - \frac{2.405 \times \dfrac{94.79}{100}}{2.592}\right) = 12.0\%$$

d. The air voids or voids in total mixture (V_A or VTM) is determined using the formula shown below:

$$VFA = 100\left(\frac{VMA - VTM}{VMA}\right)$$

$\Rightarrow$

$$VTM \text{ or } V_a = VMA\left(1 - \frac{VFA}{100}\right) \qquad (12.45)$$

$\Rightarrow$

$$VTM \text{ or } V_a = 12.0\left(1 - \frac{68.0}{100}\right) = 3.9\%$$

e. The theoretical maximum specific gravity of the mixture (G_{mm}) is determined using the formula below:

$$VTM = 100\left(1 - \frac{G_{mb}}{G_{mm}}\right)$$

$\Rightarrow$

$$G_{mm} = \frac{G_{mb}}{\left(1 - \dfrac{VTM}{100}\right)} \qquad (12.46)$$

$\Rightarrow$

$$G_{mm} = \frac{2.405}{\left(1 - \dfrac{3.9}{100}\right)} = 2.502$$

The MS Excel worksheet used to perform the computations to determine all mixture properties in this problem is shown in Figure 12.13.

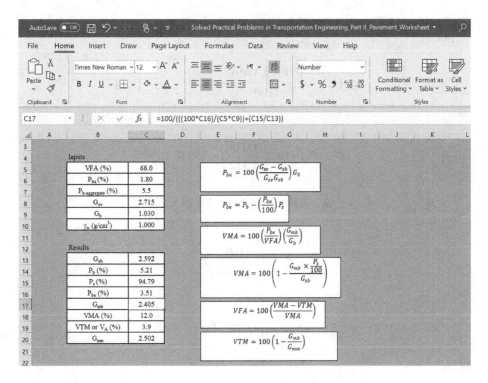

FIGURE 12.13 Image of the MS Excel worksheet used for the volumetric analysis of the asphalt mixture in Problem 12.11.

12.12 If the voids filled with asphalt of a Superpave gyratory specimen by total volume of the specimen is 10.0%, the bulk specific gravity of the specimen is 2.322, the effective specific gravity of the aggregate is 2.654, and the specific gravity of the asphalt binder is 1.025, and the asphalt binder content by total weight of mixture is 5.2%, determine the voids in mineral aggregate (VMA) and percentage of air voids (V_A or VTM) in the specimen.

Solution:

According to the formula in Equation 12.41 derived in Problem 12.9 that is shown below:

$$\text{Voids filled with asphalt, percent by total volume of mixture} = \frac{(\text{VFA})(\text{VMA})}{100}$$

$$10.0 = \frac{(\text{VFA})(\text{VMA})}{100}$$

And using the formula in Equation 12.22 derived in Problem 12.2 and shown below:

$$\text{VMA} = 100\left(\frac{P_{be}}{\text{VFA}}\right)\left(\frac{G_{mb}}{G_b}\right)$$

$\Rightarrow$

$$P_{be} = \left(\frac{(\text{VMA})(\text{VFA})}{100}\right)\left(\frac{G_b}{G_{mb}}\right)$$

(Equation 12.43 in Problem 12.9).
But:

$$\frac{(VFA)(VMA)}{100} = 10.0$$

$\Rightarrow$

$$P_{be} = 10.0 \left(\frac{1.025}{2.322} \right) = 4.41\%$$

$$P_s = 100 - P_b = 100 - 5.2 = 94.8\%$$

$$P_{be} = P_b - \left(\frac{P_{ba}}{100} \right) P_s$$

$\Rightarrow$

$$P_{ba} = 100 \left(\frac{P_b - P_{be}}{P_s} \right) \tag{12.47}$$

$\Rightarrow$

$$P_{ba} = 100 \left(\frac{5.2 - 4.41}{94.8} \right) = 0.83\%$$

But:

$$P_{ba} = 100 \left(\frac{G_{se} - G_{sb}}{G_{se} G_{sb}} \right) G_b$$

$\Rightarrow$

$$G_{sb} = \frac{1}{\left(\dfrac{P_{ba}}{100 G_b} + \dfrac{1}{G_{se}} \right)} \tag{12.48}$$

$\Rightarrow$

$$G_{sb} = \frac{1}{\left(\dfrac{0.83}{100(1.025)} + \dfrac{1}{2.654} \right)} = 2.598$$

Now the voids in mineral aggregate (VMA) can be determined using the VMA formula shown below:

$$VMA = 100 \left(1 - \frac{G_{mb} \times \dfrac{P_s}{100}}{G_{sb}} \right)$$

$\Rightarrow$

$$VMA = 100\left(1 - \frac{2.322 \times \dfrac{94.8}{100}}{2.598}\right) = 15.3\%$$

The percentage of air voids (V_A or VTM) is also determined using the VFA formula as follows:

$$VFA = 100\left(\frac{VMA - VTM}{VMA}\right)$$

$\Rightarrow$

$$VMA - VTM = \left(\frac{(VFA)(VMA)}{100}\right) \tag{12.49}$$

But again:

$$\frac{(VFA)(VMA)}{100} = 10.0$$

$\Rightarrow$

$$VMA - VTM = 10.0$$

$\Rightarrow$

$$VTM \text{ or } V_A = VMA - 10.0 = 15.3 - 10.0 = 5.3\%$$

Or:

One can think of this in another way. The volume of voids in mineral aggregate is simply equal to the volume of air voids and the volume of voids filled with asphalt. If all these percentages are computed as a percent by total volume of asphalt mixture, then it can be written as:

$$VMA = VFA_{\% \text{ by volume of mixture}} + V_A \tag{12.50}$$

Where:

VMA = voids in mineral aggregate, percent by total volume of asphalt mixture
$VFA_{\% \text{by volume of mixture}}$ = voids filled with asphalt, percent by total volume of asphalt mixture
V_A or VTM = air voids, percent by total volume of asphalt mixture

But a subscript is added to the VFA because typically the VFA is determined as a percentage by volume of total voids in the asphalt mixture and not by total volume of asphalt mixture. So, in this case, it is determined by total volume of mixture. Therefore, it can be added directly to the V_A (or VTM), which is also determined as a percentage by total volume of asphalt mixture. The result is the VMA as a percentage by total volume of asphalt mixture.

$$VMA = VFA_{\% \text{ by volume of mixture}} + V_A$$

$\Rightarrow$

$$15.3 = 10.0 + V_A$$

$\Rightarrow$

$V_A = 5.3\%$ (This is the same answer as before).

The MS Excel worksheet used to perform the computations of this problem and which does the computations in an easy and efficient way, is shown in Figure 12.14.

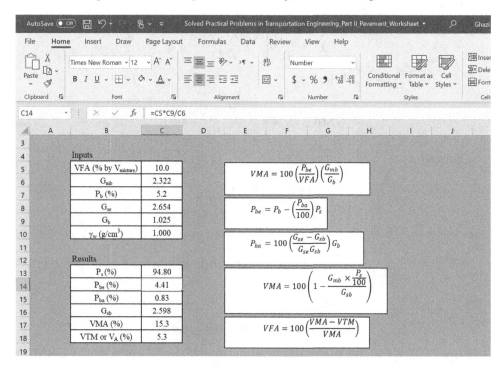

FIGURE 12.14 Image of the MS Excel worksheet used for the volumetric analysis of the asphalt mixture in Problem 12.12.

Another solution for this problem is explained below:

$$G_{mm} = \frac{P_{mm}}{\dfrac{P_s}{G_{se}} + \dfrac{P_b}{G_b}} \tag{12.51}$$

Where:

G_{mm} = theoretical maximum specific gravity of asphalt mixture

P_{mm} = total loose asphalt mixture, percentage by total weight of asphalt mixture = 100%

P_s = percentage of aggregate by total weight of asphalt mixture

G_{se} = effective specific gravity of aggregate

P_b = percentage of asphalt binder by total weight of asphalt mixture

G_b = specific gravity of asphalt binder

$\Rightarrow$

$$G_{mm} = \frac{100}{\dfrac{94.8}{2.654} + \dfrac{5.2}{1.025}} = 2.451$$

$$\text{VTM} = 100\left(1 - \frac{G_{mb}}{G_{mm}}\right)$$

$\Rightarrow$

$$\text{VTM or } V_A = 100\left(1 - \frac{2.322}{2.451}\right) = 5.3\%$$

$$\text{VMA} = \text{VFA}_{\%\text{ by volume of mixture}} + V_A$$

$\Rightarrow$

$$\text{VMA} = 10.0 + 5.3 = 15.3\%$$

The MS Excel worksheet used to perform the computations of the second solution of this problem is shown in Figure 12.15.

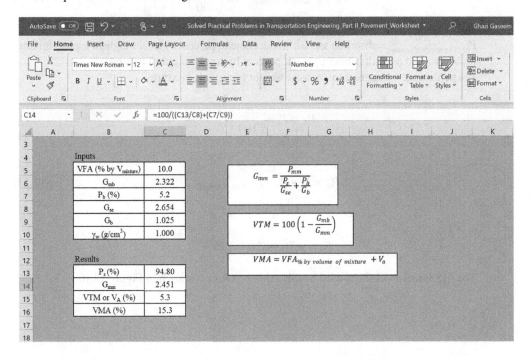

FIGURE 12.15   Image of the MS Excel worksheet used for the volumetric analysis of the asphalt mixture in Problem 12.12 (second solution).

12.13 In a Marshall mixture design test, if the absorbed asphalt binder by total weight of mixture is 1.50%, and the asphalt binder content by total weight of aggregate is 5.0%, determine the effective asphalt binder content by weight of asphalt mixture (P_{be}).

Solution:

Since the given asphalt binder content is a percentage by total weight of aggregate, the asphalt binder content as a percent by total weight of asphalt mixture (P_b) is determined first as shown below:

$$P_s = \frac{100}{\left(1 + \dfrac{P_{b\text{-aggregate}}}{100}\right)}$$

Or:

$$P_b = \frac{P_{b\text{-aggregate}}}{\left(1 + \dfrac{P_{b\text{-aggregate}}}{100}\right)}$$

$\Rightarrow$

$$P_b = \frac{5.0}{\left(1 + \dfrac{5.0}{100}\right)} = 4.76\%$$

The absorbed asphalt binder is given by total weight of mixture and the typical value is normally given as a percentage by total weight of aggregate (P_{ba}). Therefore,

$$P_{ba\text{-\% by weight of mixture}} = P_{ba}\left(\frac{P_s}{100}\right) \tag{12.52}$$

Where:

$P_{ba\text{-\% by weight of mixture}}$ = absorbed asphalt binder content, percent by total weight of asphalt mixture
P_{ba} = absorbed asphalt binder content, percent by total weight of aggregate
P_s = percentage of aggregate, by total weight of asphalt mixture

$$P_{be} = P_b - \left(\frac{P_{ba}}{100}\right)P_s$$

$\Rightarrow$

$$P_{be} = P_b - P_{ba\text{-\% by weight of mixture}} \tag{12.53}$$

$\Rightarrow$

$$P_{be} = 4.76 - 1.50 = 3.26\%$$

The MS Excel worksheet used to solve this problem is shown in Figure 12.16.

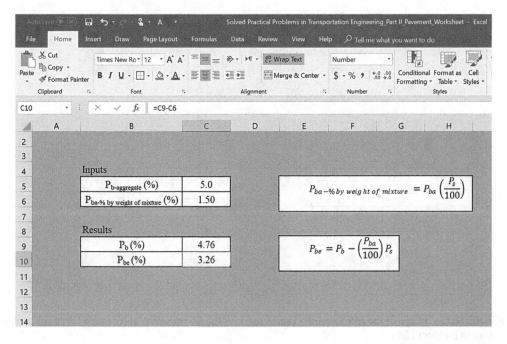

FIGURE 12.16 Image of the MS Excel worksheet used for the computation of P_{be} of the asphalt mixture in Problem 12.13.

12.14 If the voids in mineral aggregate (VMA) of a Superpave gyratory specimen is 16.0% and 25% of these voids are air voids, determine the voids filled with asphalt (VFA).

Solution:

Since 25% of the VMA is air voids, and knowing that both the air voids content (V_A) and the VMA are determined as a percent of the total volume of the asphalt mixture, then:

$$V_A = \frac{25}{100}(\text{VMA}) = 0.25(16.0) = 4.0\%$$

$$\text{VFA} = 100\left(\frac{\text{VMA} - \text{VTM}}{\text{VMA}}\right)$$

$\Rightarrow$

$$\text{VFA} = 100\left(\frac{16.0 - 4.0}{16.0}\right) = 75.0\%$$

The MS Excel worksheet used to solve this problem is shown in Figure 12.17.

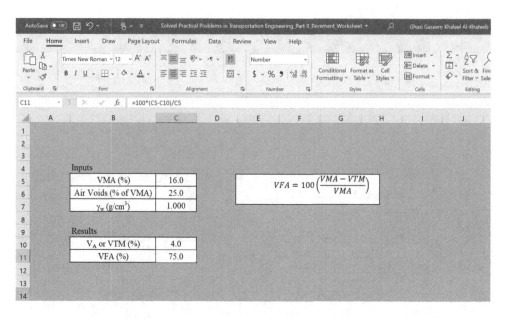

FIGURE 12.17 Image of the MS Excel worksheet used for the Computation of V_A and VFA of the asphalt mixture in Problem 12.14.

12.15 If the total volume of a Marshall specimen is 515.0 cm³, the theoretical maximum specific gravity of the loose mixture (G_{mm}) is 2.566, and the voids in total mixture (VTM) by total volume of mixture is 4.0%, determine the weight of the specimen (in grams).

Solution:

$$VTM = 100\left(1 - \frac{G_{mb}}{G_{mm}}\right)$$

⇒

$$G_{mb} = G_{mm}\left(1 - \frac{VTM}{100}\right) \qquad (12.54)$$

⇒

$$G_{mb} = 2.566\left(1 - \frac{4.0}{100}\right) = 2.463$$

$$G_{mb} = \frac{W_{mb}}{V_{mb}\gamma_w} = \frac{W_T}{V_T\gamma_w}$$

(Equation 12.11 used in Problem 12.1).

⇒

$$W_T = G_{mb}V_{mb}\gamma_w = 2.463\,(515)\,(1.000) = 1268.6\ g$$

The MS Excel worksheet used to solve this problem is shown in Figure 12.18.

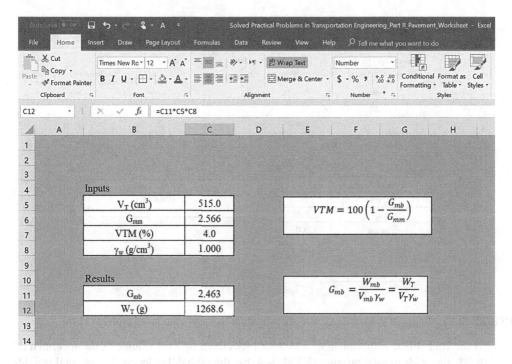

FIGURE 12.18   Image of the MS Excel worksheet used for the computations of Problem 12.15.

12.16 If the voids filled with asphalt of a Superpave gyratory specimen by total volume of the specimen is 7.0%, the specific gravity of the asphalt binder is 1.015, and the effective asphalt binder content (P_{be}) by total weight of specimen is 3.00%, determine the bulk specific gravity of the specimen (G_{mb}).

Solution:

The formula of Equation 12.22 derived in Problem 12.2 for VMA will be used to solve for G_{mb} as shown below:

$$VMA = 100\left(\frac{P_{be}}{VFA}\right)\left(\frac{G_{mb}}{G_b}\right)$$

$\Rightarrow$

$$G_{mb} = \left(\frac{(VFA)(VMA)}{100}\right)\left(\frac{G_b}{P_{be}}\right)$$

According to the formula in Equation 12.41 derived in Problem 12.9 and shown below:

$$\text{Voids filled with asphalt, percent by total volume of mixture} = \frac{(VFA)(VMA)}{100}$$

$\Rightarrow$

$$\frac{(VFA)(VMA)}{100} = 7.0$$

$\Rightarrow$

$$G_{mb} = 7.0\left(\frac{1.015}{3.00}\right) = 2.368$$

The MS Excel worksheet used to solve this problem is shown in Figure 12.19.

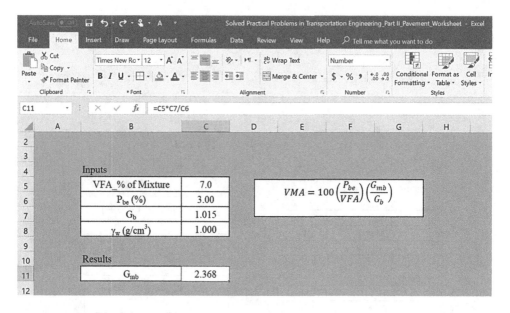

FIGURE 12.19 Image of the MS Excel worksheet used for the computations of Problem 12.16.

12.17 If the voids filled with asphalt of a Marshall specimen by total volume of the specimen is 6.0%, the voids in total mixture (VTM) is 4.0%, the asphalt binder content by total weight of specimen is 4.8%, the theoretical maximum specific gravity of the mixture (G_{mm}) is 2.488, and the specific gravity of the asphalt binder is 1.020, determine the bulk specific gravity of the aggregate (G_{sb}), the effective binder content by weight of specimen (P_{be}), and the voids filled with asphalt as a percent by volume of total voids (VFA).

Solution:

The formula that relates the VMA, air voids, and VFA as a percent of total mixture will be used:

$$VMA = VFA_{\% \text{ by volume of mixture}} + V_A \text{ or VTM}$$

$\Rightarrow$

$$VMA = 6.0 + 4.0 = 10.0\%$$

$$VTM = 100\left(1 - \frac{G_{mb}}{G_{mm}}\right)$$

$\Rightarrow$

$$G_{mb} = G_{mm}\left(1 - \frac{VTM}{100}\right)$$

$\Rightarrow$

$$G_{mb} = 2.488\left(1 - \frac{4.0}{100}\right) = 2.388$$

$$P_s = 100 - P_b = 100 - 4.8 = 95.2\%$$

$$VMA = 100\left(1 - \frac{G_{mb} \times \dfrac{P_s}{100}}{G_{sb}}\right)$$

$\Rightarrow$

$$G_{sb} = \frac{\left(\dfrac{G_{mb}P_s}{100}\right)}{\left(1 - \dfrac{VMA}{100}\right)} \qquad (12.55)$$

$\Rightarrow$

$$G_{sb} = \frac{\left(\dfrac{2.388 \times 95.2}{100}\right)}{\left(1 - \dfrac{10.0}{100}\right)} = 2.526$$

Again, the formula of Equation 12.22 derived in Problem 12.2 for VMA will be used to solve for P_{be} as shown below:

$$VMA = 100\left(\frac{P_{be}}{VFA}\right)\left(\frac{G_{mb}}{G_b}\right)$$

$\Rightarrow$

$$P_{be} = \left(\frac{(VMA)(VFA)}{100}\right)\left(\frac{G_b}{G_{mb}}\right)$$

But:

Voids filled with asphalt, percent by total volume of mixture $= \dfrac{(VFA)(VMA)}{100}$

$\Rightarrow$

$$\frac{(VFA)(VMA)}{100} = 6.0$$

$\Rightarrow$

$$P_{be} = 6.0\left(\frac{1.020}{G_{mb}}\right) = 2.56\%$$

$$VFA = 100\left(\frac{VMA - VTM}{VMA}\right)$$

$\Rightarrow$

$$\text{VFA} = 100\left(\frac{10.0 - 4.0}{10.0}\right) = 60.0\%$$

The MS Excel worksheet used to perform all the computations of this problem to determine the volumetric properties of the Marshall specimen easily and efficiently is shown in Figure 12.20.

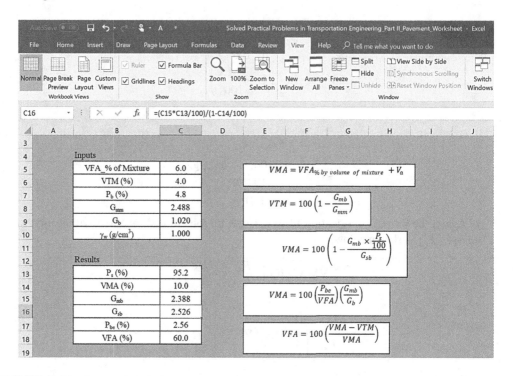

FIGURE 12.20 Image of the MS Excel worksheet used for the volumetric analysis of the asphalt mixture in Problem 12.17.

12.18 In a Marshall asphalt mixture design test, if the voids in total mixture (VTM) is 4.2%, the bulk specific gravity of the mixture is 2.444, the specific gravity of the asphalt binder is 1.030, the absorbed asphalt binder by total weight of aggregate is 1.20%, and the asphalt binder content by total weight of mixture is 4.0%, determine the voids in mineral aggregate (VMA), and the voids filled with asphalt (VFA).

Solution:

The percentage of aggregate by the total weight of the specimen is determined simply using the formula below:

$$P_s = 100 - P_b = 100 - 4.0 = 96.0\%$$

The effective asphalt binder content by the total weight of the specimen (P_{be}) is also determined using the formula for P_{be} as follows:

$$P_{be} = P_b - \left(\frac{P_{ba}}{100}\right)P_s$$

⇒

$$P_{be} = 4.0 - \left(\frac{1.20}{100}\right)96.0 = 2.85\%$$

The expression of Equation 12.14 derived in Problem 12.1 for VMA will be used herein to determine VMA as shown below:

$$\text{VMA} = \text{VTM} + P_{be}\left(\frac{G_{mb}}{G_b}\right)$$

⇒

$$\text{VMA} = 4.2 + 2.85\left(\frac{2.444}{1.030}\right) = 10.96\%$$

$$\text{VFA} = 100\left(\frac{\text{VMA} - \text{VTM}}{\text{VMA}}\right)$$

⇒

$$\text{VFA} = 100\left(\frac{11.0 - 4.2}{11.0}\right) = 61.7\%$$

The MS Excel worksheet used to perform all the computations of this problem and to determine the VMA and VFA of the specimen easily and efficiently is shown in Figure 12.21.

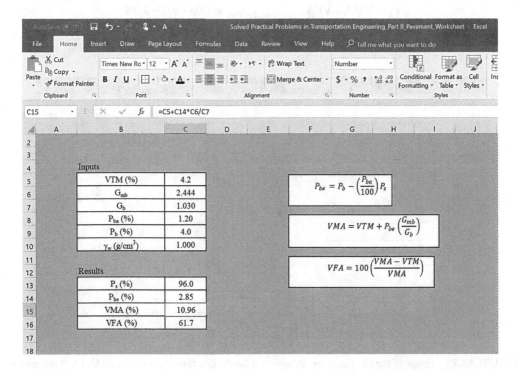

FIGURE 12.21   Image of the MS Excel worksheet used for the computations of Problem 12.18.

12.19 If the voids filled with asphalt of a Superpave gyratory specimen by total volume of the specimen is 9.0%, and 25.0% of the voids in mineral aggregate (VMA) is air voids, determine the VMA and the VTM (V_a = the percentage of air voids by total volume of specimen).

Solution:

The formula that relates the VMA, air voids, and VFA as a percent of total mixture will be used (Equation 12.50 derived in Problem 12.12):

$$VMA = VFA_{\% \text{ by volume of mixture}} + V_A \text{ or VTM}$$

$$VFA_{\% \text{ by volume of mixture}} = 9.0\%$$

But the air voids content is equal to 25.0% of the VMA, therefore:

$$V_A = VTM = 0.25 \text{ VMA}$$

$\Rightarrow$

$$VMA = 9.0 + 0.25 \text{ VMA}$$

$\Rightarrow$

$$VMA = 12.0\%$$

$\Rightarrow$

$$V_A = VTM = 0.25 \text{ VMA} = 0.25(12.0) = 3.0\%$$

Or:

$$V_A = VMA - VFA_{\% \text{ by volume of mixture}} = 12.0 - 9.0 = 3.0\%$$

See the MS Excel worksheet shown in Figure 12.22.

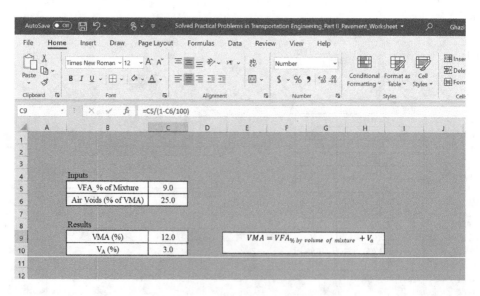

FIGURE 12.22 Image of the MS Excel worksheet used for the determination of V_A and VMA of the asphalt mixture in Problem 12.19.

12.20 In a Marshall specimen having a total volume of 510.0 cm³, the volume of the absorbed asphalt is 15.0 cm³, the volume of air voids is 30.0 cm³, and the weight of the asphalt binder is 60.0 g, determine the VMA and G_{mb} if the asphalt binder content by total weight of specimen is 4.8% and the specific gravity (G_b) of the asphalt binder is 1.030.

Solution:

The phasing system of the asphalt mixture specimen is important in this problem to understand the volumetric analysis and the relationships between the volumes of the different phases in the system and the weights of these phases (see Figure 12.23).

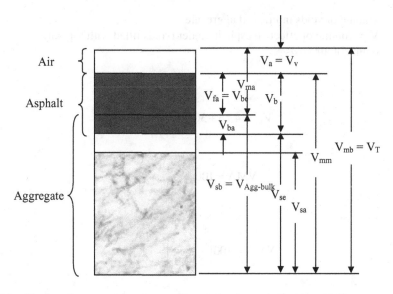

FIGURE 12.23 Phasing diagram of the compacted asphalt mixture in Problem 12.20.

The volume of asphalt binder is calculated from the weight of the asphalt binder as follows:

$$G_b = \frac{W_b}{V_b \gamma_w}$$

⇒

$$V_b = \frac{W_b}{G_b \gamma_w} = \frac{60.0}{1.030(1.000)} = 58.3 \text{ g}$$

The volume of effective asphalt binder (same as the volume of voids filled with asphalt) is simply the volume of asphalt binder minus the volume of absorbed asphalt binder based on the phasing system of the asphalt mixture specimen:

$$V_{be} = V_{fa} = V_b - V_{ba} \tag{12.56}$$

Where:
$V_{be} = V_{fa} =$ volume of effective asphalt binder (voids filled with asphalt)
$V_b =$ volume of asphalt binder
$V_{ba} =$ volume of absorbed asphalt binder

Therefore,

$$V_{be} = V_{fa} = 58.3 - 15.0 = 43.3 \text{ cm}^3$$

The volume of voids in mineral aggregate is also equal to the volume of effective asphalt binder (voids filled with asphalt) plus the volume of air voids according to the phasing system of the mixture:

$$V_{ma} = V_{be} + V_a \qquad (12.57)$$

Where:
   V_{ma} = volume of voids in mineral aggregate
   $V_{be} = V_{fa}$ = volume of effective asphalt binder (voids filled with asphalt)
   V_a = volume of air voids

$\Rightarrow$

$$V_{ma} = 43.3 + 30.0 = 73.3 \text{ cm}^3$$

$$\text{VMA} = 100 \left(\frac{V_a + V_{be}}{V_{mb}} \right)$$

$\Rightarrow$

$$\text{VMA} = 100 \left(\frac{73.3}{510.0} \right) = 14.4\%$$

$$\text{VTM} = 100 \left(\frac{V_a}{V_{mb}} \right)$$

$\Rightarrow$

$$\text{VTM} = 100 \left(\frac{30.0}{510.0} \right) = 5.9\%$$

Now to determine the G_{mb} of the specimen, the total weight of the specimen is required. It can be determined from the following formula:

$$P_b = 100 \frac{W_b}{W_T}$$

$\Rightarrow$

$$W_T = 100 \frac{W_b}{P_b} = 100 \frac{60.0}{4.8} = 1250.0 \text{ g}$$

Now, the G_{mb} is calculated using the formula below:

$$G_{mb} = \frac{W_{mb}}{V_{mb} \gamma_w}$$

⇒

$$G_{mb} = \frac{1250.0}{510.0(1.000)} = 2.451$$

The MS Excel worksheet used to perform the computations of the volumetric analysis of the Marshall specimen in this problem in a rapid and efficient way is shown in Figure 12.24.

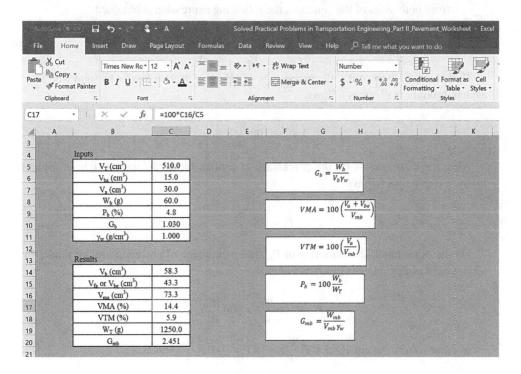

FIGURE 12.24 Image of the MS Excel worksheet used for the volumetric analysis of the asphalt mixture in Problem 12.20.

12.21 Prove that $P_{be} \neq P_b - P_{ba}$ despite the fact that from the phasing system of the asphalt mixture $V_{be} = V_b - V_{ba}$.

And show that:

$$P_{be} = P_b - P_{ba}\left(\frac{P_s}{100}\right)$$

Solution:

$$V_{be} = V_b - V_{ba}$$

The relationship between volume phase, weight, and specific gravity is given as below:

$$G_b = \frac{W_b}{V_b \gamma_w} = \frac{W_{be}}{V_{be} \gamma_w} = \frac{W_{ba}}{V_{ba} \gamma_w}$$

Therefore, $V_{be} = V_b - V_{ba}$ becomes:

$$\frac{W_{be}}{G_b \gamma_w} = \frac{W_b}{G_b \gamma_w} - \frac{W_{ba}}{G_b \gamma_w} \tag{12.58}$$

By multiplying both sides of the equation by the term $\dfrac{100}{W_T}$ and cancelling the term

$\dfrac{1}{G_b \gamma_w}$ from both sides of the equation, the following expression is obtained:

$$100 \frac{W_{be}}{W_T} = 100 \frac{W_b}{W_T} - 100 \frac{W_{ba}}{W_T} \tag{12.59}$$

Now the definition of P_{be}, the percentage of effective asphalt binder by total weight of asphalt mixture is given as follows:

$$P_{be} = 100 \frac{W_{be}}{W_T}$$

Also, the definition of P_b, the percentage of asphalt binder by total weight of asphalt mixture is given as follows:

$$P_b = 100 \frac{W_b}{W_T}$$

On the other hand, the definition of P_{ba}, the percentage of absorbed asphalt binder by total weight of aggregate is given as follows:

$$P_{ba} = 100 \frac{W_{ba}}{W_s} \neq 100 \frac{W_{ba}}{W_T}$$

In other words, the equation $100 \dfrac{W_{be}}{W_T} = 100 \dfrac{W_b}{W_T} - 100 \dfrac{W_{ba}}{W_T}$ becomes:

$$P_{be} = P_b - 100 \frac{W_{ba}}{W_T} \tag{12.60}$$

And since $P_{ba} = 100 \dfrac{W_{ba}}{W_s} \neq 100 \dfrac{W_{ba}}{W_T}$, therefore:

$$P_{be} \neq P_b - P_{ba}$$

But rather:

$$P_{be} = P_b - 100 \frac{W_{ba}}{W_T}$$

By multiplying the term $100 \dfrac{W_{ba}}{W_T}$ in the above equation by the term $\dfrac{W_s}{W_s}$, it becomes:

$$P_{be} = P_b - 100 \frac{W_{ba}}{W_T} \frac{W_s}{W_s}$$

But:

$$P_{ba} = 100 \frac{W_{ba}}{W_s} \tag{12.61}$$

And:

$$P_s = 100 \frac{W_s}{W_T} \tag{12.62}$$

Or:

$$\frac{P_s}{100} = \frac{W_s}{W_T}$$

Therefore, the equation $P_{be} = P_b - 100 \dfrac{W_{ba}}{W_T} \dfrac{W_s}{W_s}$ becomes:

$$P_{be} = P_b - P_{ba}\left(\frac{P_s}{100}\right)$$

12.22 Prove that $VFA \neq VMA - VTM$ despite the fact that from the phasing system of the asphalt mixture $V_{fa} = V_{be} = V_{ma} - V_a$.
And show that:

$$VFA = 100\left(\frac{VMA - VTM}{VMA}\right)$$

Solution:

$$V_{fa} = V_{be} = V_{ma} - V_a.$$

By multiplying both sides of the equation by the term $\dfrac{100}{V_T}$, the following expression is obtained:

$$100\frac{V_{fa}}{V_T} = 100\frac{V_{ma}}{V_T} - 100\frac{V_a}{V_T} \tag{12.63}$$

Now the definition of VMA, the voids in mineral aggregate as a percent by total volume of asphalt mixture is given as follows:

$$VMA = 100\frac{V_{ma}}{V_T}$$

Also, the definition of VTM, the air voids as a percent by total volume of asphalt mixture is given as follows:

$$VTM = 100\frac{V_a}{V_T}$$

On the other hand, the definition of VFA, the voids filled with asphalt as a percent by volume of total voids in the asphalt mixture is given as follows:

$$VFA = 100 \frac{V_{fa}}{V_{ma}}$$

In other words, the equation $100 \frac{V_{fa}}{V_T} = 100 \frac{V_{ma}}{V_T} - 100 \frac{V_a}{V_T}$ becomes:

$$100 \frac{V_{fa}}{V_T} = VMA - VTM \qquad (12.64)$$

But since:

$$VFA = 100 \frac{V_{fa}}{V_{ma}} \neq 100 \frac{V_{fa}}{V_T}$$

$\Rightarrow$

$$VFA \neq VMA - VTM$$

$$100 \frac{V_{fa}}{V_T} = V\dot{M}A - VTM$$

By dividing both sides of the above equation by the term $100 \frac{V_{ma}}{V_T}$, it becomes:

$$\frac{\left(100 \frac{V_{fa}}{V_T}\right)}{\left(100 \frac{V_{ma}}{V_T}\right)} = \frac{VMA - VTM}{\left(100 \frac{V_{ma}}{V_T}\right)}$$

Or:

$$\left(100 \frac{V_{fa}}{V_{ma}}\right)\left(\frac{V_T}{100 V_T}\right) = \frac{VMA - VTM}{\left(100 \frac{V_{ma}}{V_T}\right)}$$

Multiplying both sides of the above equation by 100 provides:

$$\left(100 \frac{V_{fa}}{V_{ma}}\right) = 100 \frac{VMA - VTM}{\left(100 \frac{V_{ma}}{V_T}\right)} \qquad (12.65)$$

Knowing that:

$$VMA = 100 \frac{V_{ma}}{V_T}$$

And:

$$\text{VFA} = 100 \frac{V_{\text{fa}}}{V_{\text{ma}}}$$

The equation $\left(100 \dfrac{V_{\text{fa}}}{V_{\text{ma}}} \right) = 100 \dfrac{\text{VMA} - \text{VTM}}{\left(100 \dfrac{V_{\text{ma}}}{V_T} \right)}$ becomes:

$$\text{VFA} = 100 \left(\frac{\text{VMA} - \text{VTM}}{\text{VMA}} \right)$$

13 Design and Analysis of Asphalt Pavements

Chapter 13 highlights the design and analysis of asphalt pavements which is an important part of pavement engineering that deals with structural-empirical as well as empirical-mechanistic designs of asphalt pavements and the analysis of these pavements for stresses, strains, and deflections. The main inputs for empirical design of asphalt pavements include traffic loading, material properties, reliability, performance indicators, and pavement drainage quality. The mechanistic-empirical (M-E) design, on the other hand, takes into consideration the mechanistic analysis of the pavement structure and its performance over time regarding fatigue, rutting, and thermal cracking based on the available traffic, material, and environmental inputs. Consequently, the availability of this data to input in the M-E design defines three levels of M-E design. Level 3 provides the least level of accuracy in M-E design. This level may be used for less important roads or highways such as low-volume roads where the consequences of early failure are minimum. In this level, normally typical default values (such as average values) for the design inputs in the area are used and selected by the user. Level 2 offers an intermediate level of accuracy and uses values for the design inputs from available data sources of the agency, from limited testing, or based on correlations. The user selects these values to be used in the design. On the other hand, Level 1 provides the highest level of accuracy, and, therefore, has the least uncertainty (or error) level. This level, due to its high reliability and accuracy, is used for major highways and important roads with heavy traffic where the safety and economic consequences of early failure are high. In this level, the values of the design inputs are based on laboratory or field testing that requires more resources and time than other levels. The mechanistic analysis of asphalt pavements focuses on the determination of stresses, strains, and deflection on critical locations in the pavement structure under traffic loading. The analysis of asphalt pavements is composed of three different types: (1) the first one deals with one-layered pavement systems where the pavement structure is considered as one layer with infinite thickness, (2) the second type deals with two-layered pavement systems where the asphalt pavement structure is considered as two layers; one with finite thickness over another layer with infinite thickness, and (3) the third type considers asphalt pavements as three-layered systems; the first two layers with finite thickness over the last underneath layer with an infinite thickness. The second and third types are very close to reality and the real-life situation. The two-layered pavement systems represent full-depth asphalt pavements where there is an asphalt pavement layer over a subgrade layer. The three-layered pavement systems represent the conventional (traditional) asphalt pavements where there is an asphalt layer over base layers over a subgrade layer. The analysis of pavements goes in parallel with design due to its high correlation with design, particularly M-E design. The mechanistic analysis of an asphalt pavement provides an idea of how the different pavement layers would dissipate the high stresses and strains on the surface of the pavement as it gets deeper and deeper in the pavement structure. Pavement structures with high-quality materials or high-thickness layers would dissipate more stresses and strains than layers with low-quality materials or low-thickness layers. Therefore, reliability and cost play a big role in the entire design process. This section will provide questions and practical problems that deal with design as well as analysis of asphalt pavements. The analysis part will focus on multilayered pavement systems because they represent the real-world situation.

13.1 Select a typical or reasonable value from the set in the second column for each item (or property) of the group in the first column in Table 13.1.

TABLE 13.1

Typical Values and Flexible Pavement Properties for Problem 13.1

Property	Typical Value
Modulus of Elasticity for HMA	80 psi
Resilient Modulus for Base Materials	12 inches
Vertical Stress on top of subgrade	8,000 psi
Dual Wheel Spacing	40 ksi
Resilient Modulus for Subgrade	250 ksi
Applied Single Wheel Loading	9,000 lb
Applied Wheel Pressure for Trucks	10 psi

Solution:

See Table 13.2.

TABLE 13.2

Matching of Typical Values for Flexible Pavement Properties for Problem 13.1

Property	Answer
Modulus of Elasticity for HMA	250 ksi
Resilient Modulus for Base Materials	40 ksi
Vertical Stress on top of subgrade	10 psi
Dual Wheel Spacing	12 inches
Resilient Modulus for Subgrade	8,000 psi
Applied Single Wheel Loading	9,000 lb
Applied Wheel Pressure for Trucks	80 psi

13.2 Indicate the effect of the increase in the value of each of the following design inputs on the design thickness of the hot-mix asphalt (HMA) layer of a flexible pavement (see Table 13.3).

TABLE 13.3

Effect of Design Inputs on HMA Design Thickness for Problem 13.2

Item	Increase	Decrease	No Effect
Base Resilient Modulus (M_R)			
Δpsi			
Reliability Level, R (%)			
S_0			
HMA Modulus			
Subgrade CBR			
Terminal Serviceability Index, p_t			
Traffic ESALs			

Solution:

See Table 13.4.

TABLE 13.4

Identifying the Effect of Design Inputs on HMA Design Thickness for Problem 13.2

Item	Increase	Decrease	No Effect
Base Resilient Modulus (M_R)		X	
Δpsi		X	
Reliability Level, R (%)	X		
S_0	X		
HMA Modulus		X	
Subgrade CBR		X	
Terminal Serviceability Index, p_t	X		
Traffic ESALs	X		

13.3 Order or rank the asphalt pavement structures subjected to different traffic loadings as shown in Figure 13.1 according to fatigue performance from best to worst assuming that the environmental conditions are the same.

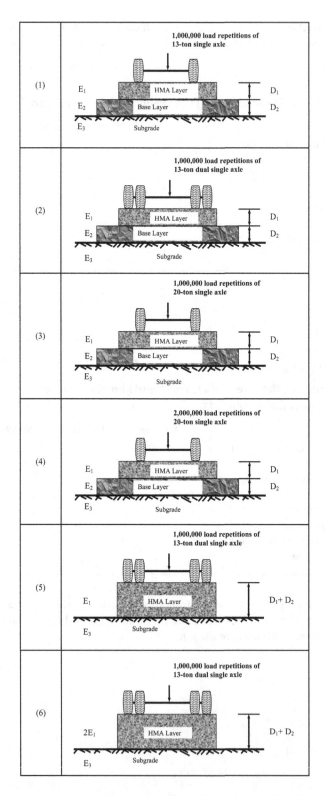

FIGURE 13.1 Flexible pavement structures under traffic loading to be ranked for Problem 13.3.

Solution:

All the pavement structures have the same subgrade layer with the same modulus (E_3). Therefore, the ranking will be based on the thickness of the HMA layer, the traffic loading, and the thickness of the base layer taking into consideration the following points:
- Higher traffic loading results in lower pavement performance.
- Higher number of load repetitions leads to lower performance.
- Dual loads lead to higher pavement performance.
- Higher HMA layer thickness leads to higher pavement performance.
- Higher base-layer thickness also leads to higher pavement performance.
- Higher modulus results in higher pavement performance.
- The replacement of a base layer with an asphalt layer with a higher modulus leads to higher pavement performance.

Based on the above points, the following ranking is obtained (see Table 13.5).

TABLE 13.5

Ranking of the Given Flexible Pavement Structures in Problem 13.3

Order or Ranking of Pavement Structures from Best to Worst

6
5
2
1
3
4

Pavement Structure #6: load (13 ton), load repetitions (1000000), modulus of HMA layer ($2E_1$), HMA layer thickness (D_1+D_2), dual wheels.

Pavement Structure #5: same as #6 except for modulus: load (13 ton), load repetitions (1000000), modulus of HMA layer (E_1), HMA layer thickness (D_1+D_2), dual wheels.

Pavement Structure #2: load (13 ton), load repetitions (1000000), modulus of HMA and base layers (E_1 and E_2, respectively), two layers (HMA and base) with total thickness (D_1+D_2), dual wheels.

Pavement Structure #1: same as pavement structure #2 except for dual wheels: load (13 ton), load repetitions (1000000), modulus of HMA and base layers (E_1 and E_2, respectively), two layers (HMA and base) with total thickness (D_1+D_2), single wheels.

Pavement Structure #3: same as pavement structure #1 except for the load: load (20 ton), load repetitions (1000000), modulus of HMA and base layers (E_1 and E_2, respectively), two layers (HMA and base) with total thickness (D_1+D_2), single wheels.

Pavement Structure #4: same as pavement structure #3 except for the load repetitions: load (20 ton), load repetitions (2000000), modulus of HMA and base layers (E_1 and E_2, respectively), two layers (HMA and base) with total thickness (D_1+D_2), single wheels.

13.4 State the mistake in each of the following flexible pavement structure designs and correct it (see Figure 13.2).

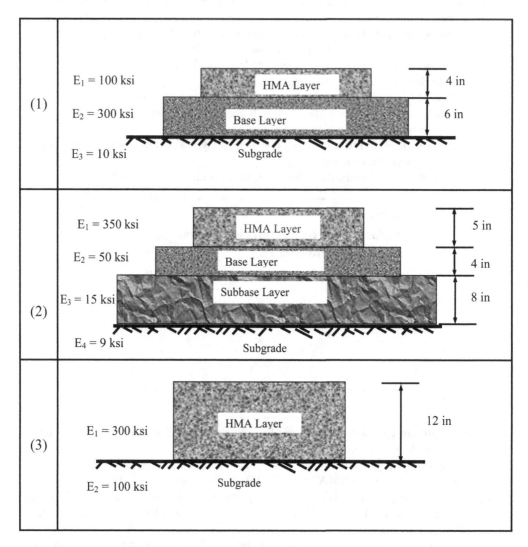

FIGURE 13.2 Mistakes in the design of flexible pavement structures for Problem 13.4.

Solution:

Pavement Structure Design #1: the modulus of the HMA layer is smaller than the modulus of the base layer. A correction can be: $E_1 = 300$ ksi and $E_2 = 100$ ksi.

Pavement Structure Design #2: the thickness of the HMA layer is higher than the thickness of the base layer. A correction can be: $D_1 = 4$ in and $D_2 = 5$ in.

Pavement Structure Design #3: the modulus of the subgrade layer is very high (not typical). A correction can be: $E_3 = 10$ ksi (see Table 13.6).

TABLE 13.6
Identifying the Mistakes in the Design of Flexible
Pavement Structures for Problem 13.4

Pavement Design	Mistake	Correction
1	$E_1 = 100$ ksi $< E_2 = 300$ ksi	$E_1 = 300$ ksi $> E_2 = 100$ ksi
2	$D_1 = 5$ in $> D_2 = 4$ in	$D_1 = 4$ in $< D_2 = 5$ in
3	$E_2 = 100$ ksi (too high)	$E_2 = 10$ ksi (typical)

13.5 Compute the predicted total number of equivalent single-axle loads (ESALs) for a highway flexible pavement for a new six-lane rural highway with a first-year annual average daily traffic (AADT) of 3000 veh/day for both directions. The design lane factor (f_d) is 0.4, the traffic growth rate is 4%, the design period is 10 years, and the traffic mix is as shown in Figure 13.3. Ignore the ESALs of passenger cars. What is the truck factor of the truck in the third category? Which load is considered the critical load?

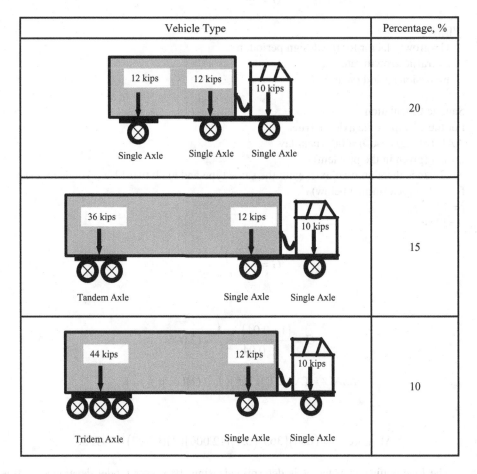

FIGURE 13.3   Truck traffic mix and loadings for the six-lane rural highway pavement in Problem 13.4.

Solution:

The total ESALs is calculated as shown in the following procedure:

$$\text{ESAL}_i = (\text{AADT}_i)(365)(f_d)(G)(N_i)(f_{\text{LE}-i}) \tag{13.1}$$

Where:
ESAL$_i$ = equivalent single-axle loads for traffic/axle category i
AADT$_i$ = annual average daily traffic for category i
365 = number of days per year
f_d = lane distribution factor for the highway
G = growth factor
N_i = number of axles with the same type and load
$f_{\text{LE-i}}$ = load equivalency factor for axle category i

The growth factor is calculated using the formula below:

$$G = \frac{(1+r)^n - 1}{r} \tag{13.2}$$

Where:
G = growth factor for the design period, n
r = traffic growth rate
n = design period (years)

Sample Calculation:
For the 12-kip single axle in Truck 1:
AADT=0 .20 (3000)=600 truck/day
f_d=0.4 (given in the problem)
N=2 (since there are two axles with the same type and load; two 12-kip single axles)
f_{IE}=0.197 (determined below)
r=4%
n= 10 years

$$G = \frac{(1+r)^n - 1}{r}$$

$\Rightarrow$

$$G = \frac{(1+0.04)^{10} - 1}{0.04} = 12.006$$

$$\text{ESAL}_i = (\text{AADT}_i)(365)(f_d)(G)(N_i)(f_{\text{LE}-i})$$

$\Rightarrow$

$$\text{ESAL}_{\text{12-kip Single}} = (600)(365)(0.4)(12.006)(2)(0.197) = 397556$$

The load equivalency factor is determined using the power model developed in this book for single-axle loads based on numbers in the table of the load equivalency factors from the Asphalt Institute's *Thickness Design Manual (MS-1)*, 9th Edition, 1982 as shown in the following relationship (see Figure 13.4).

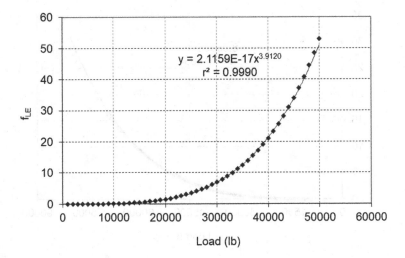

FIGURE 13.4 Load versus load equivalency factors for flexible pavements (single-axle). Based on numbers in the table of the load equivalency factors from the Asphalt Institute's *Thickness Design Manual (MS-1)*, 9th edition, 1982.

The best-fit model that describes the relationship between the load and the load equivalency factor for single-axle loads is a power model with high coefficient of determination (r^2) as shown below:

$$f_{LE} = 2.1159 \times 10^{-17} \left(Load \right)^{3.9120} \tag{13.3}$$

Where:
f_{LE} = load equivalency factor for single-axle loads
Load = single-axle load in (lb)

$\Rightarrow$

$$f_{LE} = 2.1159 \times 10^{-17} \left(12000 \right)^{3.9120} = 0.192$$

For tandem-axle loads and tridem-axle loads, the following relationships are used, respectively.

The load equivalency factor is determined using the power model developed in this book for tandem-axle loads based on numbers in the table of the load equivalency factors from the Asphalt Institute's *Thickness Design Manual (MS-1)*, 9th Edition, 1982 as shown in the following relationship (see Figure 13.5).

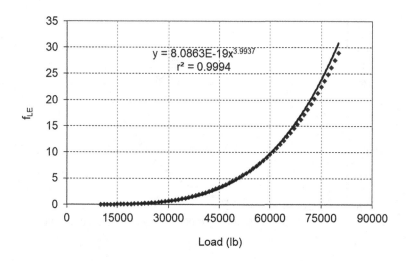

FIGURE 13.5 Load versus load equivalency factors for flexible pavements (tandem-axle). Based on numbers in the table of the load equivalency factors from the Asphalt Institute's *Thickness Design Manual (MS-1)*, 9th edition, 1982.

The best-fit model that describes the relationship between the load and the load equivalency factor for tandem-axle loads is a power model with high coefficient of determination (r^2) as shown below:

$$f_{LE} = 8.0863 \times 10^{-19} \left(\text{Load}\right)^{3.9937}$$ (13.4)

Where:

f_{LE} = load equivalency factor for tandem-axle loads
Load = tandem-axle load in (lb)

For tridem-axle loads, the load equivalency factor is determined using the power model developed in this book based on numbers in the table of the load equivalency factors from the Asphalt Institute's *Thickness Design Manual (MS-1)*, 9th Edition, 1982 as shown in the following relationship (see Figure 13.6).

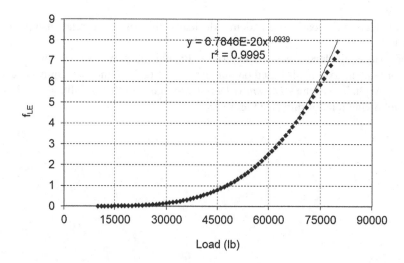

FIGURE 13.6 Load versus load equivalency factors for flexible pavements (tridem-axle). Based on numbers in the table of the load equivalency factors from the Asphalt Institute's *Thickness Design Manual (MS-1)*, 9th edition, 1982.

The best-fit model that describes the relationship between the load and the load equivalency factor for tridem-axle loads is a power model with high coefficient of determination (r^2) as shown below:

$$f_{LE} = 6.7846 \times 10^{-20} \left(\text{Load} \right)^{4.0939} \tag{13.5}$$

Where:
f_{LE} = load equivalency factor for tridem-axle loads
Load = tridem-axle load in (lb)

The ESAL results of the other axle types are summarized in Table 13.7 based on the same formula.

TABLE 13.7
ESALs Results for Problem 13.5

	Axle Type	AADT	365	f_d	G	N	f_{LE}	ESALs
	Single Axle	600	365	0.4	12.01	1	0.094	98,947
	Single Axle					2	0.192	403,821
	Single Axle	450				1	0.094	74,210
	Single Axle					1	0.192	151,433
	Tandem Axle					1	1.271	1,002,819
	Single Axle	300				1	0.094	49,473
	Single Axle					1	0.192	100,955
	Tridem Axle					1	0.694	364,948
Total ESALs								2,246,606

The truck factor is calculated using the following formula:

$$f_T = \left(N_i \right)\left(f_{LE-i} \right) \tag{13.5}$$

Where:
Load_i = number of axles for axle category i
f_{LE-i} = the load equivalency factor for axle category i

$\Rightarrow$

$$f_T = 1 \times 0.094 + 1 \times 0.192 + 1 \times 0.694 = 0.980$$

The critical load is the load with the highest load equivalency factor. Therefore, in this case, the critical load is the 44-kip tridem axle.

The MS Excel worksheet used to compute the ESALs in this problem is shown in Figure 13.7.

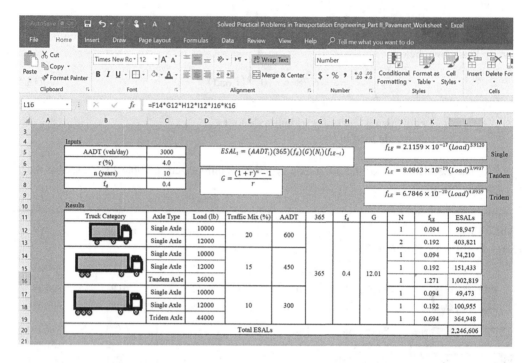

FIGURE 13.7 Image of the MS Excel worksheet used for the computations of ESALs for Problem 13.5.

13.6 A highway flexible pavement is designed for a design period of 20 years. The traffic that is expected to use the highway pavement is of the type shown in Figure 13.8. If the first-year annual average daily traffic (AADT) of this type of trucks is 200 truck/day in both directions, the annual traffic growth rate is 2%, and the design lane factor is 0.45, determine the total equivalent single-axle loads (ESALs).

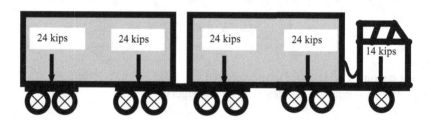

FIGURE 13.8 Type of truck traffic for the highway pavement of Problem 13.6.

Solution:

The total ESALs is calculated as shown in the following procedure:

$$\text{ESAL}_i = (\text{AADT}_i)(365)(f_d)(G)(N_i)(f_{\text{LE}-i})$$

The growth factor is calculated using the formula below:

$$G = \frac{(1+r)^n - 1}{r}$$

Sample Calculation:
For the 24-kip tandem axle:
AADT=200 truck/day
f_d=0.45 (given in the problem)
N=4 (since there are two axles with the same type and load; two 24-kip tandem axles)
f_{lE}=0.252 (determined below)
r=2%
n=20 years

$$G = \frac{(1+r)^n - 1}{r}$$

⇒

$$G = \frac{(1+0.02)^{20} - 1}{0.02} = 24.30$$

$$ESAL_i = (AADT_i)(365)(f_d)(G)(N_i)(f_{LE-i})$$

⇒

$$ESAL_{24\text{-kip Tandem}} = (200)(365)(0.45)(24.30)(4)(0.252) = 803811$$

The power model developed earlier for the load equivalency factor of tandem-axle loads will be used to compute f_{LE} for the 24-kip tandem-axle load as shown below:

$$f_{LE} = 8.0863 \times 10^{-19} (Load)^{3.9937}$$

⇒

$$f_{LE} = 8.0863 \times 10^{-19} (24000)^{3.9937} = 0.252$$

The ESAL results of the other axle types are summarized in Table 13.8 based on the same formula.

TABLE 13.8
ESALs Computations and Results for Problem 13.6

Axle Type	AADT	365	f_d	G	N	f_{LE}	ESALs
Single Axle	200	365	0.45	24.30	1	0.351	280,055
Tandem Axle					4	0.252	803,811
Total ESALs							1,083,866

The MS Excel worksheet used to compute the ESALs in this problem is shown in Figure 13.9.

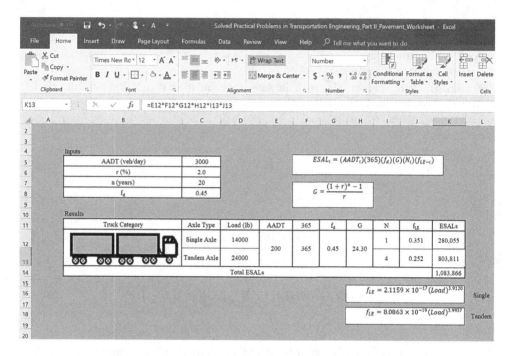

FIGURE 13.9 Image of the MS Excel worksheet used for the computations of ESALs for Problem 13.6.

13.7 Design a flexible pavement for a new six-lane rural highway that connects two cities using the AASHTO design guide procedure to carry the traffic mix shown in Table 13.9. The base and subbase layers will be constructed using the same crushed stone material and therefore one combined thickness could be used for both layers. The material properties are also provided in Table 13.10.

TABLE 13.9

Truck Traffic Mix and Loadings for Problem 13.7

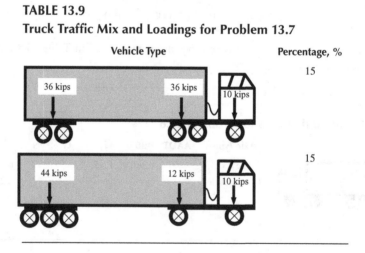

Vehicle Type	Percentage, %
	15
	15

The annual average daily traffic (AADT) that is expected to use the highway in both directions is 5000 veh/day in the first year, the traffic growth rate is 5%, and the design life of the pavement is 10 years. Assume that the traffic growth rate will remain the same during the design life of the pavement.

Given Data:

$S_o = 0.5$

$p_i = 4.5$

$p_t = 2.5$

Drainage data: it is expected that it will take one week for water to be removed from the pavement, and the pavement structure will be exposed to moisture levels approaching saturation 20% of the time (see Figure 13.10).

TABLE 13.10

Material Properties for Designing the Highway Flexible Pavement in Problem 13.7

Material	E (ksi)
HMA	400
Base Course	20
Subbase Course	20
Subgrade	10

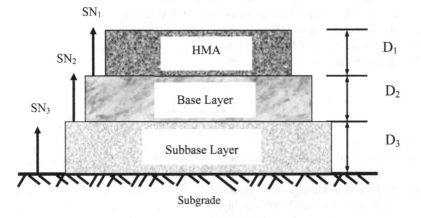

FIGURE 13.10   Sketch of the flexible pavement structure for Problem 13.7.

Solution:

The following formula is used for the design of asphalt pavements following the AASHTO design method:

$$\log_{10} W_{18} = Z_R S_0 + 9.36 \log_{10}(SN+1) - 0.20 + \frac{\log_{10}\left[\dfrac{\Delta PSI}{(4.2-1.5)}\right]}{0.40 + \left[\dfrac{1094}{(SN+1)^{5.19}}\right]} \qquad (13.7)$$

$$+ 2.32 \log_{10} M_R - 8.07$$

Where:

W_{18} = predicted number of 18,000 lb (80 kN) single-axle load applications
Z_R = standard normal deviation for a given reliability
S_0 = overall standard deviation
SN = structural number of the pavement
$\Delta PSI = p_i - p_t$
p_i = initial serviceability index
p_t = terminal serviceability index
M_R = resilient modulus (psi)

For this pavement structure, there are three structural numbers (SN_1, SN_2, and SN_3) defined as shown below:

$$SN_1 = a_1 D_1 \qquad (13.8)$$

Where:

SN_1 = Structural Number 1 above the base layer
a_1 = coefficient of Layer 1 (hot-mix asphalt (HMA) layer)
D_1 = thickness of Layer 1 (HMA layer)

$$SN_2 = a_1 D_1 + a_2 m_2 D_1 \qquad (13.9)$$

Where:

SN_2 = Structural Number 2 above the subbase layer
a_1 = coefficient of Layer 1 (hot-mix asphalt (HMA) layer)
a_2 = coefficient of Layer 2 (base layer)
D_1 = thickness of Layer 1 (HMA layer)
D_2 = thickness of Layer 2 (base layer)

$$SN_3 = a_1 D_1 + a_2 m_2 D_1 + a_3 m_3 D_3 \qquad (13.10)$$

Where:

SN_3 = Structural Number 3 above the subgrade layer
a_1 = coefficient of Layer 1 (hot-mix asphalt (HMA) layer)
a_2 = coefficient of Layer 2 (base layer)
a_3 = coefficient of Layer 3 (subbase layer)
D_1 = thickness of Layer 1 (HMA layer)
D_2 = thickness of Layer 2 (base layer)
D_3 = thickness of Layer 3 (subbase layer)

A polynomial of the third degree is used to fit the data of the elastic modulus of the HMA layer versus the layer coefficient (a_1). The data used to develop this relationship is extracted from the "Chart for Estimating Structural Layer Coefficient of Dense-Graded Asphalt Concrete Based on the Elastic (Resilient) Modulus" in the *AASHTO Guide for Design of Pavement Structures* (1993). This model can be used to determine the layer coefficient (a_1) at any modulus value (see Figure 13.11).

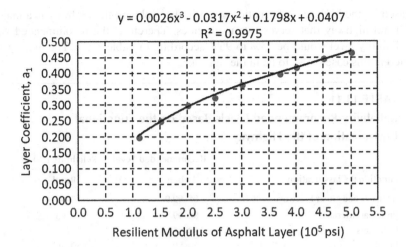

FIGURE 13.11   Resilient modulus versus layer coefficient of asphalt layer.

Therefore,

$$a_1 = 0.0026M_R^3 - 0.0317M_R^2 + 0.1798M_R + 0.0407 \qquad (13.11)$$

Where:
a_1 = coefficient of Layer 1 (HMA layer)
M_R = resilient or elastic modulus of Layer 1 or HMA layer (10^5 psi)

$\Rightarrow$

$$a_1 = 0.0026(4.0)^3 - 0.0317(4.0)^2 + 0.1798(4.0) + 0.0407 = 0.42$$

The coefficient of Layer 2 (the base layer) is given by the following logarithmic formula (Source: *AASHTO Guide for Design of Pavement Structures*, 1993):

$$a_2 = 0.249\log(E_2) - 0.977 \qquad (13.12)$$

Where:
a_2 = coefficient of Layer 2 (base layer)
E_2 = elastic modulus of Layer 2 or base layer (psi)

$\Rightarrow$

$$a_2 = 0.249\log(20000) - 0.977 = 0.094$$

The coefficient of Layer 3 (the subbase layer) is also given by the following logarithmic formula (Source: *AASHTO Guide for Design of Pavement Structures*, 1993):

$$a_3 = 0.227\log(E_3) - 0.839 \qquad (13.13)$$

Where:
a_3 = coefficient of Layer 3 (subbase layer)
E_3 = elastic modulus of Layer 3 or subbase layer (psi)

$\Rightarrow$

$$a_3 = 0.227\log(20000) - 0.839 = 0.137$$

Based on the highway information given in the problem, the highway is a major principal rural highway that serves two main cities. Therefore, the recommended value for the reliability level would be 75% to 95% according to Table 13.11. The highest value is selected in this case (95%) to be in the safe side.

TABLE 13.11
Reliability Recommended Levels Based on the Functional Classification of the Highway

Functional Classification	Recommended Level of Reliability (%)	
	Urban	Rural
Interstate and Other Freeways	85–99.9	80–99.9
Principal Arterials	80–99	75–95
Collectors	80–95	75–95
Local	50–80	50–80

Reproduced with Permission from *AASHTO Guide for Design of Pavement Structures*, 1993, by AASHTO, Washington, D.C, USA.

For a reliability (R) level of 95%, the Z-value is equal to −1.645 according to Table 13.12.

TABLE 13.12
Standard Normal Deviates (Z-Values) at Various Levels of Reliability, R

Reliability, R (%)	Z_R
50	0.000
60	−0.253
70	−0.524
75	−0.674
80	−0.841
85	−1.037
90	−1.282
95	−1.645
97	−1.881
99	−2.327
99.9	−3.090

Reproduced with Permission from *AASHTO Guide for Design of Pavement Structures*, 1993, by AASHTO, Washington, DC, USA.

$S_0 = 0.50$

$$\Delta PSI = p_i - p_t \tag{13.14}$$

Where:

p_i = initial serviceability index

p_t = terminal serviceability index

$\Rightarrow$

$$\Delta PSI = 4.5 - 2.5 = 2.0$$

Now the total ESALs should be determined as shown in the following procedure:

$$ESAL_i = (AADT_i)(365)(f_d)(G)(N_i)(f_{LE-i})$$

The growth factor is calculated using the formula below:

$$G = \frac{(1+r)^n - 1}{r}$$

Sample Calculation for the ESALs:
For the 36-kip tandem axle in Truck 1:
AADT = 0.15 (5000) = 750 truck/day

Since the highway is a six-lane highway, the lane distribution factor is 40% (0.40) according to the Asphalt Institute (AI) values in Table 13.13. And based on the AASHTO recommended values in Table 13.14, for three lanes in each direction, $f_d = 60\text{-}80\%$ in one direction. The traffic in this problem is given for both directions; therefore, a value of 0.40 will be used for f_d.

TABLE 13.13

Percentage of Total Truck Traffic in Design Lane (f_d)

Number of Traffic Lanes in Two Directions	% of Trucks in Design Lane
2	50
4	45 (35–48)
≥ 6	40 (25–48)

Reproduced with Permission from "Thickness Design-Asphalt Pavements for Highways and Streets MS-1", 1999, by Asphalt Institute, Lexington, KY, USA.

Or:

TABLE 13.14

Lane Distribution Factor (f_d)

Number of Lanes in Each Direction	% of 18-kip ESAL in Design Lane
1	100
2	80–100
3	60–80
4	50–75

Reproduced with Permission from *AASHTO Guide for Design of Pavement Structures*, 1993, by AASHTO, Washington, DC, USA.

$f_d = 0.40$

$N = 2$ (since there are two axles with the same type and load; two 36-kip tandem axles)

$f_{IE} = 1.271$ (determined below)

The power model for the load equivalency factor of tandem-axle loads developed above in Problem 13.5 will be used:

$$f_{LE} = 8.0863 \times 10^{-19} \left(\text{Load} \right)^{3.9937}$$

$\Rightarrow$

$$f_{LE} = 8.0863 \times 10^{-19} \left(36000 \right)^{3.9937} = 1.271$$

$$r = 5\%$$

$$n = 10 \text{ years}$$

$$G = \frac{\left(1+r\right)^n - 1}{r}$$

$\Rightarrow$

$$G = \frac{\left(1+0.05\right)^{10} - 1}{0.05} = 12.58$$

$$\text{ESAL}_i = (\text{AADT}_i)(365)(f_d)(G)(N_i)(f_{LE-i})$$

$\Rightarrow$

$$\text{ESAL}_{36\text{-kip Tandem}} = (750)(365)(0.4)(12.58)(2)(1.271) = 3801291$$

The results of the other axle types are summarized in Table 13.15 based on the same formula.

TABLE 13.15

ESALs Computations and Results for Problem 13.7

Truck Type	Axle Type	AADT	365	f_d	G	N	f_{LE}	ESALs
	Single	750	365	0.40	12.58	1	0.094	129,574
	Tandem					2	1.271	3,501,927
	Single	750				1	0.094	129,574
	Single					1	0.192	264,408
	Tridem					1	0.694	955,822
Total ESALs								4,981,305

Now using the resilient (elastic) modulus of the subgrade layer (10000 psi) and based on the design ESALs that the pavement structure should be capable of carrying, the structural number (SN_3) is determined using the formula below:

$$\log_{10} W_{18} = Z_R S_0 + 9.36 \log_{10}(SN+1) - 0.20 + \frac{\log_{10}\left[\dfrac{\Delta PSI}{(4.2-1.5)}\right]}{0.40 + \dfrac{1094}{(SN+1)^{5.19}}}$$

$$+ 2.32 \log_{10} M_R - 8.07$$

$\Rightarrow$

$$\log_{10}(4981305) = -1.282(0.5) + 9.36 \log_{10}(SN+1) - 0.20$$

$$+ \frac{\log_{10}\left[\dfrac{2.0}{(4.2-1.5)}\right]}{0.40 + \dfrac{1094}{(SN+1)^{5.19}}} + 2.32 \log_{10}(10000) - 8.07$$

$\Rightarrow$

$$6.6973 = -1.282(0.5) + 9.36 \log_{10}(3.32+1) - 0.20$$

$$+ \frac{\log_{10}\left[\dfrac{2.0}{(4.2-1.5)}\right]}{0.40 + \dfrac{1094}{(3.32+1)^{5.19}}} + 2.32 \log_{10}(10000) - 8.07$$

The MS Excel Solver tool is used as shown in the procedure below:
The left side of the equation is basically set to:

$$\text{Left Side} = \text{Design ESALs} \tag{13.15}$$

In this case,

$$\text{Left Side} = \text{Design ESALs} = 4981305$$

And the right side of the equation is set to:

$$\text{Right Side} = 10^{\left[-1.282(0.5)+9.36\log_{10}(SN+1)-0.20+\dfrac{\log_{10}\left[\dfrac{2.0}{(4.2-1.5)}\right]}{0.40+\dfrac{1094}{(SN+1)^{5.19}}}+2.32\log_{10}(10000)-8.07\right]} \tag{13.16}$$

An error is defined as the difference between the left side and the right side; and in this case the error is defined as follows:

$$\text{Error} = \left[\log(\text{Left Side}) - \log(\text{right side})\right]^2 \tag{13.17}$$

When the error is zero or minimum, that is when SN is equal to the correct value. Therefore, the Excel Solver will perform iterative approach by changing SN value until the error converges to a value of zero or minimum value possible. That will be the solution of SN. The MS Excel Solver is used three times; each time, only the resilient (elastic) modulus is changed to determine the SN. For SN_3, the resilient modulus of the subgrade is used. For SN_2, the resilient modulus of the subbase is used. And for SN_1, the resilient modulus of the base is used. The total ESALs is the same for the three cases (see Figure 13.12).

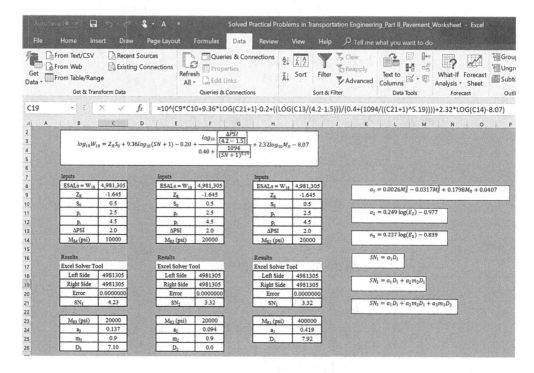

FIGURE 13.12 Image of the MS Excel worksheet used for the determination of SN values and design thicknesses for Problem 13.7.

Based on that, the following results are obtained (see Table 13.16).

TABLE 13.16

Values of SN_1, SN_2, and SN_3 for the Flexible Pavement of Problem 13.7

Structural Number	Value
SN_1	3.32
SN_2	3.32
SN3	4.23

Now the thickness of each layer is determined as explained in the following procedure:

$$SN_1 = a_1 D_1$$

$\Rightarrow$

$$3.32 = 0.42D_1$$

$\Rightarrow$

$$D_1 = 7.92 \text{ in} \cong 8 \text{ in}$$

The drainage coefficients for the base and subbase layers (m_2 and m_3) are determined from Table 13.17 based on the given drainage data (for the condition "water removed within one week", the quality of drainage is "Fair", and for the condition "20% of the time the pavement structure is exposed to moisture levels approaching saturation", the drainage coefficients value is 1.00–0.80. Therefore,

$m_2 = m_3 = 0.90$ (an average value in the range has been selected) (see Table 13.17).

TABLE 13.17

Recommended Values of Drainage Coefficients (m) for Untreated Bases and Subbases in Flexible Pavements

Quality of Drainage		Percentage of Time Pavement Structure Is Exposed to Moisture Levels Approaching Saturation			
Rating	Water removed within	< 1%	1–5%	5–25%	> 25%
Excellent	2 hours	1.40–1.35	1.35–1.30	1.30–1.20	1.20
Good	1 day	1.35–1.25	1.25–1.15	1.15–1.00	1.00
Fair	1 week	1.25–1.15	1.15–1.05	1.00–0.80	0.80
Poor	1 month	1.15–1.05	1.05–0.80	0.80–0.60	0.60
Very Poor	Never drain	1.05–0.95	0.95–0.75	0.75–0.40	0.40

Reproduced with Permission from *AASHTO Guide for Design of Pavement Structures*, 1993, by AASHTO, Washington, DC, USA.

$$SN_2 = a_1 D_1 + a_2 m_2 D_1$$

$\Rightarrow$

$$3.32 = 0.42(8) + 0.094(0.9)D_2$$

$\Rightarrow$

$$D_2 = 0$$

$$SN_3 = a_1 D_1 + a_2 m_2 D_1 + a_3 m_3 D_3$$

$\Rightarrow$

$$4.23 = 0.42(8) + 0.094(0.9)(0) + 0.137(0.9)D_3$$

$\Rightarrow$

$$D_3 = 7.1 \text{ in} \cong 8 \text{ in}$$

Because the same crushed stone material (with the same modulus) is used for the base and the subbase layers, the thickness of the subbase layer comes out to be zero. In other words, only one layer (base) will be used for the pavement structure having a total thickness of 8.0 inches. The final design will be as follows (see Table 13.18 and Figure 13.13).

TABLE 13.18

Design Thicknesses of Asphalt and Base Layers for Problem 13.7

Thickness	in (cm)
D_1	8 ($\cong$ 20)
D_2	8 ($\cong$ 20)

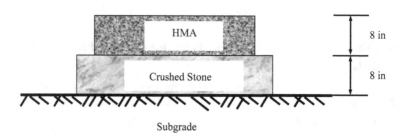

FIGURE 13.13   Flexible pavement structure and design thicknesses for Problem 13.7.

13.8   In the problem above, the pavement design engineer would like to reduce the thickness of the HMA layer and minimize the cost of the highway pavement. Therefore, another source for a base material has been searched. A new base material with an elastic modulus of 80000 psi will be used. The crushed stone material will only be used for the subbase layer. Use the same design inputs to determine the new HMA layer thickness and the thickness of the other pavement layers (see Table 13.19 and Figure 13.14).

TABLE 13.19

Material Properties for Designing the Highway Flexible Pavement in Problem 13.8

Material	E (ksi)
HMA	400
Base Course	80
Subbase Course	20
Subgrade	10

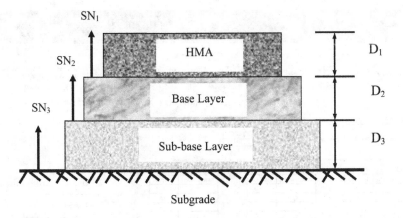

FIGURE 13.14   Sketch of the flexible pavement structure for Problem 13.8.

Given Data:
$S_o = 0.5$
$R\ (\%) = 95\ \%$
$p_i = 4.5$
$p_t = 2.5$
Drainage coefficients: $m_2 = m_3 = 0.9$
$f_d = 0.40$
$r = 5\%$
$n = 10$ years
Total predicted ESALs $= 4981305$

Solution:

Following the same procedure, and using the MS Excel Solver tool to determine the three structural numbers of the pavement (SN_1, SN_2, and SN_3), the following results are obtained (see Table 13.20).

TABLE 13.20
Values of SN_1, SN_2, and SN_3 for the
Flexible Pavement of Problem 13.8

Structural Number	Value
SN_1	1.99
SN_2	3.32
SN3	4.23

The MS Excel worksheet along with the Excel Solver tool used to solve for the results of this problem are shown in Figure 13.15.

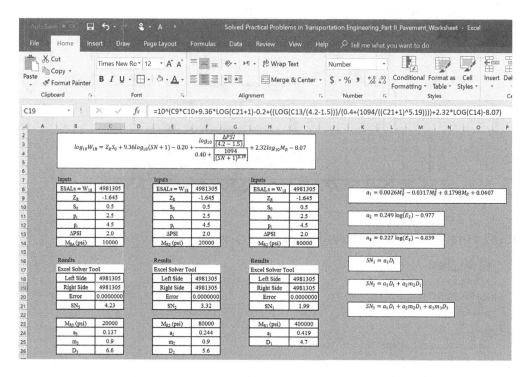

FIGURE 13.15 Image of the MS Excel worksheet used for the determination of SN values and design thicknesses for Problem 13.8.

a_1 is the same since the HMA material is the same; therefore,

$$a_1 = 0.42$$

$$a_2 = 0.249 \log(E_2) - 0.977$$

$\Rightarrow$

$$a_2 = 0.249 \log(80000) - 0.977 = 0.244$$

a_3 is the same since the subbase material is the same; therefore,

$$a_3 = 0.137$$

The thickness of each layer is now determined as explained in the following procedure:

$$SN_1 = a_1 D_1$$

$\Rightarrow$

$$1.99 = 0.42 D_1$$

$\Rightarrow$

$$D_1 = 4.7 \text{ in} \cong 5 \text{ in}$$

$$SN_2 = a_1 D_1 + a_2 m_2 D_1$$

$\Rightarrow$

$$3.32 = 0.42(5) + 0.244(0.9) D_2$$

$\Rightarrow$

$$D_2 = 5.6 \text{ in} \cong 6 \text{ in}$$

$$SN_3 = a_1 D_1 + a_2 m_2 D_1 + a_3 m_3 D_3$$

$\Rightarrow$

$$4.23 = 0.42(5) + 0.244(0.9)(6) + 0.137(0.9) D_3$$

$\Rightarrow$

$$D_3 = 6.6 \text{ in} \cong 7 \text{ in}$$

The final design will be as shown in Table 13.21.

TABLE 13.21
Final Design Thicknesses for the
Flexible Pavement of Problem 13.8

Thickness	in (cm)
D_1	5 ($\cong$ 12.5 cm)
D_2	6 ($\cong$ 15 cm)
D_3	7 ($\cong$ 18 cm)

13.9 Determine the number of equivalent single-axle loads (ESALs) that the asphalt pavement structure shown in Figure 13.16 is capable of carrying at a reliability level (R) of 90%, overall standard deviation (S_0) of 0.5, ΔPSI of 2.0, and $m_2 = 1.0$.

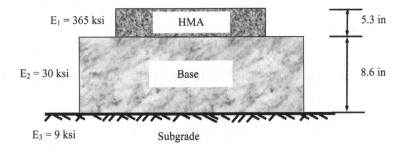

FIGURE 13.16 Flexible pavement structure for Problem 13.9.

Solution:

The following formula is used for the design of asphalt pavements:

$$\log_{10} W_{18} = Z_R S_0 + 9.36 \log_{10}(SN+1) - 0.20 + \frac{\log_{10}\left[\dfrac{\Delta PSI}{(4.2-1.5)}\right]}{0.40 + \dfrac{1094}{(SN+1)^{5.19}}}$$

$$+2.32 \log_{10} M_R - 8.07$$

For this pavement structure, there are two structural numbers (SN_1 and SN_2) as shown in Figure 13.17.

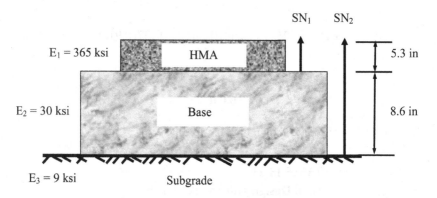

FIGURE 13.17   Sketch for SN_1 and SN_2 on the flexible pavement structure for Problem 13.9.

$$a_1 = 0.0026 M_R^3 - 0.0317 M_R^2 + 0.1798 M_R + 0.0407$$

$\Rightarrow$

$$a_1 = 0.0026(3.65)^3 - 0.0317(3.65)^2 + 0.1798(3.65) + 0.0407 = 0.401$$

$$SN_1 = a_1 D_1$$

$\Rightarrow$

$$SN_1 = 0.401(5.3) = 2.12$$

The coefficient of Layer 2 (the base layer) is given by the following logarithmic formula:

$$a_2 = 0.249 \log(E_2) - 0.977$$

$\Rightarrow$

$$a_2 = 0.249 \log(30000) - 0.977 = 0.138$$

$$SN_2 = a_1 D_1 + a_2 m_2 D_1$$

$\Rightarrow$

$$SN_2 = 0.40(5.3) + 0.138(1.0)(8.6) = 3.31$$

For a reliability (R) level of 90%, the Z-value is equal to −1.282.

Now based on the pavement structural number (SN_2), the pavement structure will be able to carry ESALs calculated based on the design formula below:

$$\log_{10} W_{18} = Z_R S_0 + 9.36 \log_{10}(SN+1) - 0.20 + \frac{\log_{10}\left[\dfrac{\Delta PSI}{(4.2-1.5)}\right]}{0.40 + \left[\dfrac{1094}{(SN+1)^{5.19}}\right]}$$

$$+ 2.32 \log_{10} M_R - 8.07$$

$\Rightarrow$

$$\log_{10} W_{18} = -1.282(0.5) + 9.36 \log_{10}(3.31+1)$$

$$-0.20 + \frac{\log_{10}\left[\dfrac{2.0}{(4.2-1.5)}\right]}{0.40 + \left[\dfrac{1094}{(3.31+1)^{5.19}}\right]} + 2.32 \log_{10}(9000) - 8.07$$

$$= 6.074$$

$\Rightarrow$

$$W_{18} = 1164377 \cong 1.2 \text{ million ESALs}$$

The MS Excel worksheet used to determine the ESALs in this problem is shown in Figure 13.18.

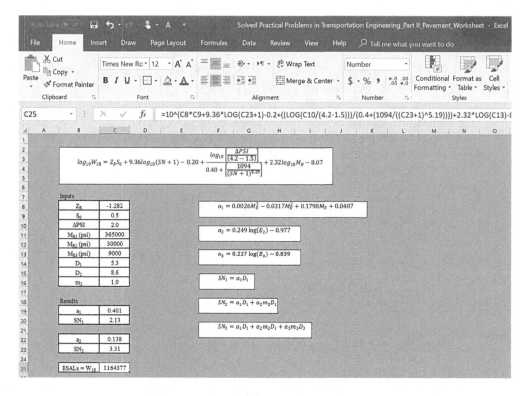

FIGURE 13.18 Image of the MS Excel worksheet used for the determination of ESALs for Problem 13.9.

13.10 A highway designer would like to set a limitation on the number of trucks that use a rural highway pavement that is designed to carry an equivalent single-axle loads (ESALs) of 0.5 million for a design period of 10 years. The truck traffic that is expected to use the highway pavement is of the type shown in Figure 13.19. Determine the maximum allowable first-year annual average daily traffic (AADT) of this type of trucks given that the traffic growth factor is 12.0 and the design lane factor is 0.45.

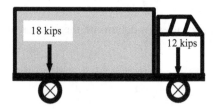

FIGURE 13.19 Type of truck traffic for Problem 13.10.

Solution:

The total number of ESALs is calculated using the formula below and based on the AADT data:

$$ESAL_i = (AADT_i)(365)(f_d)(G)(N_i)(f_{LE-i})$$

For the 12-kip single axle:
$f_d = 0.45$ (given in the problem)
$N = 1$ (since there is only one axle with the same type and load; one 12-kip single axle)

$f_{IE} = 0.192$ (determined below)

The power model developed earlier for the load equivalency factor of single-axle loads will be used to compute f_{LE} for the 12-kip single-axle load as shown below:

$$f_{LE} = 2.1159 \times 10^{-17} (\text{Load})^{3.9120}$$

$\Rightarrow$

$$f_{LE} = 2.1159 \times 10^{-17} (12000)^{3.9120} = 0.192$$

$$n = 10 \text{ years}$$

$$G = 12$$

$$\text{ESAL}_i = (\text{AADT}_i)(365)(f_d)(G)(N_i)(f_{LE-i})$$

$\Rightarrow$

$$\text{ESAL}_{12\text{-kip Single}} = (\text{AADT})(365)(0.45)(12)(1)(0.192) = 378.4 \ \text{AADT}$$

For the 18-kip single axle:

$f_d = 0.45$ (given in the problem)

$N = 1$ (since there is only one axle with the same type and load; one 18-kip single axle)

$f_{IE} = 1$

$n = 10$ years

$G = 12$

$$\text{ESAL}_i = (\text{AADT}_i)(365)(f_d)(G)(N_i)(f_{LE-i})$$

$\Rightarrow$

$$\text{ESAL}_{18\text{-kip Single}} = (\text{AADT})(365)(0.45)(12)(1)(1) = 1971 \ \text{AADT}$$

Since the highway pavement is designed to carry design ESALs of 0.5 million, therefore,

$$378.4 \ \text{AADT} + 1971 \ \text{AADT} = 500000$$

$\Rightarrow$

$$\text{AADT} = 213 \ \text{truck/day}$$

The MS Excel worksheet used to perform the computations of this problem is shown in Figure 13.20.

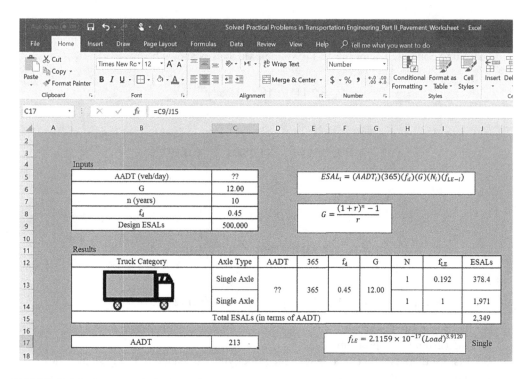

FIGURE 13.20 Image of the MS Excel worksheet used for determining the AADT of Problem 13.10.

13.11 A highway flexible pavement is designed to carry an equivalent single-axle loads (ESALs) of 795000 for a design period (n). The traffic that is expected to use the highway pavement is of the type shown in Figure 13.21. If the first-year annual average daily traffic (AADT) of this type of trucks is 200 truck/day, the annual traffic growth rate is 2%, and the design lane factor is 0.45, determine the design life for this pavement in years.

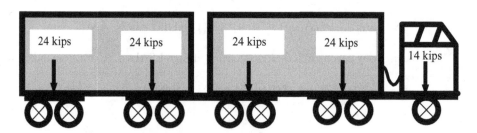

FIGURE 13.21 Type of truck traffic for Problem 13.11.

Solution:

The total number of ESALs is calculated using the formula below and based on the AADT data:

$$ESAL_i = (AADT_i)(365)(f_d)(G)(N_i)(f_{LE-i})$$

For the 14-kip single axle:
$f_d = 0.45$ (given in the problem)
$N = 1$ (since there is only one axle with the same type and load; one 14-kip single axle)
$f_{IE} = 0.351$ (calculated below)

The power model developed earlier for the load equivalency factor of single-axle loads will be used to compute f_{LE} for the 14-kip single-axle load as follows:

$$f_{LE} = 2.1159 \times 10^{-17} (\text{Load})^{3.9120}$$

$\Rightarrow$

$$f_{LE} = 2.1159 \times 10^{-17} (14000)^{3.9120} = 0.351$$

$$ESAL_i = (AADT_i)(365)(f_d)(G)(N_i)(f_{LE-i})$$

$\Rightarrow$

$$ESAL_{\text{14-kip Single}} = (200)(365)(0.45)(G)(1)(0.351) = 11526\,G$$

For the 24-kip tandem axle:
 $f_d = 0.45$ (given in the problem)
 N = 4 (since there are four axles with the same type and load; four 24-kip tandem axles)
 $f_{lE} = 0.252$ (computed below)

The power model developed earlier for the load equivalency factor of tandem-axle loads will be used to compute f_{LE} for the 24-kip tandem-axle load as follows:

$$f_{LE} = 8.0863 \times 10^{-19} (\text{Load})^{3.9937}$$

$\Rightarrow$

$$f_{LE} = 8.0863 \times 10^{-19} (24000)^{3.9937} = 0.252$$

$$ESAL_i = (AADT_i)(365)(f_d)(G)(N_i)(f_{LE-i})$$

$\Rightarrow$

$$ESAL_{\text{24-kip Tandem}} = (200)(365)(0.45)(G)(4)(0.252) = 33082\,G$$

Therefore, the estimated number of ESALs based on the AADT data is equal to:

$$\text{Estimated ESALs} = 11526\,G + 33082\,G$$

Since the highway pavement is designed to carry design ESALs of 795000, the estimated ESALs should equal to the design ESALs. Consequently,

$$11526\,G + 33082\,G = 795000$$

$\Rightarrow$

$$G = 17.82$$

$$G = \frac{(1+r)^n - 1}{r}$$

$\Rightarrow$

$$G = \frac{(1+0.02)^n - 1}{0.02} = 17.82$$

$\Rightarrow$

$$n = 15.4 \cong 16 \text{ years}$$

The MS Excel worksheet used to perform the computations of this problem is shown in Figure 13.22.

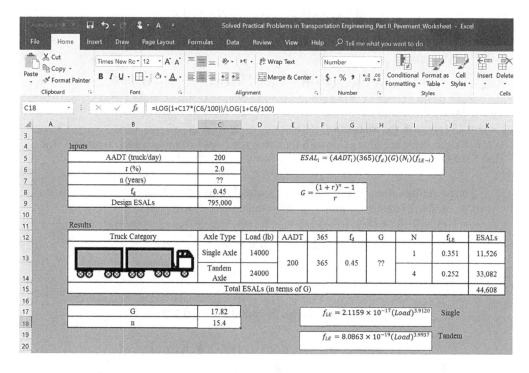

FIGURE 13.22   Image of the MS Excel worksheet used for estimating G and N of Problem 13.11.

13.12 The full-depth asphalt pavement shown in Figure 13.23 is designed to carry an equivalent single-axle loads (ESALs) of 6 million for a design period of 20 years at a reliability level of 90%. Determine the resilient modulus of the modified subgrade that should be used under the HMA layer if the overall standard deviation (S_0) is 0.5 and ΔPSI is 2.0 (see Figure 13.23).

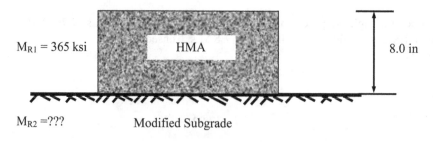

M_{R1} = 365 ksi   HMA   8.0 in

M_{R2} =???   Modified Subgrade

FIGURE 13.23   Full-depth asphalt pavement structure for Problem 13.12.

Solution:

The layer coefficient is determined for layer 1 (HMA layer) as below:

$$a_1 = 0.0026 M_R^3 - 0.0317 M_R^2 + 0.1798 M_R + 0.0407$$

$\Rightarrow$

$$a_1 = 0.0026(3.65)^3 - 0.0317(3.65)^2 + 0.1798(3.65) + 0.0407 = 0.401$$

$$SN_1 = a_1 D_1$$

$\Rightarrow$

$$SN_1 = 0.401(8.0) = 3.208$$

For a reliability (R) level of 90%, the Z-value is equal to −1.282.

$$\log_{10} W_{18} = Z_R S_0 + 9.36 \log_{10}(SN+1) - 0.20 + \frac{\log_{10}\left[\dfrac{\Delta PSI}{(4.2-1.5)}\right]}{0.40 + \left[\dfrac{1094}{(SN+1)^{5.19}}\right]}$$

$$+ 2.32 \log_{10} M_R - 8.07$$

$\Rightarrow$

$$\log_{10}(6 \times 10^6) = -1.282(0.5) + 9.36 \log_{10}(3.208+1) - 0.20$$

$$+ \frac{\log_{10}\left[\dfrac{2.0}{(4.2-1.5)}\right]}{0.40 + \left[\dfrac{1094}{(3.208+1)^{5.19}}\right]} + 2.32 \log_{10}(M_{R2}) - 8.07$$

$\Rightarrow$

$$6.778 = -1.282(0.5) + 9.36 \log_{10}(3.208+1) - 0.20$$

$$+ \frac{\log_{10}\left[\dfrac{2.0}{(4.2-1.5)}\right]}{0.40 + \left[\dfrac{1094}{(3.208+1)^{5.19}}\right]} + 2.32 \log_{10}(M_{R2}) - 8.07$$

Solving the above equation for M_{R2} provides the following solution:

$$M_{R2} = 19907 \text{ psi}$$

Therefore, a modified subgrade with a resilient (elastic) modulus of 19907 psi ($\cong$ 137 MPa) should be used under the full-depth asphalt layer so that the pavement will be capable of carrying a 20-year expected traffic level of 6 million ESALs.

The MS Excel worksheet used to determine the ESALs in this problem is shown in Figure 13.24.

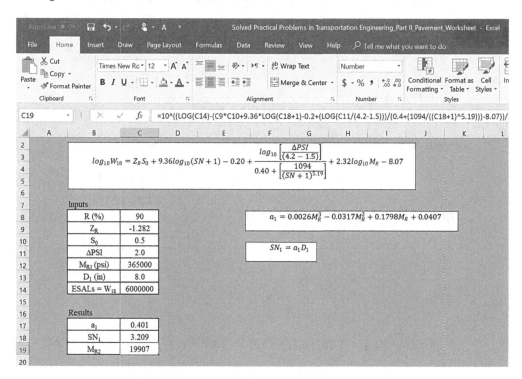

FIGURE 13.24 Image of the MS Excel worksheet used for determining M_{R2} of Problem 13.12.

13.13 The flexible pavement structure shown in Figure 13.25 is designed for a four-lane highway based on the following inputs: reliability level (R)=90%, overall standard deviation (S_0)=0.5, ΔPSI=2.0, m_2=1.0, traffic growth rate (r)=4%, design period, n=20 years, and design lane factor, f_d=0.40. The traffic mix that will use the highway consists of 10% trucks of the type shown in Figure 13.26.

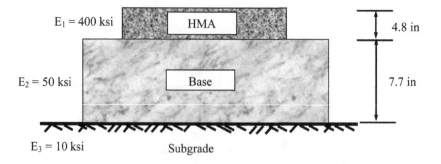

FIGURE 13.25 Flexible pavement structure for Problem 13.13.

If the annual average daily traffic (AADT) for the first year is 3000 veh/day, determine the following:

a. The ESALs after 20 years that this pavement structure can carry.
b. The load equivalency factor of the single axle of the design truck.
c. The maximum load that this single axle can carry.
    Note: ignore the effect of the passenger cars.

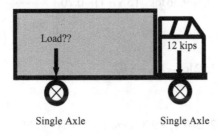

Single Axle   Single Axle

FIGURE 13.26   Type of truck traffic for Problem 13.13.

Solution:

The layer coefficients are determined:

$$a_1 = 0.0026M_R^3 - 0.0317M_R^2 + 0.1798M_R + 0.0407$$

⇒

$$a_1 = 0.0026(4.0)^3 - 0.0317(4.0)^2 + 0.1798(4.0) + 0.0407 = 0.419$$

$$SN_1 = a_1D_1$$

⇒

$$SN_1 = 0.419(4.8) = 2.01$$

The coefficient of Layer 2 (the base layer) is given by the following logarithmic formula:

$$a_2 = 0.249\log(E_2) - 0.977$$

⇒

$$a_2 = 0.249\log(50000) - 0.977 = 0.193$$

$$SN_2 = a_1D_1 + a_2m_2D_1$$

⇒

$$SN_2 = 0.419(4.8) + 0.193(1.0)(7.7) = 3.50$$

For a reliability (R) level of 90%, the Z-value is equal to −1.282.

Now based on the pavement structural number (SN_2), the pavement structure will be able to carry ESALs calculated based on the design formula below:

$$\log_{10} W_{18} = Z_R S_0 + 9.36 \log_{10}(SN+1) - 0.20 + \frac{\log_{10}\left[\dfrac{\Delta PSI}{(4.2-1.5)}\right]}{0.40 + \left[\dfrac{1094}{(SN+1)^{5.19}}\right]}$$

$$+ 2.32 \log_{10} M_R - 8.07$$

$\Rightarrow$

$$\log_{10} W_{18} = -1.282(0.5) + 9.36 \log_{10}(3.50+1) - 0.20$$

$$+ \frac{\log_{10}\left[\dfrac{2.0}{(4.2-1.5)}\right]}{0.40 + \left[\dfrac{1094}{(3.50+1)^{5.19}}\right]} + 2.32 \log_{10}(9000) - 8.07 = 6.327$$

$\Rightarrow$

$$W_{18} = 2125163 \text{ ESALs}$$

But the estimated number of ESALs based on the AADT is calculated as below:

$$ESAL_i = (AADT_i)(365)(f_d)(G)(N_i)(f_{LE-i})$$

For the 12-kip single axle:
$f_d = 0.40$ (given in the problem)
$N = 1$ (since there is only one axle with the same type and load; one 12-kip single axle)
$f_{IE} = 0.192$ (computed below)

The power model developed earlier for the load equivalency factor of single-axle loads will be used to compute f_{LE} for the 12-kip single-axle load as follows:

$$f_{LE} = 2.1159 \times 10^{-17}(\text{Load})^{3.9120}$$

$\Rightarrow$

$$f_{LE} = 2.1159 \times 10^{-17}(12000)^{3.9120} = 0.192$$

$$ESAL_i = (AADT_i)(365)(f_d)(G)(N_i)(f_{LE-i})$$

$\Rightarrow$

$$ESAL_{12\text{-kip Single}} = (3000 \times 0.10)(365)(0.40)(29.78)(1)(0.192) = 250393$$

For the other single axle:
$f_d = 0.40$ (given in the problem)
$N = 1$
$f_{IE} = ?$

$$ESAL_i = (AADT_i)(365)(f_d)(G)(N_i)(f_{LE-i})$$

$\Rightarrow$

$$ESAL_{x\text{-Single}} = (3000 \times 0.10)(365)(0.40)(29.78)(1)(f_{LE-x}) = 1304280\, f_{LE-x}$$

The predicted number of ESALs based on AADT should be equal to the total number of design ESALs the pavement structure can carry in the design life; therefore,

$$W_{18} = 2125163 = 250393 + 1304280\, f_{LE-x}$$

$\Rightarrow$

$$f_{LE-x} = 1.437$$

Using the same power model developed earlier for the load equivalency factor of single-axle loads, the single-axle load that corresponds to a f_{LE} of 1.433 will be determined as shown below:

$$f_{LE} = 2.1159 \times 10^{-17} (\text{Load})^{3.9120}$$

$\Rightarrow$

$$1.437 = 2.1159 \times 10^{-17} (\text{Load})^{3.9120}$$

$$\text{Load} = 20076\ \text{lb}$$

The MS Excel worksheet used to determine the ESALs in this problem is shown in Figure 13.27.

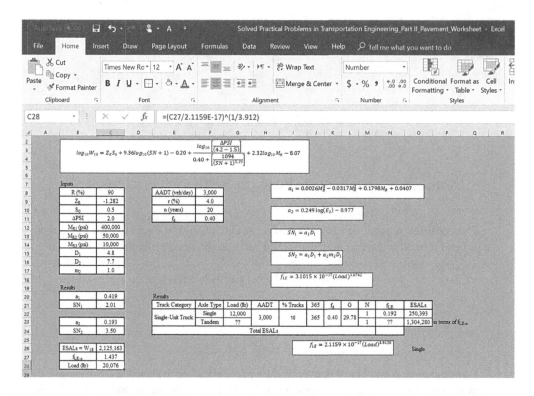

FIGURE 13.27 Image of the MS Excel worksheet used for determining the load of the rear single axle of the truck for Problem 13.13.

13.14 The full-depth asphalt pavement shown in Figure 13.28 is designed to carry an equivalent single-axle loads (ESALs) of 4.5 million for a design period of 20 years at a reliability level of 90%. Determine the terminal serviceability index (p_t) that this asphalt pavement is designed for so that it will carry the design ESALs knowing that the overall standard deviation (S_0) is 0.5 and the initial serviceability index (p_i) is 4.2.

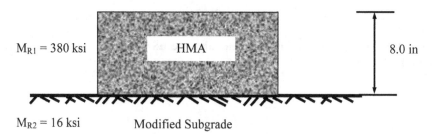

$M_{R1} = 380$ ksi   HMA   8.0 in

$M_{R2} = 16$ ksi   Modified Subgrade

FIGURE 13.28 Full-depth asphalt pavement structure for Problem 13.14.

Solution:

The layer coefficient is determined for layer 1 (HMA layer) as below:

$$a_1 = 0.0026M_R^3 - 0.0317M_R^2 + 0.1798M_R + 0.0407$$

$\Rightarrow$

$$a_1 = 0.0026(3.8)^3 - 0.0317(3.8)^2 + 0.1798(3.8) + 0.0407 = 0.409$$

$$SN_1 = a_1 D_1$$

$\Rightarrow$

$$SN_1 = 0.409(8.0) = 3.271$$

For a reliability (R) level of 90%, the Z-value is equal to -1.282.

$$\log_{10} W_{18} = Z_R S_0 + 9.36 \log_{10}(SN + 1) - 0.20 + \frac{\log_{10}\left[\dfrac{\Delta PSI}{(4.2 - 1.5)}\right]}{0.40 + \left[\dfrac{1094}{(SN + 1)^{5.19}}\right]}$$

$$+ 2.32 \log_{10} M_R - 8.07$$

$\Rightarrow$

$$\log_{10}(4.5 \times 10^6) = -1.282(0.5) + 9.36 \log_{10}(3.271 + 1) - 0.20$$

$$+ \frac{\log_{10}\left[\dfrac{\Delta PSI}{(4.2 - 1.5)}\right]}{0.40 + \left[\dfrac{1094}{(3.271 + 1)^{5.19}}\right]} + 2.32 \log_{10}(16000) - 8.07$$

$\Rightarrow$

MS Excel worksheet is used to solve the above equation for ΔPSI. The following solution is obtained:

$$\Delta PSI = 2.2$$

$$\Delta PSI = p_t - p_i$$

$\Rightarrow$

$$p_i = p_t - \Delta PSI = 4.2 - 2.2 = 2.0$$

Therefore, a terminal serviceability index (p_t) of 2.0 is used so that the given full-depth asphalt pavement structure with the given design inputs will be able to carry the design ESALs of 4.5 million.

The MS Excel worksheet used to determine ΔPSI and the terminal serviceability index (p_t) is shown in Figure 13.29.

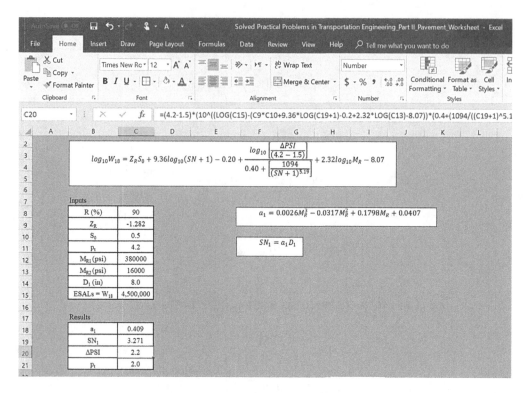

FIGURE 13.29 Image of the MS Excel worksheet used for determining ΔPSI and p_t for Problem 13.14.

13.15 A rural highway asphalt pavement with a total structural number (SN) of 3.50 is designed for a design period of 20 years. The truck traffic that is expected to use the highway pavement is of the type shown in Figure 13.30. If the first-year annual average daily traffic (AADT) of this type of trucks is 350 truck/day, determine the load equivalency factor and the maximum allowable load for the tridem axle shown below given the following design inputs (note: ignore the effect of passenger cars):

<u>Given Data:</u>
R (%)=95%
G=33.0
f_d=0.45
S_0=0.5
ΔPSI=2.0
$M_{\text{R-subgrade}}$=12000 psi (see Figure 13.30).

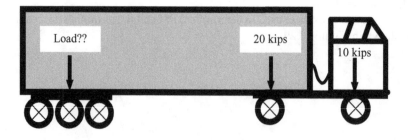

FIGURE 13.30 Type of truck traffic for Problem 13.15.

Solution:

$$SN = 3.50$$

For a reliability (R) level of 95%, the Z-value is equal to -1.645.

Now based on the pavement structural number (SN), the pavement structure will be able to carry ESALs calculated based on the design formula below:

$$\log_{10} W_{18} = Z_R S_0 + 9.36 \log_{10}(SN+1) - 0.20 + \frac{\log_{10}\left[\dfrac{\Delta PSI}{(4.2-1.5)}\right]}{0.40 + \dfrac{1094}{(SN+1)^{5.19}}}$$

$$+ 2.32 \log_{10} M_R - 8.07$$

$\Rightarrow$

$$\log_{10} W_{18} = -1.645(0.5) + 9.36 \log_{10}(3.50+1) - 0.20$$

$$+ \frac{\log_{10}\left[\dfrac{2.0}{(4.2-1.5)}\right]}{0.40 + \dfrac{1094}{(3.50+1)^{5.19}}} + 2.32 \log_{10}(12000) - 8.07$$

$\Rightarrow$

$$W_{18} = 4141294 \text{ ESALs}$$

This number represents the design ESALs that the pavement structure can carry through the design life based on the empirical design formula shown above.

But the estimated number of ESALs based on the AADT is calculated as below:

$$ESAL_i = (AADT_i)(365)(f_d)(G)(N_i)(f_{LE-i})$$

Sample Calculation for the ESALs:
For the 10-kip single axle:
$f_d = 0.45$ (given in the problem)
$N = 1$
$f_{IE} = 0.094$ (calculated below)

The power model developed earlier for the load equivalency factor of single-axle loads will be used to compute f_{LE} for the 10-kip single-axle load as follows:

$$f_{LE} = 2.1159 \times 10^{-17} (\text{Load})^{3.9120}$$

$\Rightarrow$

$$f_{LE} = 2.1159 \times 10^{-17} (10000)^{3.9120} = 0.094$$

$$ESAL_i = (AADT_i)(365)(f_d)(G)(N_i)(f_{LE-i})$$

$\Rightarrow$

$$ESAL_{10\text{-kip Single}} = (350)(365)(0.45)(33.0)(1)(0.094) = 178477$$

The results of the other axle loads are summarized in Table 13.22.

TABLE 13.22

Computations of the ESALs for Problem 13.15

Truck Type	Axle Type	AADT	365	f_d	G	N	f_{LE}	ESALs
	Single	350	365	0.45	33.0	1	0.094	178,477
	Single					1	1.416	2,686,655
	Tridem					1	??	1,897,088 $f_{LE\text{-}x}$

Total ESALs

The predicted number of ESALs should be equal to the design ESALs the pavement structure can carry in the design life; therefore,

$$W_{18} = 4141294 = 178477 + 2686655 + 1897088\, f_{LE-x}$$

$\Rightarrow$

$$f_{LE-x} = 0.673$$

Using the same power model developed earlier for the load equivalency factor of tridem-axle loads, the tridem-axle load that corresponds to a f_{LE} of 0.673 will be determined as shown below:

$$f_{LE} = 6.7846 \times 10^{-20} (\text{Load})^{4.0939}$$

$\Rightarrow$

$$0.673 = 6.7846 \times 10^{-20} (\text{Load})^{4.0939}$$

$$\text{Load} = 43666\ \text{lb}$$

Therefore, the maximum allowable load for the tridem axle of the truck that will use this highway pavement with the given design data is about 44000 lb.

The MS Excel worksheet, used to calculate the estimated ESALs based on the AADT data, the design ESALs based on the empirical design formula using the given design inputs, and the maximum allowable load for the tridem axle of the truck that is expected to use the highway pavement, is shown in Figure 13.31.

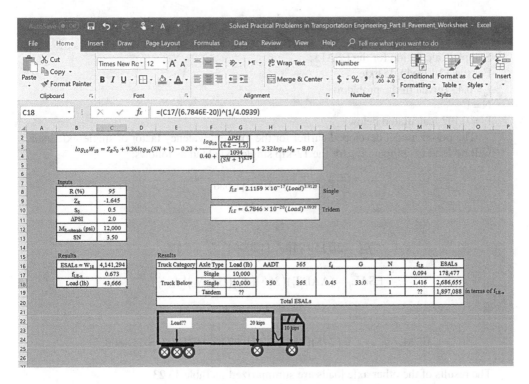

FIGURE 13.31 Image of the MS Excel worksheet used for determining the load of the tridem axle of the truck for Problem 13.15.

13.16 A rural highway flexible pavement is designed to carry an equivalent single-axle loads (ESALs) of 2×10^6 for a design period of 20 years. The traffic that is expected to use the highway pavement is of the type shown in Figure 13.32. If the first-year annual average daily traffic (AADT) of this type of trucks is 275 trucks/day (the transportation agency in the area is planning to put a limitation on the maximum allowable load for the tridem axle in the trucks using the highway), determine the load equivalency factor and the maximum allowable load of the tridem axle given that the traffic growth factor is 29.8 and the design lane factor is 0.45.

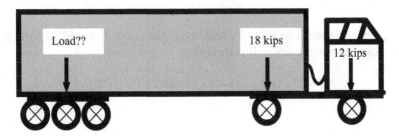

FIGURE 13.32 Type of truck traffic for Problem 13.16.

Solution:

The asphalt pavement structure given in this problem is able to carry design ESALs of 2 million for the design period of 20 years.

The estimated number of ESALs based on the AADT is calculated as below:

$$\text{ESAL}_i = (\text{AADT}_i)(365)(f_d)(G)(N_i)(f_{\text{LE}-i})$$

Sample Calculation for the ESALs:
For the 12-kip single axle:
$f_d = 0.45$ (given in the problem)
$N = 1$
$f_{1E} = 0.192$ (calculated below)

The power model developed earlier for the load equivalency factor of single-axle loads will be used to compute f_{LE} for the 12-kip single-axle load as follows:

$$f_{\text{LE}} = 2.1159 \times 10^{-17} (\text{Load})^{3.9120}$$

$\Rightarrow$

$$f_{\text{LE}} = 2.1159 \times 10^{-17} (12000)^{3.9120} = 0.192$$

$$\text{ESAL}_i = (\text{AADT}_i)(365)(f_d)(G)(N_i)(f_{\text{LE}-i})$$

$\Rightarrow$

$$\text{ESAL}_{12\text{-kip Single}} = (275)(365)(0.45)(29.8)(1)(0.192) = 258408$$

The results of the other axle loads are summarized in Table 13.23.

TABLE 13.23
Computations of the ESALs for Problem 13.16

Truck Type	Axle Type	AADT	365	f_d	G	N	f_{LE}	ESALs
	Single	275	365	0.45	33.0	1	0.192	258,408
	Single					1	1.000	1,346,029
	Tridem					1	??	1,346,029 f_{LE-x}

Total ESALs

The predicted number of ESALs should be equal to the design ESALs the pavement structure can carry in the design life; therefore,

$$W_{18} = 2000000 = 258408 + 1346029 + 1346029\, f_{\text{LE}-x}$$

$\Rightarrow$

$$f_{\text{LE}-x} = 0.294$$

Using the same power model developed earlier for the load equivalency factor of tridem-axle loads, the tridem-axle load that corresponds to a f_{LE} of 0.294 will be determined as shown below:

$$f_{\text{LE}} = 6.7846 \times 10^{-20} (\text{Load})^{4.0939}$$

$$\Rightarrow$$

$$0.294 = 6.7846 \times 10^{-20} \left(\text{Load}\right)^{4.0939}$$

$$\text{Load} = 35669 \text{ lb}$$

Therefore, the maximum allowable load for the tridem axle of the truck that will use this highway pavement with the given design data is about 36 kips.

The MS Excel worksheet used to perform all the computations to determine the maximum allowable load for the tridem axle of the truck that is expected to use the highway pavement is shown in Figure 13.33.

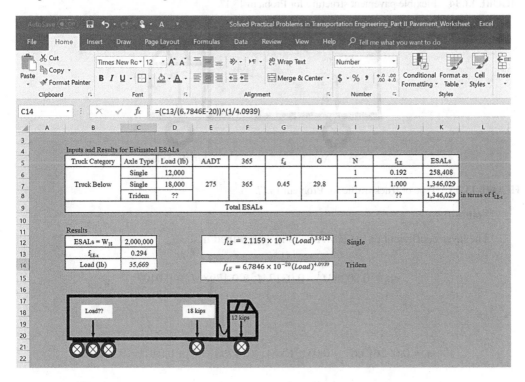

FIGURE 13.33 Image of the MS Excel worksheet used for determining the load of the tridem axle of the truck for Problem 13.16.

13.17 A highway designer would like to set a limitation on the first-year average annual daily traffic (AADT) for the asphalt pavement design shown in Figure 13.34 if the design is to be used for a multi-lane highway with a traffic mix consisting of 60% passenger cars and 40% trucks of the type shown in Figure 13.35. Determine the AADT given the following design data (note: ignore the effect of passenger cars):

Given Design Data:

R = 90%

Overall standard deviation $(S_0) = 0.5$

$\Delta \text{PSI} = 2.0$

$m_2 = 1.0$

$f_d = 0.40$

r = 4%

n = 10 years

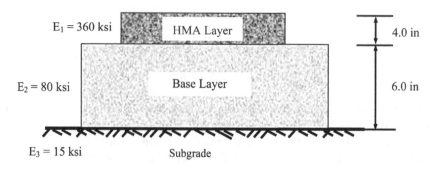

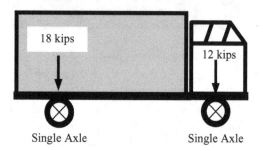

FIGURE 13.34 Flexible pavement structure for Problem 13.17.

FIGURE 13.35 Type of truck traffic for Problem 13.17.

Solution:

The layer coefficient of the HMA layer is determined first:

$$a_1 = 0.0026M_R^3 - 0.0317M_R^2 + 0.1798M_R + 0.0407$$

$\Rightarrow$

$$a_1 = 0.0026(3.6)^3 - 0.0317(3.6)^2 + 0.1798(3.6) + 0.0407 = 0.398$$

$$SN_1 = a_1D_1$$

$\Rightarrow$

$$SN_1 = 0.398(4.0) = 1.59$$

The coefficient of Layer 2 (the base layer) is given by the following logarithmic formula:

$$a_2 = 0.249\log(E_2) - 0.977$$

$\Rightarrow$

$$a_2 = 0.249\log(80000) - 0.977 = 0.244$$

$$SN_2 = a_1D_1 + a_2m_2D_1$$

$\Rightarrow$

$$SN_2 = 0.419(4.0) + 0.244(1.0)(6.0) = 3.06$$

For a reliability (R) level of 90%, the Z-value is equal to -1.282.

$$\log_{10} W_{18} = Z_R S_0 + 9.36 \log_{10}(SN+1) - 0.20 + \frac{\log_{10}\left[\dfrac{\Delta PSI}{(4.2-1.5)}\right]}{0.40 + \left[\dfrac{1094}{(SN+1)^{5.19}}\right]}$$

$$+ 2.32 \log_{10} M_R - 8.07$$

$\Rightarrow$

$$\log_{10} W_{18} = -1.282(0.5) + 9.36 \log_{10}(3.06+1) - 0.20$$

$$+ \frac{\log_{10}\left[\dfrac{\Delta PSI}{(4.2-1.5)}\right]}{0.40 + \left[\dfrac{1094}{(3.06+1)^{5.19}}\right]} + 2.32 \log_{10}(16000) - 8.07$$

$$SN = 3.50$$

For a reliability (R) level of 95%, the Z-value is equal to -1.645.

Now based on the pavement structural number (SN), the pavement structure will be able to carry ESALs calculated based on the design formula below:

$$\log_{10} W_{18} = Z_R S_0 + 9.36 \log_{10}(SN+1) - 0.20 + \frac{\log_{10}\left[\dfrac{\Delta PSI}{(4.2-1.5)}\right]}{0.40 + \left[\dfrac{1094}{(SN+1)^{5.19}}\right]}$$

$$+ 2.32 \log_{10} M_R - 8.07$$

$\Rightarrow$

$$\log_{10} W_{18} = -1.645(0.5) + 9.36 \log_{10}(3.50+1) - 0.20$$

$$+ \frac{\log_{10}\left[\dfrac{2.0}{(4.2-1.5)}\right]}{0.40 + \left[\dfrac{1094}{(3.50+1)^{5.19}}\right]} + 2.32 \log_{10}(15000) - 8.07$$

$\Rightarrow$

$$W_{18} = 2281793 \text{ ESALs}$$

This number represents the design ESALs that the pavement structure can carry through the design life based on the empirical design formula shown above.

But the estimated number of ESALs based on the AADT is calculated as below:

$$\text{ESAL}_i = (\text{AADT}_i)(365)(f_d)(G)(N_i)(f_{\text{LE}-i})$$

Sample Calculation for the ESALs:
For the 12-kip single axle:
$f_d = 0.40$

$$G = \frac{(1+r)^n - 1}{r}$$

$\Rightarrow$

$$G = \frac{(1+0.04)^{10} - 1}{0.04} = 12.01$$

$N = 1$
$f_{\text{LE}} = 0.192$ (calculated below)

The power model developed earlier for the load equivalency factor of single-axle loads will be used to compute f_{LE} for the 12-kip single-axle load as follows:

$$f_{\text{LE}} = 2.1159 \times 10^{-17} (\text{Load})^{3.9120}$$

$\Rightarrow$

$$f_{\text{LE}} = 2.1159 \times 10^{-17} (12000)^{3.9120} = 0.192$$

$$\text{ESAL}_i = (\text{AADT}_i)(365)(f_d)(G)(N_i)(f_{\text{LE}-i})$$

$\Rightarrow$

$$\text{ESAL}_{10\text{-kip Single}} = (\text{AADT})(365)(0.40)(12.01)(1)(0.192) = 134.6 \text{ AADT}$$

The results of the other axle loads are summarized in Table 13.24.

TABLE 13.24

Computations of the ESALs for Problem 13.17

Truck Type	Axle Type	AADT	% Trucks	365	f_d	G	N	f_{LE}	ESALs
	Single	??	40	365	0.40	12.01	1	0.192	134.6
	Single						1	1.000	701.2
Total ESALs (in terms of AADT)									835.8 AADT

The predicted number of ESALs should be equal to the design ESALs the pavement structure can carry in the design life; therefore,

$$W_{18} = 2281793 = 835.8 \text{ AADT}$$

$$\Rightarrow$$

$$\text{AADT} \cong 2730 \text{ veh/day}$$

Therefore, the annual average daily traffic of the first year for this pavement structure is expected to be at maximum 2730 veh/day so that the pavement structure will be able to carry the given traffic mix under the existing conditions.

The MS Excel worksheet used to calculate the computations of this problem is shown in Figure 13.36.

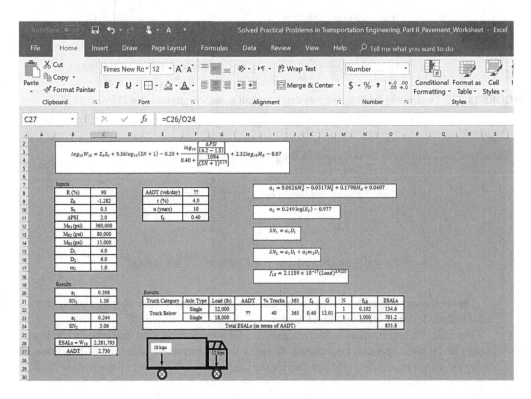

FIGURE 13.36   Image of the MS Excel worksheet used for determining the AADT for Problem 13.17.

13.18 A full-depth asphalt pavement is subjected to a circular single-wheel load of 5 kips with a contact pressure of 88 psi that corresponds to a single-axle with dual wheels as shown in Figure 13.37. If the maximum allowable horizontal tensile strain at the bottom of the hot-mix asphalt (HMA) layer is located at the mid-point between the two wheels as shown in the figure and its value is 220 micro-strains, then based on a mechanistic-empirical pavement design method and given the following design inputs, determine the minimum thickness of the HMA layer. Use the Asphalt Institute (AI) distress model (transfer function) for fatigue cracking shown below to perform the mechanistic analysis of the pavement.

$$N_f = 0.0796 \left(\varepsilon_t \right)^{-3.291} \left(E_1 \right)^{-0.854} \tag{13.18}$$

Where:

N_f = number of load repetitions to fatigue
ε_t = critical tensile strain at the bottom of the HMA layer
E_1 = elastic modulus of the HMA layer
<u>Given Design Data:</u>
R = 90%
Overall standard deviation (S_0) = 0.5
ΔPSI = 2.0

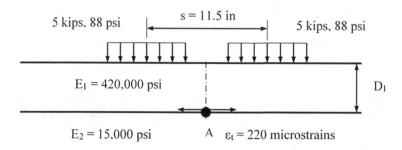

FIGURE 13.37   Dual-wheel loading on a full-depth asphalt pavement structure for Problem 13.18.

Solution:

Based on the AI distress model given in the problem:

$$N_f = 0.0796(\varepsilon_t)^{-3.291}(E_1)^{-0.854}$$

$\Rightarrow$

$$N_f = 0.0796(220\times10^{-6})^{-3.291}(420000)^{-0.854} \cong 1366927$$

This is the number of load repetitions that the pavement structure will be able to carry until fatigue failure based on the AI model.

The load of a single wheel is 5 kips, therefore, for a single-axle with dual wheels, the load would be 20 kips. The load equivalency factor of the 20-kip single-axle load is determined using the power model developed earlier for the load equivalency factor of single-axle loads:

$$f_{LE} = 2.1159\times10^{-17}(Load)^{3.9120}$$

$\Rightarrow$

$$f_{LE} = 2.1159\times10^{-17}(20000)^{3.9120} = 1.416$$

$$N_f = 1366927$$

This is the number of 20-kip load repetitions the pavement structure can carry until fatigue failure. Therefore,

$$\text{ESALs} = (N_f)(f_{\text{LE}}) \tag{13.19}$$

$\Rightarrow$

$$\text{ESALs} = (1366926)(1.416) = 1935841$$

Using the empirical design formula for asphalt pavements shown below, the structural number (SN) of the pavement is determined:

$$\log_{10} W_{18} = Z_R S_0 + 9.36 \log_{10}(SN+1) - 0.20 + \frac{\log_{10}\left[\dfrac{2.0}{(4.2-1.5)}\right]}{0.40 + \left[\dfrac{1094}{(SN+1)^{5.19}}\right]}$$

$$+ 2.32 \log_{10}(12000) - 8.07$$

Z_R that corresponds to a reliability level (R) of 90% is equal to −1.282.

W_{18} is the ESALs that corresponds to the load repetitions determined using the mechanistic analysis above. In other words, W_{18} is equal to 138743. Therefore,

$$\log_{10} W_{18} = Z_R S_0 + 9.36 \log_{10}(SN+1) - 0.20 + \frac{\log_{10}\left[\dfrac{\Delta PSI}{(4.2-1.5)}\right]}{0.40 + \left[\dfrac{1094}{(SN+1)^{5.19}}\right]}$$

$$+ 2.32 \log_{10} M_R - 8.07$$

$\Rightarrow$

$$\log_{10}(1935841) = (-1.282)(0.5) + 9.36 \log_{10}(SN+1) - 0.20$$

$$+ \frac{\log_{10}\left[\dfrac{\Delta PSI}{(4.2-1.5)}\right]}{0.40 + \left[\dfrac{1094}{(SN+1)^{5.19}}\right]} + 2.32 \log_{10}(15000) - 8.07$$

Using the MS Excel Solver tool to solve for the SN in the above equation, the solution is obtained as follows:

$$SN = 2.98$$

The layer coefficient of the HMA layer is determined using the formula below:

$$a_1 = 0.0026 M_R^3 - 0.0317 M_R^2 + 0.1798 M_R + 0.0407$$

$\Rightarrow$

$$a_1 = 0.0026(4.2)^3 - 0.0317(4.2)^2 + 0.1798(4.2) + 0.0407 = 0.429$$

$$SN_1 = a_1 D_1$$

$\Rightarrow$

$$D_1 = \frac{SN_1}{a_1} = \frac{2.98}{0.429} = 6.9 \text{ in} \cong 7.0 \text{ in} \left(\cong 18 \text{ cm}\right)$$

The MS Excel worksheet used to perform the mechanistic analysis of the asphalt pavement and the empirical design based on the mechanistic analysis is shown in Figure 13.38.

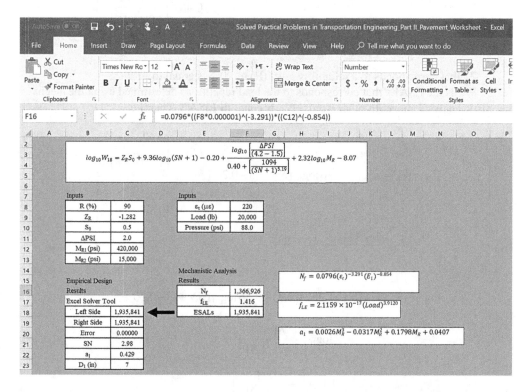

FIGURE 13.38 Image of the MS Excel worksheet used for the mechanistic-empirical pavement design for Problem 13.18.

13.19 A circular loading of 4500 lb with a tire pressure of 60 psi is applied on a flexible pavement as shown in Figure 13.39.

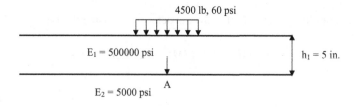

FIGURE 13.39 Single wheel circular loading on a two-layered flexible pavement for Problem 13.19.

Determine the following:
a. The vertical compressive stress at point A
b. The tensile strain at point A

Solution:

The WinJULEA (WinLEA) program is used to perform the analysis of stresses and strains for this pavement.
The following inputs and assumptions are used in the program:

(1) A slip value of 0 is used for full-friction layers or rough surfaces. A slip value of 100000 is used for full slip when there is no friction between layers.
(2) The thickness of the last layer is entered as 0.
(3) The Poisson ratio (PR) is assumed for each layer. A value of 0.35 is typically used for the HMA layer. A value of 0.40 is normally used for base and subbase layers. And a value of 0.45 is used for the subgrade layer.
(4) The values of the applied loads are entered with the X-coordinate and Y-coordinate.
(5) The contact area for each load is calculated using the formula below and entered in the program.
(6) The X-coordinate and Y-coordinate along with the depth for the evaluation points are entered.

The values of the Poisson ratio for the asphalt layer and the subgrade layer are assumed to be 0.35 and 0.45, respectively. The contact area is calculated as follows:

$$A_c = \frac{P}{q} \qquad (13.20)$$

Whe re:
A_c = contact area
P = applied loading
q = applied pressure

$\Rightarrow$

$$A_c = \frac{4500}{60} = 75 \text{ in}^2$$

The inputs needed for the program are shown in Figure 13.40.

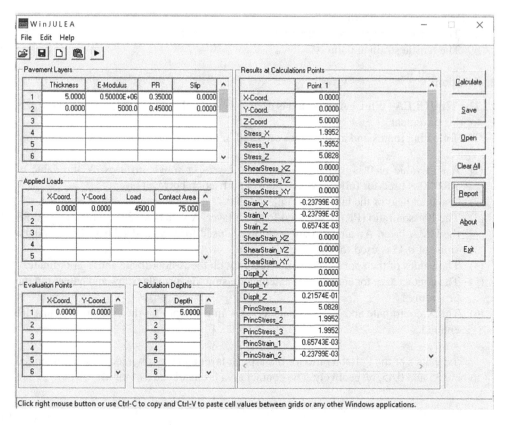

FIGURE 13.40   Inputs for the WinJULEA program for Problem 13.19.

The evaluation point is located at x = 0, y = 0, and z = 5 in. The results of the analysis after conducting the calculations by hitting "Calculate" are summarized below (see Table 13.25).

TABLE 13.25
WinJULEA Flexible Pavement Analysis Results for Problem 13.19

```
*** RESULTS
-----------------------------------------------------------------------------------------
-----------------------------------------------------------------------------------------
-----------------------------------------------------------------------------------------
-----------------------------------------------------------------------------------------
-----------------------------------------------------------------------------------------

DEPTH         X-COORD       Y-COORD       STRESS-X      STRESS-Y      STRESS-Z      SHEAR-XZ
SHEAR-YZ      SHEAR-XY      STRAIN-X      STRAIN-Y      STRAIN-Z      SHEAR-XZ      SHEAR-YZ
SHEAR-XY      DISPLT-X      DISPLT-Y      DISPLT-Z      P.STRESS1     P.STRESS2     P.STRESS3
P.STRAIN1     P.STRAIN2     P.STRAIN3     MAX.SHEAR     OCT.STRESS    OCT.SHEAR
-----------------------------------------------------------------------------------------
-----------------------------------------------------------------------------------------
-----------------------------------------------------------------------------------------
-----------------------------------------------------------------------------------------

0.500E+01     0.000E+00     0.000E+00     0.200E+01     0.200E+01     0.508E+01     0.000E+00
0.000E+00     0.000E+00    -0.238E-03    -0.238E-03     0.657E-03     0.000E+00     0.000E+00
0.000E+00     0.000E+00     0.000E+00     0.216E-01     0.508E+01     0.200E+01     0.200E+01
0.657E-03    -0.238E-03    -0.238E-03     0.154E+01     0.302E+01     0.146E+01
```

From the above results, the vertical stress at point A (at a depth, $z = 5$ in) is stress-z $= 5.08$ psi, and the tensile strain at point A is the horizontal strain-x $= -0.238 \times 10^{-3}$ in/in $= 238$ µε. +ve is used for compression and −ve is used for tension.

a. The vertical compressive stress at point A is equal to 5.08 psi ($\cong$ 35 kPa)
b. The tensile strain at point A is equal to 238 µε.

13.20 A full-depth asphalt pavement is subjected to a circular single-wheel load of 10.2 kips with a contact pressure of 90.0 psi as shown in Figure 13.41. The elastic modulus of the hot-mix asphalt (HMA) layer is 150000 psi and the elastic modulus of the subgrade is 7500 psi. If the maximum allowable horizontal tensile strain at the bottom of the asphalt layer under the centerline of the wheel load is 360 microstrains, determine the following:

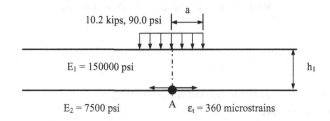

FIGURE 13.41 Single wheel circular loading on a full-depth asphalt pavement (two-layered system) for Problem 13.20.

a. The radius of the contact area
b. The thickness of the asphalt layer (h_1) based on a mechanistic-empirical pavement design method
c. The number of load repetitions (applications) until fatigue failure using the Asphalt Institute (AI) distress model for fatigue cracking $N_f = 0.0796 (\varepsilon_t)^{-3.291} (E_1)^{-0.854}$

Solution:

a. The radius of the contact area is determined as follows:

$$A_c = \pi a^2 \qquad (13.21)$$

$\Rightarrow$

$$a = \sqrt{\frac{A_c}{\pi}}$$

Where:
   A_c = contact area for applied loading
   a = radius of contact area

$$A_c = \frac{P}{q}$$

$\Rightarrow$

$$A_c = \frac{10200}{90} = 113.3 \text{ in}^2 \left(\cong 731 \text{ cm}^2\right)$$

⇒

$$a = \sqrt{\frac{113.3}{\pi}} = 6.0 \text{ in} \left(15.2 \text{ cm}\right)$$

b. The thickness of the asphalt layer (h_1) is determined as shown in the procedure below:
    The ratio E_1/E_2 is determined:

$$\frac{E_1}{E_2} = \frac{150000}{7500} = 20$$

Using the critical tensile strain formula shown below:

$$\varepsilon = \frac{q}{E_1} F_\varepsilon \qquad\qquad (13.22)$$

Where:
   ε = critical tensile strain at the bottom of the asphalt layer
   q = applied wheel pressure
   E_1 = elastic modulus of asphalt layer
   F_ε = tensile strain factor (determined from Figure 13.42)

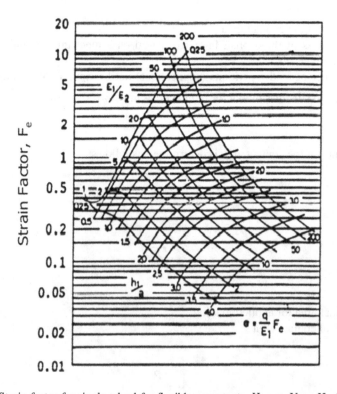

FIGURE 13.42 Strain factor for single wheel for flexible pavements. Huang, Yang H., *Pavement Analysis and Design*, 2nd edition ©2004. Reprinted by permission of Pearson Education, Inc., New York, New York, USA.

$\Rightarrow$

$$360 \times 10^{-6} = \frac{90}{150000} F_\varepsilon$$

$\Rightarrow$

$$F_\varepsilon = 0.60$$

Using this value (0.60) for the strain factor and the value of 20 for the ratio E_1/E_2, the chart in Figure 13.42 provides the following value for h_1/a:

$$\frac{h_1}{a} = 1.75$$

$\Rightarrow$

$$h_1 = 1.75a$$

$\Rightarrow$

$$h_1 = 1.75(6.0) = 10.5 \text{ in} (26.7 \text{ cm})$$

c. The number of load repetitions (applications) until fatigue failure is determined using the AI distress model:

$$N_f = 0.0796(\varepsilon_t)^{-3.291}(E_1)^{-0.854}$$

$\Rightarrow$

$$N_f = 0.0796(360 \times 10^{-6})^{-3.291}(150000)^{-0.854} = 651238$$

The MS Excel worksheet used to solve this problem is shown in Figure 13.43.

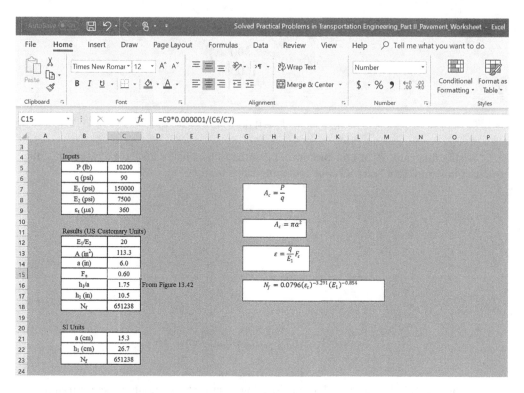

FIGURE 13.43 Image of the MS Excel worksheet used for the computations of Problem 13.20.

13.21 A full-depth asphalt pavement is subjected to a circular single-wheel load of 10 kips with a contact pressure of 88 psi as shown in Pavement System #1 in Figure 13.44. The elastic modulus of the hot-mix asphalt (HMA) layer is 200000 psi and the elastic modulus of the subgrade is 4000 psi. Determine the critical tensile strain at the bottom of the asphalt layer. Based on a mechanistic-empirical pavement design method, what is the number of load repetitions (applications) the pavement structure can carry until fatigue failure using the Asphalt Institute (AI) distress model for fatigue cracking shown in the expression below.

$$N_f = 0.0796 \left(\varepsilon_t \right)^{-3.291} \left(E_1 \right)^{-0.854} \tag{13.23}$$

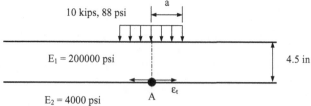

FIGURE 13.44 Single wheel circular loading on a full-depth asphalt pavement for Problem 13.21 (Pavement System #1).

Solution:

The radius of the contact area is determined as follows:

$$A_c = \pi a^2$$

$\Rightarrow$

$$a = \sqrt{\frac{A_c}{\pi}}$$

But:

$$A_c = \frac{P}{p}$$

$\Rightarrow$

$$A_c = \frac{10000}{88} = 113.6 \text{ in}^2 \left(\cong 733 \text{ cm}^2\right)$$

$\Rightarrow$

$$a = \sqrt{\frac{113.6}{\pi}} = 6.0 \text{ in}\left(15.2 \text{ cm}\right)$$

The ratio E_1/E_2 is determined:

$$\frac{E_1}{E_2} = \frac{200000}{4000} = 50$$

The ratio h_1/a is determined:

$$\frac{4.5}{6.0} = 0.75$$

Using $E_1/E_2 = 50$ and $h_1/a = 0.75$, the chart in Figure 13.42 provides the following value for the strain factor (F_ε):

$$F_\varepsilon = 2.25$$

Using the critical tensile strain formula shown below:

$$\varepsilon = \frac{q}{E_1} F_\varepsilon$$

$\Rightarrow$

$$\varepsilon = \frac{88}{200000}\left(2.25\right) = 990 \times 10^{-6} \text{ in/in} = 990 \ \mu\varepsilon$$

The number of load repetitions (applications) until fatigue failure is determined using the AI distress model:

$$N_f = 0.0796\left(\varepsilon_t\right)^{-3.291}\left(E_1\right)^{-0.854}$$

$\Rightarrow$

$$N_f = 0.0796\left(990 \times 10^{-6}\right)^{-3.291}\left(200000\right)^{-0.854} \cong 18248$$

The MS Excel worksheet used to solve this problem is shown in Figure 13.45.

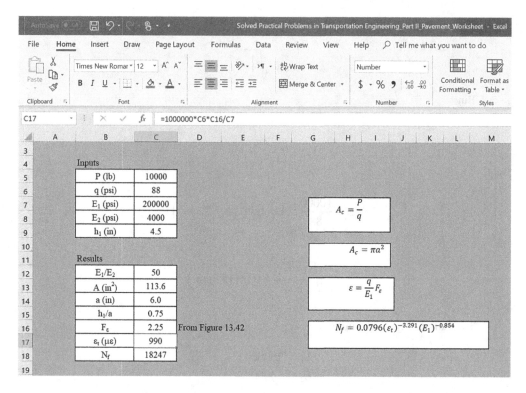

FIGURE 13.45   Image of the MS Excel worksheet used for the computations of Problem 13.21.

13.22 If the load in the pavement system of Problem 13.21 above is to be applied over a set of dual wheels with a center-to-center spacing between the dual wheels of 11.5 inches as shown in Pavement System #2 in Figure 13.46, determine the critical tensile strain at the bottom of the asphalt layer. In this case, what is the number of load repetitions (applications) the pavement structure can carry until fatigue failure using the Asphalt Institute (AI) distress model for fatigue cracking $N_f = 0.0796(\varepsilon_t)^{-3.291}(E_1)^{-0.854}$. What is the increase in the design life of the pavement based on the mechanistic analysis?

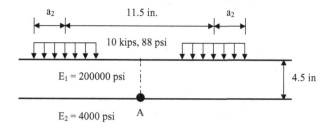

FIGURE 13.46   Dual wheel loading on a full-depth asphalt pavement for Problem 13.22 (Pavement System #2).

Solution:

The WinJULEA (WinLEA) program is used to perform the analysis of stresses and strains for this pavement.

The inputs used in the program are shown in Figure 13.47.

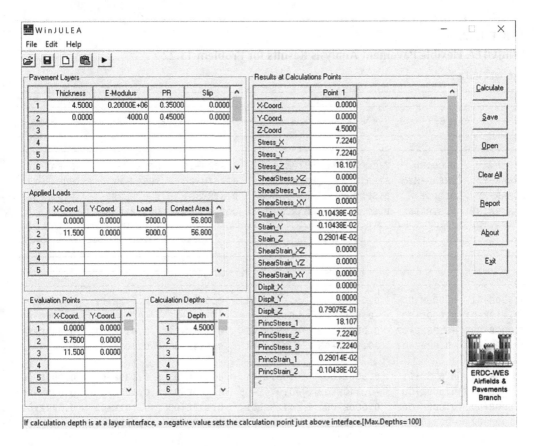

FIGURE 13.47   Inputs for the WinJULEA program for Problem 13.22.

The results of the WinLEA program run are summarized in the report below. Based on these results, the tensile strain at the bottom of the asphalt layer is calculated at three different locations with the coordinates (0, 0, 6), (5.75, 0, 6), and (11.5, 0, 6), where the first number is the X-coordinate, the second number corresponds to the Y-coordinate, and the third number is the Z-coordinate (depth) as shown in Table 13.26. In this case, the first point (0, 0, 6) is simply under the centerline of the first wheel load at a depth of 6 in (at the bottom of the asphalt layer), the second point (5.75, 0, 6) is at the mid-point between the two wheel loads at a depth of 6 in (at the bottom of the asphalt layer), and the third point (11.5, 0, 6) is under the centerline of the second wheel load at a depth of 6 in (at the bottom of the asphalt layer). The tensile strain results for these three points are summarized in Tables 13.27 and 13.28.

TABLE 13.26

The Coordinates of the Three Points in Problem 13.22

Point	Location (X, Y, Z)
1	(0, 0, 6)
2	(5.75, 0, 6)
3	(11.5, 0, 6)

TABLE 13.27

WinJULEA Flexible Pavement Analysis Results for Problem 13.22

```
*** RESULTS
```


DEPTH	X-COORD	Y-COORD	STRESS-X	STRESS-Y	STRESS-Z	SHEAR-XZ
SHEAR-YZ	SHEAR-XY	STRAIN-X	STRAIN-Y	STRAIN-Z	SHEAR-XZ	SHEAR-YZ
SHEAR-XY	DISPLT-X	DISPLT-Y	DISPLT-Z	P.STRESS1	P.STRESS2	P.STRESS3
P.STRAIN1	P.STRAIN2	P.STRAIN3	MAX.SHEAR	OCT.STRESS	OCT.SHEAR	


```
0.450E+01   0.000E+00    0.000E+00    0.632E+01   0.542E+01    0.142E+02   -0.119E+01
-0.119E+01   0.000E+00   -0.625E-03   -0.949E-03   0.222E-02   -0.864E-03   0.000E+00
0.000E+00   -0.281E-02    0.000E+00    0.693E-01   0.143E+02    0.614E+01   0.542E+01
0.229E-02   -0.689E-03   -0.949E-03    0.446E+01   0.864E+01    0.405E+01

0.450E+01   0.575E+01    0.000E+00    0.751E+01   0.577E+01    0.139E+02   0.000E+00
0.000E+00    0.000E+00   -0.340E-03   -0.970E-03   0.199E-02    0.000E+00   0.000E+00
0.000E+00    0.000E+00    0.000E+00    0.723E-01   0.139E+02    0.751E+01   0.577E+01
0.199E-02   -0.340E-03   -0.970E-03    0.408E+01   0.907E+01    0.351E+01

0.450E+01   0.115E+02    0.000E+00    0.632E+01   0.542E+01    0.142E+02   0.119E+01
0.119E+01    0.000E+00   -0.625E-03   -0.949E-03   0.222E-02    0.864E-03   0.000E+00
0.000E+00    0.281E-02    0.000E+00    0.693E-01   0.143E+02    0.614E+01   0.542E+01
0.229E-02   -0.689E-03   -0.949E-03    0.446E+01   0.864E+01    0.405E+01
```

TABLE 13.28

Tensile Strain Results for the Three Locations in Problem 13.22 using WinLEA Program

Point	Location	Tensile Strain-X ($\mu\varepsilon$)	Tensile Strain-Y ($\mu\varepsilon$)
$(0, 0, 6)$	Bottom of asphalt layer, centerline of 1st wheel load	625	949
$(5.75, 0, 6)$	Bottom of asphalt layer, mid-point between the two wheel loads	340	970
$(11.5, 0, 6)$	Bottom of asphalt layer, centerline of 2nd wheel load	625	949

As shown in Table 13.28, the critical tensile strain at the bottom of the asphalt layer is at the middle distance (midpoint) between the two wheels. The value is equal to 970 $\mu\varepsilon$. Comparing the critical tensile strain obtained in Pavement System #2 when the load is applied over dual wheels (970 $\mu\varepsilon$) with that obtained in Pavement System #1 when the load is applied on a single wheel (990 $\mu\varepsilon$). The value is decreased. If the same WinLEA program is used to perform the analysis for Pavement System #1, the critical tensile strain is obtained as 1040 $\mu\varepsilon$. In other words, the reduction in the tensile strain between the two cases is equal to:

$$\text{Reduction in } \varepsilon_t = \frac{1040 - 970}{1040} \times 100 = 6.7\%$$

For dual wheels:

$$N_f = 0.0796(\varepsilon_t)^{-3.291}(E_1)^{-0.854}$$

$\Rightarrow$

$$N_f = 0.0796(970 \times 10^{-6})^{-3.291}(200000)^{-0.854} \cong 19515$$

For single wheel:

$$N_f = 0.0796(\varepsilon_t)^{-3.291}(E_1)^{-0.854}$$

$\Rightarrow$

$$N_f = 0.0796(1040 \times 10^{-6})^{-3.291}(200000)^{-0.854} \cong 15516$$

$\Rightarrow$

$$\text{Increase in } N_f = \frac{19515 - 15516}{15516} \times 100 = 25.8\%$$

In conclusion, using an axle with dual wheels instead of single wheel will increase the fatigue design life of the asphalt pavement by about 26%.

The MS Excel worksheet used to perform the computations of this problem is shown in Figure 13.48.

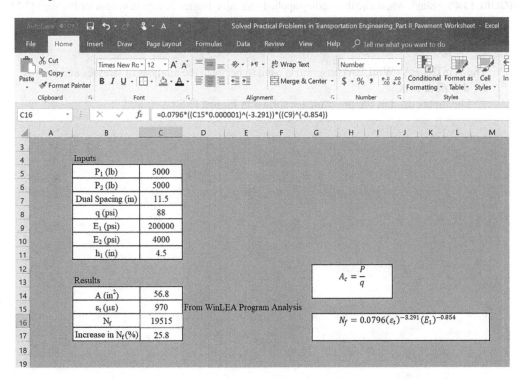

FIGURE 13.48 Image of the MS Excel worksheet used for the computations of Problem 13.22.

13.23 A circular loading with a radius of 4.0 inches is applied to the three-layer asphalt pavement shown in Figure 13.49. Calculate the critical pavement responses for fatigue cracking and for rutting in this pavement. Using the Asphalt Institute (AI) distress models for fatigue cracking and rutting as transfer functions in a mechanistic-empirical pavement design method, determine the number of load applications the pavement can sustain until fatigue failure and the number of load applications the pavement can carry until rutting failure.

$$N_f = 0.0796 \left(\varepsilon_t \right)^{-3.291} \left(E_1 \right)^{-0.854} \qquad \text{for fatigue cracking}$$

$$N_f = 1.365 \times 10^{-9} \left(\varepsilon_c \right)^{-4.477} \qquad \text{for rutting} \qquad (13.24)$$

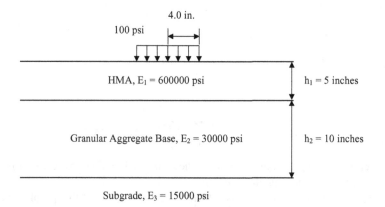

FIGURE 13.49   Single wheel circular loading applied on a three-layered pavement system for Problem 13.23.

Solution:

The WinJULEA (WinLEA) program is used to perform the analysis of stresses and strains for this pavement.

The inputs used in the program are shown in Figure 13.50.

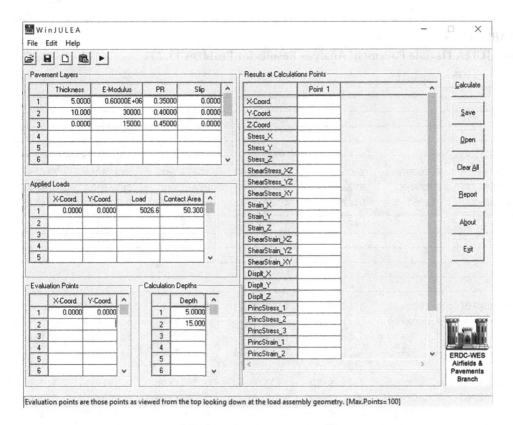

FIGURE 13.50   Inputs for the WinJULEA program for Problem 13.23.

The results of the WinLEA program run are summarized in the report below. Based on these results, the critical pavement responses are calculated at two different locations with the coordinates (0, 0, 5) and (0, 0, 15), where the first number is the X-coordinate, the second number corresponds to the Y-coordinate, and the third number is the Z-coordinate (depth). In this case, the first point (0, 0, 5) is simply under the centerline of the wheel load at a depth of 5 in (at the bottom of the asphalt layer), and the second point (0, 0, 15) is at the centerline of the wheel load at a depth of 15 in (at the top of the subgrade layer). The critical pavement response at the bottom of the asphalt layer is the horizontal tensile strain that is responsible for fatigue cracking. On the other hand, the critical pavement response at the top of the subgrade layer is the vertical compressive strain that is responsible for rutting in the asphalt pavement. The results are summarized in Tables 13.29 and 13.30.

TABLE 13.29

WinJULEA Flexible Pavement Analysis Results for Problem 13.23

```
      *** RESULTS
-----------------------------------------------------------------------------
-----------------------------------------------------------------------------
-----------------------------------------------------------------------------
-----------------------------------------------------------------------------
```

DEPTH	X-COORD	Y-COORD	STRESS-X	STRESS-Y	STRESS-Z	SHEAR-XZ
SHEAR-YZ	SHEAR-XY	STRAIN-X	STRAIN-Y	STRAIN-Z	SHEAR-XZ	SHEAR-YZ
SHEAR-XY	DISPLT-X	DISPLT-Y	DISPLT-Z	P.STRESS1	P.STRESS2	P.STRESS3
P.STRAIN1	P.STRAIN2	P.STRAIN3	MAX.SHEAR	OCT.STRESS	OCT.SHEAR	

```
-----------------------------------------------------------------------------
-----------------------------------------------------------------------------
-----------------------------------------------------------------------------
-----------------------------------------------------------------------------
```

0.500E+01	0.000E+00	0.000E+00	0.139E+01	0.139E+01	0.148E+02	0.000E+00
0.000E+00	0.000E+00	-0.170E-03	-0.170E-03	0.457E-03	0.000E+00	0.000E+00
0.000E+00	0.000E+00	0.000E+00	0.948E-02	0.148E+02	0.139E+01	0.139E+01
0.457E-03	-0.170E-03	-0.170E-03	0.672E+01	0.587E+01	0.634E+01	
0.150E+02	0.000E+00	0.000E+00	0.312E+00	0.312E+00	0.397E+01	0.000E+00
0.000E+00	0.000E+00	-0.108E-03	-0.108E-03	0.246E-03	0.000E+00	0.000E+00
0.000E+00	0.000E+00	0.000E+00	0.670E-02	0.397E+01	0.312E+00	0.312E+00
0.246E-03	-0.108E-03	-0.108E-03	0.183E+01	0.153E+01	0.172E+01	

TABLE 13.30

Critical Pavement Response Results for the Two Locations using WinJULEA Program for Problem 13.23

Point	Location	Tensile Strain-X=Tensile Strain-Y (με)	Compressive Strain-Z (με)
(0, 0, 5)	Bottom of asphalt layer, centerline of wheel load	170	
(0, 0, 15)	Top of subgrade layer, centerline of wheel load		246

As shown in Table 13.30, the critical tensile strain at the bottom of the asphalt layer is equal to 170 με, and the vertical co pressive strain at the top of the subgrade layer is equal to 246 με.

Using the AI distress model for fatigue cracking shown below:

$$N_f = 0.0796 (\varepsilon_t)^{-3.291} (E_1)^{-0.854}$$

$\Rightarrow$

$$N_f = 0.0796 (170 \times 10^{-6})^{-3.291} (600000)^{-0.854} = 2354857$$

In other words, the asphalt pavement can carry up to approximately 2.4 million load applications before fatigue failure based on the AI fatigue distress model.

And using the AI distress model for rutting shown below:

$$N_f = 1.365 \times 10^{-9} (\varepsilon_c)^{-4.477}$$

$\Rightarrow$

$$N_f = 1.365 \times 10^{-9} (246 \times 10^{-6})^{-4.477} = 19629814$$

This means that the asphalt pavement can sustain about 19.6 million load applications before rutting failure based on the AI rutting distress model.

The computations in this problem are performed using the MS Excel worksheet shown in Figure 13.51.

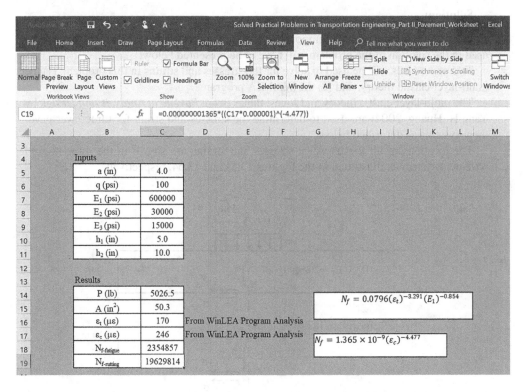

FIGURE 13.51   Image of the MS Excel worksheet used of the computations for Problem 13.23.

13.24 For the three-layer asphalt pavement system shown in Figure 13.52, determine all the stresses and strains at the axis of symmetry at the interfaces of the pavement layers using the multi-layer elastic theory and the WinLEA program.

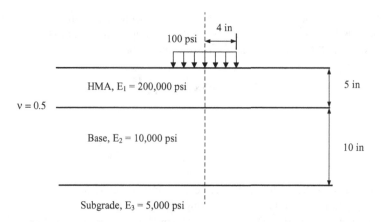

FIGURE 13.52 Single wheel circular loading on a three-layered pavement system for Problem 13.24.

Solution:

Prior to solving the problem, important theory and derivations for formulas that will be used in the solution of the problem will have to be presented below:

Two procedures will be used to solve for the stresses and strains in this problem: (1) the multi-layer elastic theory, and (2) the WinLEA program.

The multi-layer elastic theory and Jones' tables (from *Pavement Analysis and Design* by Yang H. Huang, 2004) are used to determine the stresses and strain for three-layer pavement systems as described below:

The locations of all the interface stresses and strains are shown in the three-layered pavement system as illustrated in the Figures 13.53 and 13.54, respectively:

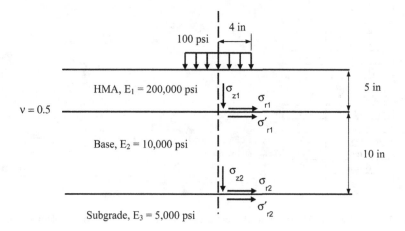

FIGURE 13.53 Location of the interface stresses in the three-layered pavement system for Problem 13.24.

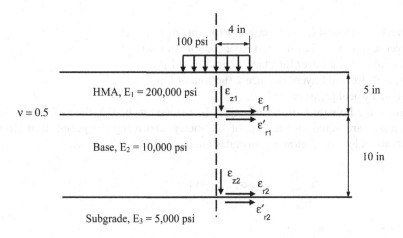

FIGURE 13.54   Location of the interface strains in the three-layered pavement system for Problem 13.24.

The following ratios are determined:

$$k_1 = \frac{E_1}{E_2} \tag{13.25}$$

$$k_2 = \frac{E_2}{E_3} \tag{13.26}$$

$$A = \frac{a}{h_2} \tag{13.27}$$

$$H = \frac{h_1}{h_2} \tag{13.28}$$

Where:
   E_1 = elastic modulus of Layer #1 (asphalt layer)
   E_2 = elastic modulus of Layer #2 (base layer)
   E_3 = elastic modulus of Layer #3 (subgrade layer)
   a = radius of contact area of applied load
   h_1 = thickness of Layer #1
   h_2 = thickness of Layer #2
   • Using the above ratios and Jones' tables, the following are determined:
     ZZ1, ZZ2, ZZ1–RR1, and ZZ2–RR2
     These are stress factors used to calculate the following pavement stresses:

$$\sigma_{z1} = q(ZZ1) \tag{13.29}$$

$$\sigma_{z2} = q(ZZ2) \tag{13.30}$$

$$\sigma_{z1} - \sigma_{r1} = q(ZZ1 - RR1) \tag{13.31}$$

$$\sigma_{z2} - \sigma_{r2} = q(ZZ2 - RR2) \tag{13.32}$$

Where:

σ_{z1} = vertical stress at Layer Interface 1 (bottom of Layer 1)
σ_{z2} = vertical stress at Layer Interface 2 (bottom of Layer 2)
σ_{r1} = radial stress at Layer Interface 1 (bottom of Layer 1)
σ_{r2} = radial stress at Layer Interface 2 (bottom of Layer 2)
q = applied wheel pressure

- Since the Poisson ratio for all layers is assumed to be 0.5, the radial and tangential stresses are equal on the axis of symmetry, and using the generalized Hook's law shown below, the following formulas are obtained for the strains:

$$\varepsilon_{x,y,\text{ or }z} = \frac{1}{E}\left[\sigma_{x,y,\text{ or }z} - v\left(\sigma_{y,x,\text{ or }x} + \sigma_{z,z,\text{ or }y}\right)\right]$$ (13.33)

Where:

ε_x = strain in x-direction
σ_x = stress in x-direction
σ_y = stress in y-direction
σ_z = stress in z-direction
E = elastic modulus

Since the Poisson ratio (v) is equal to 0.5 and σ_r and σ_t are equal on the axis of symmetry, therefore:

$$\varepsilon_z = \frac{1}{E}\left(\sigma_z - \sigma_r\right)$$ (13.34)

$$\varepsilon_r = \frac{1}{2E}\left(\sigma_r - \sigma_z\right)$$ (13.35)

$$\varepsilon_z = -2\varepsilon_r$$ (13.36)

Where:

ε_z = vertical strain
ε_r = radial strain
σ_z = vertical stress
σ_r = radial stress
E = elastic modulus

Using these formulas, the tangential strains and radial strains at Interface 1 and Interface 2 are determined as follows:

At Interface 1 (bottom of Layer #1), and since the point is located in Layer #1:

$$\varepsilon_z = \frac{1}{E}\left(\sigma_z - \sigma_r\right)$$

$\Rightarrow$

$$\varepsilon_{z1} = \frac{1}{E_1}\left(\sigma_{z1} - \sigma_{r1}\right)$$ (13.37)

$$\varepsilon_r = \frac{1}{2E}\left(\sigma_r - \sigma_z\right)$$

⇒

$$\varepsilon_{r1} = \frac{1}{2E_1}\left(\sigma_{r1} - \sigma_{z1}\right) \tag{13.38}$$

Where:
ε_{z1} = vertical strain at Interface 1 and bottom of Layer #1
ε_{r1} = radial strain at Interface 1 and bottom of Layer #1
σ_{z1} = vertical stress at Interface 1 and bottom of Layer #1
σ_{r1} = radial stress at Interface 1 and bottom of Layer #1
E_1 = elastic modulus of Layer #1

At Interface 2 (bottom of Layer #2), and since the point is located in Layer #2:

$$\varepsilon_z = \frac{1}{E}\left(\sigma_z - \sigma_r\right)$$

⇒

$$\varepsilon_{z2} = \frac{1}{E_2}\left(\sigma_{z2} - \sigma_{r2}\right) \tag{13.39}$$

$$\varepsilon_r = \frac{1}{2E}\left(\sigma_r - \sigma_z\right)$$

⇒

$$\varepsilon_{r2} = \frac{1}{2E_2}\left(\sigma_{r2} - \sigma_{z2}\right) \tag{13.40}$$

Where:
ε_{z2} = vertical strain at Interface 2 and bottom of Layer #2
ε_{r2} = radial strain at Interface 2 and bottom of Layer #2
σ_{z2} = vertical stress at Interface 2 and bottom of Layer #2
σ_{r2} = radial stress at Interface 2 and bottom of Layer #2
E_2 = elastic modulus of Layer #2

At Interface 1 (top of Layer #2), and since the point is located in Layer #2:

$$\varepsilon_z = \frac{1}{E}\left(\sigma_z - \sigma_r\right)$$

⇒

$$\acute{\varepsilon}_{z1} = \frac{1}{E_2}\left(\acute{\sigma}_{z1} - \acute{\sigma}_{r1}\right) \tag{13.41}$$

$$\varepsilon_r = \frac{1}{2E}\left(\sigma_r - \sigma_z\right)$$

$\Rightarrow$

$$\acute{\varepsilon}_{r1} = \frac{1}{2E_2}\left(\acute{\sigma}_{r1} - \acute{\sigma}_{z1}\right) \qquad (13.42)$$

Where:

$\acute{\varepsilon}_{z1}$ = vertical strain at Interface 1 and top of Layer #2

$\acute{\varepsilon}_{r1}$ = radial strain at Interface 1 and top of Layer #2

$\acute{\sigma}_{z1}$ = vertical stress at Interface 1 and top of Layer #2

$\acute{\sigma}_{r1}$ = radial stress at Interface 1 and top of Layer #2

E_2 = elastic modulus of Layer #2

At Interface 2 (top of Layer #3), and since the point is located in Layer #3:

$$\varepsilon_z = \frac{1}{E}\left(\sigma_z - \sigma_r\right)$$

$\Rightarrow$

$$\acute{\varepsilon}_{z2} = \frac{1}{E_3}\left(\acute{\sigma}_{z2} - \acute{\sigma}_{r2}\right) \qquad (13.43)$$

$$\varepsilon_r = \frac{1}{2E}\left(\sigma_r - \sigma_z\right)$$

$\Rightarrow$

$$\acute{\varepsilon}_{r2} = \frac{1}{2E_3}\left(\acute{\sigma}_{r2} - \acute{\sigma}_{z2}\right) \qquad (13.44)$$

Where:

$\acute{\varepsilon}_{z2}$ = vertical strain at Interface 2 and top of Layer #3

$\acute{\varepsilon}_{r2}$ = radial strain at Interface 2 and top of Layer #3

$\acute{\sigma}_{z2}$ = vertical stress at Interface 2 and top of Layer #3

$\acute{\sigma}_{r2}$ = radial stress at Interface 2 and top of Layer #3

E_3 = elastic modulus of Layer #3

- Using the continuity conditions at the interfaces, it implies that the radial strain at the bottom of one layer is equal to the radial strain at the top of the next layer. Therefore, the following formulas are obtained:

$$\varepsilon_{r1} = \acute{\varepsilon}_{r1} \qquad (13.45)$$

$$\varepsilon_{r2} = \acute{\varepsilon}_{r2} \qquad (13.46)$$

Where:

ε_{r1} = radial strain at Interface 1 and bottom of Layer #1

ε_{r2} = radial strain at Interface 2 and bottom of Layer #2

$\acute{\varepsilon}_{r1}$ = radial strain at Interface 1 and top of Layer #2

$\acute{\varepsilon}_{r2}$ = radial strain at Interface 2 and top of Layer #3

$$\varepsilon_{r1} = \acute{\varepsilon}_{r1}$$

$\Rightarrow$

$$\frac{1}{2E_1}\left(\sigma_{r1} - \sigma_{z1}\right) = \frac{1}{2E_2}\left(\acute{\sigma}_{r1} - \acute{\sigma}_{z1}\right)$$

$\Rightarrow$

$$\left(\acute{\sigma}_{r1} - \acute{\sigma}_{z1}\right) = \frac{\left(\sigma_{r1} - \sigma_{z1}\right)}{\left(\dfrac{E_1}{E_2}\right)} = \frac{\left(\sigma_{r1} - \sigma_{z1}\right)}{k_1}$$

Or:

$$\left(\acute{\sigma}_{r1} - \acute{\sigma}_{z1}\right) = \frac{\left(\sigma_{r1} - \sigma_{z1}\right)}{k_1} \tag{13.47}$$

$$\varepsilon_{r2} = \acute{\varepsilon}_{r2}$$

$\Rightarrow$

$$\frac{1}{2E_2}\left(\sigma_{r2} - \sigma_{z2}\right) = \frac{1}{2E_3}\left(\acute{\sigma}_{r2} - \acute{\sigma}_{z2}\right)$$

$\Rightarrow$

$$\left(\acute{\sigma}_{r2} - \acute{\sigma}_{z2}\right) = \frac{\left(\sigma_{r2} - \sigma_{z2}\right)}{\left(\dfrac{E_2}{E_3}\right)} = \frac{\left(\sigma_{r2} - \sigma_{z2}\right)}{k_2}$$

Or:

$$\left(\acute{\sigma}_{r2} - \acute{\sigma}_{z2}\right) = \frac{\left(\sigma_{r2} - \sigma_{z2}\right)}{k_2} \tag{13.48}$$

• Based on the continuity conditions, also the following equations are valid:

$$\acute{\sigma}_{z1} = \sigma_{z1} \tag{13.49}$$

$$\acute{\sigma}_{z2} = \sigma_{z2} \tag{13.50}$$

$$\acute{\varepsilon}_{z1} = \varepsilon_{z1} \tag{13.51}$$

$$\acute{\varepsilon}_{z2} = \varepsilon_{z2} \tag{13.52}$$

- The sign convention used in the solution for the three-layer pavement systems is as follows: −ve for tension and +ve for compression.

$$k_1 = \frac{E_1}{E_2}$$

⇒

$$k_1 = \frac{200000}{10000} = 20$$

$$k_2 = \frac{E_2}{E_3}$$

⇒

$$k_2 = \frac{10000}{5000} = 2$$

$$A = \frac{a}{h_2}$$

⇒

$$A = \frac{4}{10} = 0.4$$

$$H = \frac{h_1}{h_2}$$

⇒

$$H = \frac{5}{10} = 0.5$$

Using these values and Jones' tables, the following stress factors are obtained:

$$ZZ1 = 0.13480$$

$$ZZ2 = 0.03998$$

$$ZZ1 - RR1 = 1.89817$$

$$ZZ2 - RR2 = 0.06722$$

$$\sigma_{z1} = q(ZZ1)$$

$$\sigma_{z2} = q(ZZ2)$$

$$\sigma_{z1} - \sigma_{r1} = q\left(ZZ1 - RR1\right)$$

$$\sigma_{z2} - \sigma_{r2} = q\left(ZZ2 - RR2\right)$$

$\Rightarrow$

$$\sigma_{z1} = 100\left(0.13480\right) = 13.48 \text{ psi} \left(92.9 \text{ kPa}\right)$$

$$\sigma_{z2} = 100\left(0.03998\right) = 4.00 \text{ psi} \left(27.6 \text{ kPa}\right)$$

$$\sigma_{z1} - \sigma_{r1} = 100\left(1.89817\right) = 189.8 \text{ psi}$$

$\Rightarrow$

$$\sigma_{r1} = -176.3 \text{ psi} \left(1215.8 \text{ kPa}\right)$$

$$\sigma_{z2} - \sigma_{r2} = 100\left(0.06722\right) = 6.72 \text{ psi}$$

$\Rightarrow$

$$\sigma_{r2} = -2.72 \text{ psi} \left(-18.8 \text{ kPa}\right)$$

$$\varepsilon_{z1} = \frac{1}{E_1}\left(\sigma_{z1} - \sigma_{r1}\right)$$

$\Rightarrow$

$$\varepsilon_{z1} = \frac{1}{200000}\left(189.8\right) = 949 \times 10^{-6} = 949 \ \mu\varepsilon$$

$$\varepsilon_{z2} = \frac{1}{E_2}\left(\sigma_{z2} - \sigma_{r2}\right)$$

$\Rightarrow$

$$\varepsilon_{z2} = \frac{1}{10000}\left(6.72\right) = 672 \times 10^{-6} = 672 \ \mu\varepsilon$$

$$\varepsilon_{r1} = \frac{1}{2E_1}\left(\sigma_{r1} - \sigma_{z1}\right)$$

$\Rightarrow$

$$\varepsilon_{r1} = \frac{1}{2\left(200000\right)}\left(-189.8\right) = 474.5 \times 10^{-6} = 474.5 \ \mu\varepsilon$$

$$\varepsilon_{r2} = \frac{1}{2E_2}\left(\sigma_{r2} - \sigma_{z2}\right)$$

$\Rightarrow$

$$\varepsilon_{r2} = \frac{1}{2\left(10000\right)}\left(-6.72\right) = 336 \times 10^{-6} = 336 \ \mu\varepsilon$$

$$\left(\acute{\sigma}_{r1} - \acute{\sigma}_{z1}\right) = \frac{\left(\sigma_{r1} - \sigma_{z1}\right)}{k_1}$$

$\Rightarrow$

$$\left(\acute{\sigma}_{r1}-\acute{\sigma}_{z1}\right)=\frac{-189.8}{20}=-9.49 \text{ psi}$$

But due to continuity conditions at layer interfaces:

$$\acute{\sigma}_{z1}=\sigma_{z1}=13.48 \text{ psi}\left(92.9 \text{ kPa}\right)$$

$\Rightarrow$

$$\acute{\sigma}_{r1}=3.99 \text{ psi}\left(27.5 \text{ kPa}\right)$$

$$\left(\acute{\sigma}_{r2}-\acute{\sigma}_{z2}\right)=\frac{\left(\sigma_{r2}-\sigma_{z2}\right)}{k_2}$$

$\Rightarrow$

$$\left(\acute{\sigma}_{r2}-\acute{\sigma}_{z2}\right)=\frac{-6.72}{2}=-3.36 \text{ psi}$$

But due to continuity conditions at layer interfaces:

$$\acute{\sigma}_{z2}=\sigma_{z2}=4.00 \text{ psi}\left(27.6 \text{ kPa}\right)$$

$\Rightarrow$

$$\acute{\sigma}_{r2}=0.64 \text{ psi}\left(4.4 \text{ kPa}\right)$$

$$\acute{\varepsilon}_{z1}=\frac{1}{E_2}\left(\acute{\sigma}_{z1}-\acute{\sigma}_{r1}\right)$$

$\Rightarrow$

$$\acute{\varepsilon}_{z1}=\frac{1}{10000}\left(9.49\right)=949\times10^{-6}=949 \ \mu\varepsilon$$

$$\acute{\varepsilon}_{z2}=\frac{1}{E_3}\left(\acute{\sigma}_{z2}-\acute{\sigma}_{r2}\right)$$

$\Rightarrow$

$$\acute{\varepsilon}_{z2}=\frac{1}{5000}\left(3.36\right)=672\times10^{-6}=672 \ \mu\varepsilon$$

$$\acute{\varepsilon}_{r1}=\frac{1}{2E_2}\left(\acute{\sigma}_{r1}-\acute{\sigma}_{z1}\right)$$

$\Rightarrow$

$$\acute{\varepsilon}_{r1}=\frac{1}{2\left(10000\right)}\left(-9.49\right)=-474.5\times10^{-6}=-474.5 \ \mu\varepsilon$$

$$\acute{\varepsilon}_{r2} = \frac{1}{2E_3}\left(\acute{\sigma}_{r2} - \acute{\sigma}_{z2}\right)$$

$\Rightarrow$

$$\acute{\varepsilon}_{r2} = \frac{1}{2(5000)}(-3.36) = -336 \times 10^{-6} = -336\ \mu\varepsilon$$

The pavement response (stress and strain) results at the two-layer interfaces are summarized in Table 13.31.

TABLE 13.31

Pavement Response (Stress and Strain) Results at the Two-Layer Interfaces Using Multi-Layer Elastic Theory for Problem 13.24

Interface	Pavement Layer	Vertical Stress (σ_z)	Radial Stress (σ_r)	Vertical Strain (ε_z)	Radial Strain (ε_r)
1	Bottom of layer #1	$\sigma_{z1} = 13.48$	$\sigma_{r1} = -176.3$	$\varepsilon_{z1} = 949$	$\varepsilon_{r1} = -474.5$
1	Top of layer #2	$\acute{\sigma}_{z1} = 13.48$	$\acute{\sigma}_{r1} = 3.99$	$\acute{\varepsilon}_{z1} = 949$	$\acute{\varepsilon}_{r1} = -474.5$
2	Bottom of layer #2	$\sigma_{z2} = 4.00$	$\sigma_{r2} = -2.72$	$\varepsilon_{z2} = 672$	$\varepsilon_{r2} = -336$
2	Top of layer #3	$\acute{\sigma}_{z2} = 4.00$	$\acute{\sigma}_{r2} = 0.64$	$\acute{\varepsilon}_{z2} = 672$	$\acute{\varepsilon}_{r2} = -336$

The MS Excel worksheet shown in Figure 13.55 is used to perform all computations related to the pavement response analysis to determine the stresses and strains at the two-layer interfaces under the centerline of the wheel load.

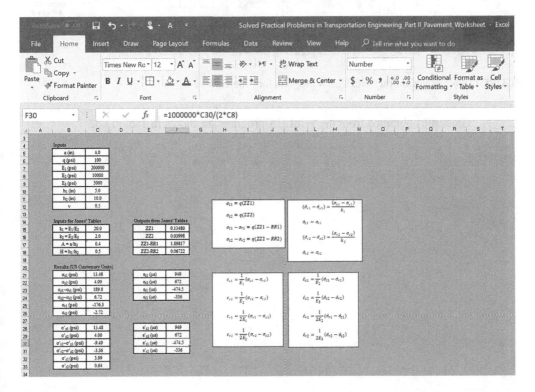

FIGURE 13.55 Image of the MS Excel worksheet used for the flexible pavement analysis of Problem 13.24.

The WinLEA program is also used to perform the analysis of the asphalt pavement structure in this problem to compare the results. The same Poisson ratio is used ($\nu = 0.5$) so that the comparison is one-to-one. The inputs used in the WinLEA program are shown in Figure 13.56.

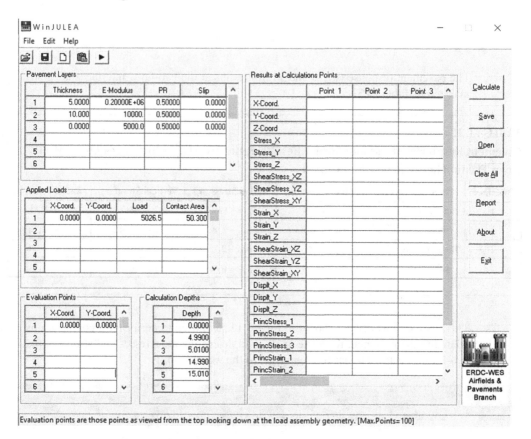

FIGURE 13.56   Inputs for the WinJULEA program for Problem 13.24.

The results obtained from the WinLEA program are shown in Table 13.32.

TABLE 13.32

WinJULEA Flexible Pavement Analysis Results for Problem 13.24

```
*** RESULTS

-----------------------------------------------------------------------------
-----------------------------------------------------------------------------
-----------------------------------------------------------------------------
-----------------------------------------------------------------------------
-----------------------------------------------------------------------------

DEPTH       X-COORD     Y-COORD     STRESS-X    STRESS-Y    STRESS-Z    SHEAR-XZ
SHEAR-YZ    SHEAR-XY    STRAIN-X    STRAIN-Y    STRAIN-Z    SHEAR-XZ    SHEAR-YZ
SHEAR-XY    DISPLT-X    DISPLT-Y    DISPLT-Z    P.STRESS1   P.STRESS2   P.STRESS3
P.STRAIN1   P.STRAIN2   P.STRAIN3   MAX.SHEAR   OCT.STRESS  OCT.SHEAR


-----------------------------------------------------------------------------
-----------------------------------------------------------------------------
-----------------------------------------------------------------------------
-----------------------------------------------------------------------------
-----------------------------------------------------------------------------

0.000E+00   0.000E+00   0.000E+00   0.236E+03   0.236E+03   0.999E+02   0.000E+00
0.000E+00   0.000E+00   0.340E-03   0.340E-03  -0.681E-03   0.000E+00   0.000E+00
0.000E+00   0.000E+00   0.000E+00   0.281E-01   0.236E+03   0.236E+03   0.999E+02
0.340E-03   0.340E-03  -0.681E-03   0.681E+02   0.191E+03   0.642E+02

0.499E+01   0.000E+00   0.000E+00  -0.175E+03  -0.175E+03   0.135E+02   0.000E+00
0.000E+00   0.000E+00  -0.472E-03  -0.472E-03   0.944E-03   0.000E+00   0.000E+00
0.000E+00   0.000E+00   0.000E+00   0.272E-01   0.135E+02  -0.175E+03  -0.175E+03
0.944E-03  -0.472E-03  -0.472E-03   0.944E+02  -0.112E+03   0.890E+02

0.501E+01   0.000E+00   0.000E+00   0.395E+01   0.395E+01   0.135E+02   0.000E+00
0.000E+00   0.000E+00  -0.479E-03  -0.479E-03   0.958E-03   0.000E+00   0.000E+00
0.000E+00   0.000E+00   0.000E+00   0.272E-01   0.135E+02   0.395E+01   0.395E+01
0.958E-03  -0.479E-03  -0.479E-03   0.479E+01   0.715E+01   0.452E+01

0.150E+02   0.000E+00   0.000E+00  -0.271E+01  -0.271E+01   0.400E+01   0.000E+00
0.000E+00   0.000E+00  -0.335E-03  -0.335E-03   0.670E-03   0.000E+00   0.000E+00
0.000E+00   0.000E+00   0.000E+00   0.200E-01   0.400E+01   0.271E+01   0.271E+01
0.670E-03  -0.335E-03  -0.335E-03   0.335E+01  -0.473E+00   0.316E+01

0.150E+02   0.000E+00   0.000E+00   0.640E+00   0.640E+00   0.399E+01   0.000E+00
0.000E+00   0.000E+00  -0.335E-03  -0.335E-03   0.669E-03   0.000E+00   0.000E+00
0.000E+00   0.000E+00   0.000E+00   0.200E-01   0.399E+01   0.640E+00   0.640E+00
0.669E-03  -0.335E-03  -0.335E-03   0.167E+01   0.176E+01   0.158E+01
```

The pavement response (stress and strain) results at the two-layer interfaces as obtained from the WinLEA program are summarized in Table 13.33.

TABLE 13.33

Pavement Response (Stress and Strain) Results at the Two-Layer Interfaces Using WinJULEA Program for Problem 13.24

Interface	Pavement Layer	Vertical Stress (σ_z)	Radial Stress (σ_r)	Vertical Strain (ε_z)	Radial Strain (ε_r)
1	Bottom of layer #1	$\sigma_{z1}=13.5$	$\sigma_{r1}=-175$	$\varepsilon_{z1}=944$	$\varepsilon_{r1}=-472$
1	Top of layer #2	$\acute{\sigma}_{z1}=13.5$	$\acute{\sigma}_{r1}=3.95$	$\acute{\varepsilon}_{z1}=958$	$\acute{\varepsilon}_{r1}=-479$
2	Bottom of layer #2	$\sigma_{z2}=4.00$	$\sigma_{r2}=-2.71$	$\varepsilon_{z2}=670$	$\varepsilon_{r2}=-335$
2	Top of layer #3	$\acute{\sigma}_{z2}=3.99$	$\acute{\sigma}_{r1}=0.64$	$\acute{\varepsilon}_{z2}=669$	$\acute{\varepsilon}_{r2}=-335$

The results obtained from the WinLEA program are compared with the results obtained from the multilayer elastic theory analysis and Jones' tables; the results look similar and very comparable.

13.25 For the three-layer asphalt pavement system shown in Figure 13.57, at the axis of symmetry, the radial tensile strain at the bottom of the HMA layer is 285 με, the vertical compressive stress on the top of the base layer is 8.1 psi, the vertical compressive stress at the bottom of the base layer is 2.4 psi, and the radial tensile strain on the top of the subgrade layer is 202 με. Determine the following pavement responses at the axis of symmetry (under the centerline of the wheel load) using the continuity conditions at the layer interfaces and the multi-layer elastic theory formulas:

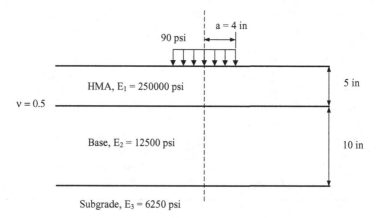

FIGURE 13.57   Three-layered flexible pavement system for Problem 13.25.

a.   The vertical compressive strain at the bottom of the asphalt layer
b.   The vertical compressive strain at the top of the base layer
c.   The vertical compressive strain at the bottom of the base layer
d.   The vertical compressive strain at the top of the subgrade layer
e.   The radial tensile strain at the top of the base layer
f.   The radial tensile strain at the bottom of the base layer

g. The vertical compressive stress at the bottom of the asphalt layer
h. The vertical compressive stress at the top of the subgrade layer
i. The radial tensile stress at the bottom of the asphalt layer
j. The radial compressive stress at the top of the base layer
k. The radial tensile stress at the bottom of the base layer
l. The radial tensile stress at the top of the subgrade layer

Solution:

First, the given pavement responses are summarized based on the description and the sign convention:

$\varepsilon_{r1} = -285 \ \mu\varepsilon$

$\acute{\sigma}_{z1} = 8.1 \ \text{psi}$
$\sigma_{z2} = 2.4 \ \text{psi}$

$\acute{\varepsilon}_{r2} = -202 \ \mu\varepsilon$

Second, the requirements are determined based on the continuity conditions and the formulas derived from the generalized Hook's law for strain:

a. The vertical compressive strain at the bottom of the asphalt layer is calculated as shown below:

This is ε_{z1} and is calculated using the formula below:

$$\varepsilon_{z1} = -2\varepsilon_{r1}$$

$\Rightarrow$

$$\varepsilon_{z1} = -2(-285) = 570 \ \mu\varepsilon$$

b. The vertical compressive strain at the top of the base layer is determined as below:

This is $\acute{\varepsilon}_{z1}$ and is calculated using the formula below:

$$\acute{\varepsilon}_{z1} = \varepsilon_{z1}$$

$\Rightarrow$

$$\acute{\varepsilon}_{z1} = 570 \ \mu\varepsilon$$

c. The vertical compressive strain at the bottom of the base layer is determined as follows:
This is ε_{z2} and is calculated using the formula below:

$$\varepsilon_{z2} = -2\varepsilon_{r2}$$

And:

$$\varepsilon_{r2} = \acute{\varepsilon}_{r2} = -202 \ \mu\varepsilon$$

$\Rightarrow$

$$\varepsilon_{z2} = -2(-202) = 404 \ \mu\varepsilon$$

d. The vertical compressive strain at the top of the subgrade layer is computed as follows:

This is $\acute{\varepsilon}_{z2}$ and is calculated using the formula below:

$$\acute{\varepsilon}_{z2} = \varepsilon_{z2}$$

$\Rightarrow$

$$\acute{\varepsilon}_{z2} = 404 \ \mu\varepsilon$$

e. The radial tensile strain at the top of the base layer is determined as below:

This is $\acute{\varepsilon}_{r1}$ and is calculated using the formula below:

$$\acute{\varepsilon}_{r1} = \varepsilon_{r1}$$

$\Rightarrow$

$$\acute{\varepsilon}_{r1} = -285 \ \mu\varepsilon$$

f. The radial tensile strain at the bottom of the base layer is calculated as follows:
This is ε_{r2} and is calculated using the formula below:

$$\varepsilon_{r2} = \acute{\varepsilon}_{r2}$$

$\Rightarrow$

$$\varepsilon_{r2} = -202 \ \mu\varepsilon$$

g. The vertical compressive stress at the bottom of the asphalt layer is computed as shown below:
This is σ_{z1} and is calculated using the formula below:

$$\sigma_{z1} = \acute{\sigma}_{z1}$$

$\Rightarrow$

$$\sigma_{z1} = 8.1 \ \text{psi} \left(55.8 \ \text{kPa} \right)$$

h. The vertical compressive stress at the top of the subgrade layer is determined as below:

This is $\acute{\sigma}_{z2}$ and is calculated using the formula below:

$$\acute{\sigma}_{z2} = \sigma_{z2}$$

$\Rightarrow$

$$\acute{\sigma}_{z2} = 2.4 \ \text{psi} \left(16.5 \ \text{kPa} \right)$$

i. The radial tensile stress at the bottom of the asphalt layer is calculated as below:
This is σ_{r1} and is calculated using the formula below:

$$\varepsilon_{r1} = \frac{1}{2E_1}(\sigma_{r1} - \sigma_{z1})$$

$\Rightarrow$

$$-285 \times 10^{-6} = \frac{1}{2(250000)}(\sigma_{r1} - 8.1)$$

$\Rightarrow$

$$\sigma_{r1} = -134.4 \text{ psi}(-926.7 \text{ kPa})$$

Or:

$$\varepsilon_{z1} = \frac{1}{E_1}(\sigma_{z1} - \sigma_{r1})$$

$\Rightarrow$

$$570 \times 10^{-6} = \frac{1}{250000}(8.1 - \sigma_{r1})$$

$\Rightarrow$

$$\sigma_{r1} = -134.4 \text{ psi}(-926.7 \text{ kPa})$$

j. The radial compressive stress at the top of the base layer is calculated as below:
This is $\acute{\sigma}_{r1}$ and is calculated using the formula below:

$$\acute{\varepsilon}_{r1} = \frac{1}{2E_2}(\acute{\sigma}_{r1} - \acute{\sigma}_{z1})$$

$\Rightarrow$

$$-285 \times 10^{-6} = \frac{1}{2(12500)}(\acute{\sigma}_{r1} - 8.1)$$

$\Rightarrow$

$$\acute{\sigma}_{r1} = 0.975 \text{ psi}(6.7 \text{ kPa})$$

Or:

$$\acute{\varepsilon}_{z1} = \frac{1}{E_2}(\acute{\sigma}_{z1} - \acute{\sigma}_{r1})$$

$\Rightarrow$

$$570 \times 10^{-6} = \frac{1}{12500}(8.1 - \acute{\sigma}_{r1})$$

$\Rightarrow$

$$\acute{\sigma}_{r1} = 0.975 \text{ psi} (6.7 \text{ kPa})$$

k. The radial tensile stress at the bottom of the base layer is computed as follows:
    This is σ_{r2} and is calculated using the formula below:

$$\varepsilon_{r2} = \frac{1}{2E_2} (\sigma_{r2} - \sigma_{z2})$$

$\Rightarrow$

$$-202 \times 10^{-6} = \frac{1}{2(12500)} (\sigma_{r2} - 2.4)$$

$\Rightarrow$

$$\sigma_{r2} = -2.7 \text{ psi} (-18.3 \text{ kPa})$$

Or:

$$\varepsilon_{z2} = \frac{1}{E_2} (\sigma_{z2} - \sigma_{r2})$$

$\Rightarrow$

$$404 \times 10^{-6} = \frac{1}{12500} (2.4 - \sigma_{r2})$$

$\Rightarrow$

$$\sigma_{r2} = -2.7 \text{ psi} (-18.3 \text{ kPa})$$

l. The radial tensile stress at the top of the subgrade layer is determined as shown below:
    This is $\acute{\sigma}_{r2}$ and is calculated using the formula below:

$$\acute{\varepsilon}_{r2} = \frac{1}{2E_3} (\acute{\sigma}_{r2} - \acute{\sigma}_{z2})$$

$\Rightarrow$

$$-202 \times 10^{-6} = \frac{1}{2(6250)} (\acute{\sigma}_{r2} - 2.4)$$

$\Rightarrow$

$$\acute{\sigma}_{r2} = -0.125 \text{ psi} (-0.9 \text{ kPa})$$

Or:

$$\acute{\varepsilon}_{z2} = \frac{1}{E_3}\left(\acute{\sigma}_{z2} - \acute{\sigma}_{r2}\right)$$

⇒

$$404 \times 10^{-6} = \frac{1}{6250}\left(2.4 - \sigma_{r2}\right)$$

⇒

$$\acute{\sigma}_{r2} = -0.125 \text{ psi}\left(-0.9 \text{ kPa}\right)$$

The MS Excel worksheet shown in Figure 13.58 is used to perform all computations related to the pavement response analysis to determine the stresses and strains at the two-layer interfaces under the centerline of the wheel load.

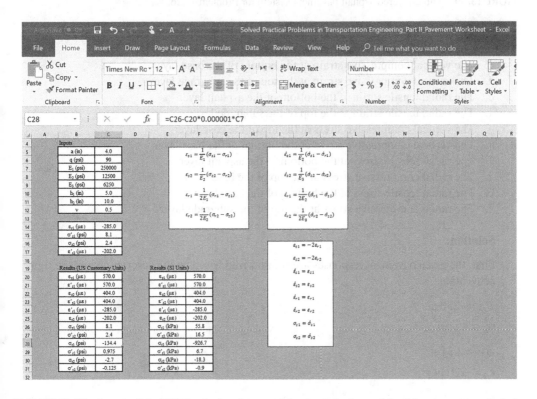

FIGURE 13.58 Image of the MS Excel worksheet used for the three-layered flexible pavement analysis for Problem 13.25.

13.26 For the three-layer asphalt pavement system shown in Figure 13.59, at the axis of symmetry, the radial tensile strain on the top of the base layer is 230 με, the vertical compressive stress at the bottom of the HMA layer is 8.5 psi, the vertical compressive stress on the top of the subgrade layer is 4.2 psi, and the radial tensile strain at the bottom of the base layer is 220 με. Determine the following pavement responses at the axis of symmetry (under the centerline of the wheel load) using the continuity conditions at the layer interfaces and the multi-layer elastic theory formulas:

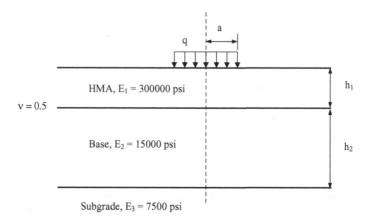

FIGURE 13.59   Three-layered asphalt pavement system for Problem 13.26.

a.   The vertical compressive strain at the bottom of the asphalt layer
b.   The vertical compressive strain at the top of the base layer
c.   The vertical compressive strain at the bottom of the base layer
d.   The vertical compressive strain at the top of the subgrade layer
e.   The radial tensile strain at the bottom of the asphalt layer
f.   The radial tensile strain at the top of the subgrade layer
g.   The vertical compressive stress at the top of the base layer
h.   The vertical compressive stress at the bottom of the base layer
i.   The radial tensile stress at the bottom of the asphalt layer
j.   The radial compressive stress at the top of the base layer
k.   The radial tensile stress at the bottom of the base layer
l.   The radial compressive stress at the top of the subgrade layer

Solution:

First, the given pavement responses are summarized based on the description and the sign convention:

$\acute{\varepsilon}_{r1} = -230\ \mu\varepsilon$

$\sigma_{z1} = 8.5\ \text{psi}$

$\acute{\sigma}_{z2} = 4.2\ \text{psi}$

$\varepsilon_{r2} = -220\ \mu\varepsilon$

Second, the requirements are determined based on the continuity conditions and the formulas derived from the generalized Hook's law for strain:

a.   The vertical compressive strain at the bottom of the asphalt layer is calculated as shown below:

   This is ε_{z1} and is calculated using the formula below:

$$\varepsilon_{z1} = -2\varepsilon_{r1}$$

But:

$$\varepsilon_{r1} = \acute{\varepsilon}_{r1} = -230 \; \mu\varepsilon$$

$\Rightarrow$

$$\varepsilon_{z1} = -2(-230) = 460 \; \mu\varepsilon$$

b. The vertical compressive strain at the top of the base layer is determined as below:

This is $\acute{\varepsilon}_{z1}$ and is calculated using the formula below:

$$\acute{\varepsilon}_{z1} = \varepsilon_{z1}$$

$\Rightarrow$

$$\acute{\varepsilon}_{z1} = 460 \; \mu\varepsilon$$

c. The vertical compressive strain at the bottom of the base layer is determined as follows:
This is ε_{z2} and is calculated using the formula below:

$$\varepsilon_{z2} = -2\varepsilon_{r2}$$

$\Rightarrow$

$$\varepsilon_{z2} = -2(-220) = 440 \; \mu\varepsilon$$

d. The vertical compressive strain at the top of the subgrade layer is computed as follows:

This is $\acute{\varepsilon}_{z2}$ and is calculated using the formula below:

$$\acute{\varepsilon}_{z2} = \varepsilon_{z2}$$

$\Rightarrow$

$$\acute{\varepsilon}_{z2} = 440 \; \mu\varepsilon$$

e. The radial tensile strain at the bottom of the asphalt layer is determined as below:
This is ε_{r1} and is calculated using the formula below:

$$\varepsilon_{r1} = \acute{\varepsilon}_{r1}$$

$\Rightarrow$

$$\varepsilon_{r1} = -230 \; \mu\varepsilon$$

f. The radial tensile strain at the top of the subgrade layer is calculated as follows:
    This is ε_{r2} and is calculated using the formula below:

$$\acute{\varepsilon}_{r2} = \varepsilon_{r2}$$

$\Rightarrow$

$$\acute{\varepsilon}_{r2} = -220 \ \mu\varepsilon$$

g. The vertical compressive stress at the top of the base layer is computed as shown below:
    This is $\acute{\sigma}_{z1}$ and is calculated using the formula below:

$$\acute{\sigma}_{z1} = \sigma_{z1}$$

$\Rightarrow$

$$\acute{\sigma}_{z1} = 8.5 \ \text{psi} \left(58.6 \ \text{kPa}\right)$$

h. The vertical compressive stress at the bottom of the base layer is determined as below:
    This is σ_{z2} and is calculated using the formula below:

$$\sigma_{z2} = \acute{\sigma}_{z2}$$

$\Rightarrow$

$$\sigma_{z2} = 4.2 \ \text{psi} \left(29.0 \ \text{kPa}\right)$$

i. The radial tensile stress at the bottom of the asphalt layer is calculated as below:
    This is σ_{r1} and is calculated using the formula below:

$$\varepsilon_{r1} = \frac{1}{2E_1}\left(\sigma_{r1} - \sigma_{z1}\right)$$

$\Rightarrow$

$$-230 \times 10^{-6} = \frac{1}{2\left(300000\right)}\left(\sigma_{r1} - 8.5\right)$$

$\Rightarrow$

$$\sigma_{r1} = -129.5 \ \text{psi} \left(892.9 \ \text{kPa}\right)$$

Or:

$$\varepsilon_{z1} = \frac{1}{E_1}\left(\sigma_{z1} - \sigma_{r1}\right)$$

$\Rightarrow$

$$460 \times 10^{-6} = \frac{1}{300000}(8.5 - \sigma_{r1})$$

$\Rightarrow$

$$\sigma_{r1} = -129.5 \text{ psi}(-892.9 \text{ kPa})$$

j. The radial compressive stress at the top of the base layer is calculated as below:

This is $\acute{\sigma}_{r1}$ and is calculated using the formula below:

$$\acute{\varepsilon}_{r1} = \frac{1}{2E_2}\left(\acute{\sigma}_{r1} - \acute{\sigma}_{z1}\right)$$

$\Rightarrow$

$$-230 \times 10^{-6} = \frac{1}{2(15000)}\left(\acute{\sigma}_{r1} - 8.5\right)$$

$\Rightarrow$

$$\acute{\sigma}_{r1} = 1.6 \text{ psi}(11.0 \text{ kPa})$$

Or:

$$\acute{\varepsilon}_{z1} = \frac{1}{E_2}\left(\acute{\sigma}_{z1} - \acute{\sigma}_{r1}\right)$$

$\Rightarrow$

$$460 \times 10^{-6} = \frac{1}{15000}\left(8.5 - \sigma_{r1}\right)$$

$\Rightarrow$

$$\acute{\sigma}_{r1} = 1.6 \text{ psi}(11.0 \text{ kPa})$$

k. The radial tensile stress at the bottom of the base layer is computed as follows:
    This is σ_{r2} and is calculated using the formula below:

$$\varepsilon_{r2} = \frac{1}{2E_2}\left(\sigma_{r2} - \sigma_{z2}\right)$$

$\Rightarrow$

$$-220 \times 10^{-6} = \frac{1}{2(15000)}\left(\sigma_{r2} - 4.2\right)$$

⇒

$$\sigma_{r2} = -2.4 \text{ psi} \left(-16.5 \text{ kPa}\right)$$

Or:

$$\varepsilon_{z2} = \frac{1}{E_2}\left(\sigma_{z2} - \sigma_{r2}\right)$$

⇒

$$440 \times 10^{-6} = \frac{1}{15000}\left(4.2 - \sigma_{r2}\right)$$

⇒

$$\sigma_{r2} = -2.4 \text{ psi} \left(-16.5 \text{ kPa}\right)$$

1. The radial compressive stress at the top of the subgrade layer is determined as shown below:

   This is $\acute{\sigma}_{r2}$ and is calculated using the formula below:

$$\acute{\varepsilon}_{r2} = \frac{1}{2E_3}\left(\acute{\sigma}_{r2} - \acute{\sigma}_{z2}\right)$$

⇒

$$-220 \times 10^{-6} = \frac{1}{2(7500)}\left(\acute{\sigma}_{r2} - 4.2\right)$$

⇒

$$\acute{\sigma}_{r2} = 0.9 \text{ psi} \left(6.2 \text{ kPa}\right)$$

Or:

$$\acute{\varepsilon}_{z2} = \frac{1}{E_3}\left(\acute{\sigma}_{z2} - \acute{\sigma}_{r2}\right)$$

⇒

$$440 \times 10^{-6} = \frac{1}{7500}\left(4.2 - \sigma_{r2}\right)$$

⇒

$$\acute{\sigma}_{r2} = 0.9 \text{ psi} \left(6.2 \text{ kPa}\right)$$

The MS Excel worksheet shown in Figure 13.60 is used to perform all computations related to the pavement response analysis to determine the stresses and strains at the two-layer interfaces under the centerline of the wheel load.

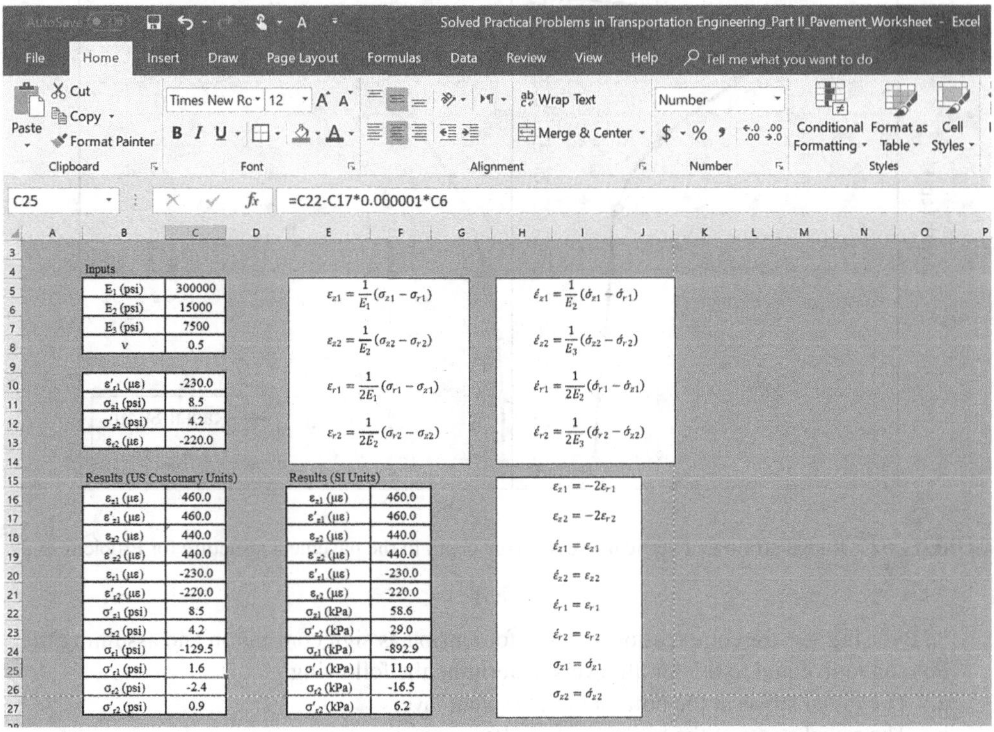

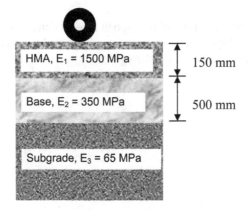

FIGURE 13.60 Image of the MS Excel worksheet used for the three-layered flexible pavement analysis for Problem 13.26.

13.27 For the pavement structure shown in Figure 13.61, the vertical stress (kPa) and radial strain (με) under the centerline of the wheel load (at the axis of symmetry) are plotted versus depth within the pavement structure as shown in Figure 13.62.

HMA, E_1 = 1500 MPa   150 mm

Base, E_2 = 350 MPa   500 mm

Subgrade, E_3 = 65 MPa

FIGURE 13.61 Single-wheel loading on an asphalt pavement structure for Problem 13.27.

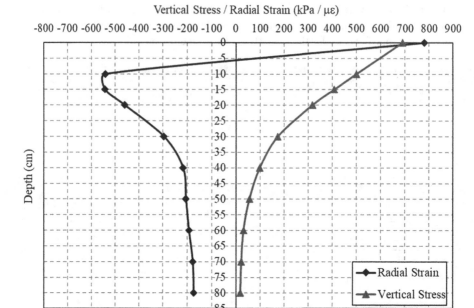

FIGURE 13.62 Radial strain and vertical stress versus depth in the pavement structure for Problem 13.27.

By using +ve for compression and −ve for tension as sign convention and assuming the Poisson ratio equal to 0.5 for all layers, determine the following:

a. The radial stress at the bottom of the asphalt layer
b. The radial stress at the top of the base layer
c. The radial stress at the bottom of the base layer
d. The radial stress at the top of the subgrade layer
e. The vertical strain at the bottom of the asphalt layer
f. The vertical strain at the bottom of the base layer
g. The applied wheel pressure
h. The radius of contact area if the applied wheel loading is 40 kN

Solution:

a. The radial stress at the bottom of the asphalt layer is determined as below:
    At the bottom of the asphalt layer, the depth is equal to 15 cm, the following pavement responses are obtained from the figure:

$$\sigma_{z1} = 410 \text{ kPa}$$

$$\varepsilon_{r1} = -550 \ \mu\varepsilon$$

$$\varepsilon_{r1} = \frac{1}{2E_1}\left(\sigma_{r1} - \sigma_{z1}\right)$$

⇒

$$-550 \times 10^{-6} = \frac{1}{2\left(1500 \times 10^3\right)}\left(\sigma_{r1} - 410\right)$$

$\Rightarrow$

$$\sigma_{r1} = -1240 \text{ kPa}$$

b. The radial stress at the top of the base layer is determined as follows:
At the top of the base layer, the depth is equal to 15 cm, the following pavement responses are obtained from the figure:

$$\acute{\sigma}_{z1} = \sigma_{z1} = 410 \text{ kPa}$$

$$\acute{\varepsilon}_{r1} = \varepsilon_{r1} = -550 \ \mu\varepsilon$$

$$\acute{\varepsilon}_{r1} = \frac{1}{2E_2}\left(\acute{\sigma}_{r1} - \acute{\sigma}_{z1}\right)$$

$\Rightarrow$

$$-550 \times 10^{-6} = \frac{1}{2\left(350 \times 10^3\right)}\left(\acute{\sigma}_{r1} - 410\right)$$

$\Rightarrow$

$$\acute{\sigma}_{r1} = 25 \text{ kPa}$$

c. The radial stress at the bottom of the base layer is calculated as shown below:
At the bottom of the base layer, the depth is equal to 65 cm, the following pavement responses are obtained from the figure:

$$\sigma_{z2} = 25 \text{ kPa}$$

$$\varepsilon_{r2} = -180 \ \mu\varepsilon$$

$$\varepsilon_{r2} = \frac{1}{2E_2}\left(\sigma_{r2} - \sigma_{z2}\right)$$

$\Rightarrow$

$$-180 \times 10^{-6} = \frac{1}{2\left(350 \times 10^3\right)}\left(\sigma_{r2} - 25\right)$$

$\Rightarrow$

$$\sigma_{r2} = -101 \text{ kPa}$$

d. The radial stress at the top of the subgrade layer is determined as below:
    At the top of the subgrade layer, the depth is equal to 65 cm, the following pavement
    responses are obtained from the figure:

$$\acute{\sigma}_{z2} = \sigma_{z2} = 25 \text{ kPa}$$

$$\acute{\varepsilon}_{r2} = \varepsilon_{r2} = -180 \ \mu\varepsilon$$

$$\acute{\varepsilon}_{r2} = \frac{1}{2E_3}\left(\acute{\sigma}_{r2} - \acute{\sigma}_{z2}\right)$$

$\Rightarrow$

$$-180 \times 10^{-6} = \frac{1}{2\left(65 \times 10^3\right)}\left(\acute{\sigma}_{r2} - 25\right)$$

$\Rightarrow$

$$\acute{\sigma}_{r2} = 1.6 \text{ kPa}$$

e. The vertical strain at the bottom of the asphalt layer is calculated as follows:

$$\varepsilon_{z1} = \frac{1}{E_1}\left(\sigma_{z1} - \sigma_{r1}\right)$$

$\Rightarrow$

$$\varepsilon_{z1} = \frac{1}{1500 \times 10^3}\left(410 - \left(-1240\right)\right) = 1100 \times 10^{-6} = 1100 \ \mu\varepsilon$$

f. The vertical strain at the bottom of the base layer is determined as below:

$$\varepsilon_{z2} = \frac{1}{E_2}\left(\sigma_{z2} - \sigma_{r2}\right)$$

$\Rightarrow$

$$\varepsilon_{z2} = \frac{1}{350 \times 10^3}\left(25 - \left(-101\right)\right) = 504 \times 10^{-6} = 504 \ \mu\varepsilon$$

g. The applied wheel pressure is determined as follows:
    It is equal to the vertical stress at the surface of the pavement (at a depth of 0 cm).
    Therefore, from the figure at a depth of 0, the vertical stress is equal to 690 kPa.

$$q = 690 \text{ kPa}$$

h. The radius of contact area if the applied wheel loading is 40 kN is determined using
    the formula below:

$$A_c = \frac{P}{q}$$

$\Rightarrow$

$$A_c = \frac{40}{690} = 0.058 \text{ m}^2 = 580 \text{ cm}^2$$

$$A_c = \pi a^2$$

$\Rightarrow$

$$a = \sqrt{\frac{A_c}{\pi}} = \sqrt{\frac{580}{\pi}} = 13.6 \text{ cm}$$

The MS Excel worksheet shown in Figure 13.63 is used to perform the computations related to the results in this problem.

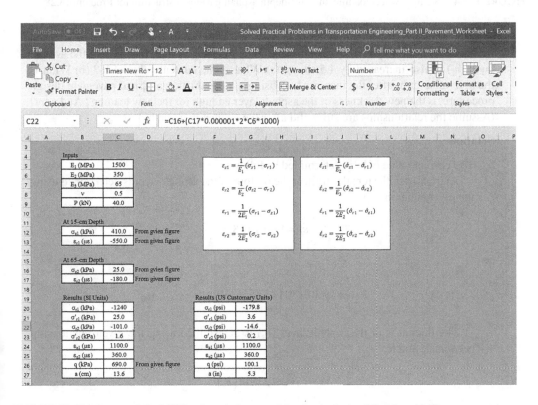

FIGURE 13.63 Image of the MS Excel worksheet used for the analysis of Problem 13.27.

13.28 For the full-depth asphalt pavement structure shown in Figure 13.64, the relationship between the critical tensile strain at the bottom of the asphalt layer (ε_t) in micro-strains and the applied truck wheel pressure or load intensity (q) in psi is given by a linear model that has the following expression:

$$\varepsilon_t = 1.9q + 178 \tag{13.53}$$

and the relationship between the tensile strain at the bottom of the asphalt layer (ε_t) in micro-strains and the thickness of the asphalt layer (h_1) in inches is given by a logarithmic model as shown in the expression below:

$$\varepsilon_t = 760 - 240\ln\left(h_1\right) \tag{13.54}$$

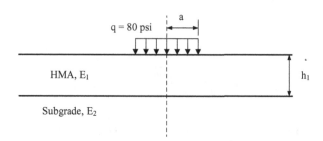

FIGURE 13.64   A single-wheel loading on a full-depth asphalt pavement structure for Problem 13.28.

If the allowable truck loading is changed such that the applied pressure (q) is increased by 1.5 times from 80 psi to 120 psi; and based on that, the design thickness of the asphalt layer (h_1) is also increased by 1.5 times, (a) What is the current value of h_1? (b) What is the new value of h_1 after the increase? (c) Using a mechanistic-empirical pavement design procedure, is the new thickness of the asphalt layer enough for the new loading? And why? (Assume the relationships are still valid after the change in q and h_1.)

Solution:

a.   The current value of h_1 is determined as follows:
        At q = 80 psi, the tensile strain at the bottom of the asphalt layer is determined using the given relationship:

$$\varepsilon_t = 1.9q + 178$$

$\Rightarrow$

$$\varepsilon_t = 1.9(80) + 178 = 330 \ \mu\varepsilon$$

The thickness of the asphalt layer (h_1) is determined using the given relationship between ε_t and h_1:

$$\varepsilon_t = 760 - 240\ln\left(h_1\right)$$

$\Rightarrow$

$$330 = 760 - 240\ln\left(h_1\right)$$

$\Rightarrow$

$$h_1 = 6.0 \text{ in}$$

b. If the thickness of the asphalt layer (h_1) is increased by 1.5 times; therefore, the new value of the thickness is equal to:

$$h_1 = 1.5(6.0) = 9.0 \text{ in}$$

c. The new thickness ($h_1 = 9.0$ in) will imply that the tensile strain at the bottom of the asphalt layer is:

$$\varepsilon_t = 760 - 240\ln(h_1)$$

$\Rightarrow$

$$\varepsilon_t = 760 - 240\ln(9.0) = 232.7 \ \mu\varepsilon \cong 233 \ \mu\varepsilon$$

Therefore, the reduction value in the tensile strain is equal to:

$$\Delta\varepsilon_t = 330 - 233 = 97 \ \mu\varepsilon$$

After the increase in loading (q) from 80 psi to 120 psi, the new tensile strain is calculated as:

$\Rightarrow$

$$\varepsilon_t = 1.9(120) + 178 = 406 \ \mu\varepsilon$$

However, the increase in loading (from 80 to 120 psi) results in an increase in the tensile strain from 330 µε to 406 µε. In other words, the change in the tensile strain is

$$\Delta\varepsilon_t = 406 - 330 = 76 \ \mu\varepsilon$$

Since the reduction in the tensile strain due to the increase in the thickness of the asphalt layer is higher than the increase in the tensile strain as a result of the increase in loading, it implies that the new design thickness of the asphalt layer (9.0 inches) would be adequate for the new loading.

The MS Excel worksheet used to conduct the computations of this problem is shown Figure 13.65.

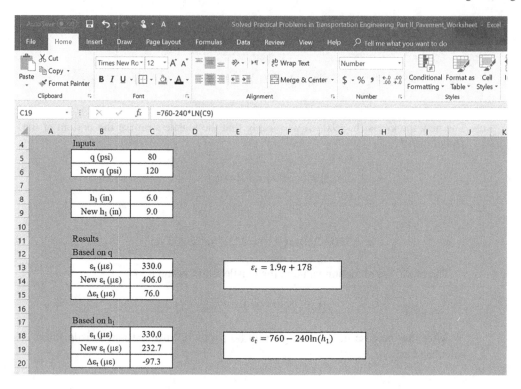

FIGURE 13.65 Image of the MS Excel worksheet used for the computations of Problem 13.28.

13.29 A full-depth asphalt pavement is expected to carry a design traffic loading of 4 million ESALs for a design life of 10 years. The pavement will be subjected to a critical circular single-wheel load of 4.5 kips with a contact pressure of 100 psi as shown in Figure 13.66. Based on a mechanistic-empirical pavement design method, determine the maximum allowable initial tensile strain at the bottom of the asphalt layer (ε_t) and the required design thickness of the full-depth asphalt layer. Use the Asphalt Institute (AI) distress model for fatigue performance shown below:

$$N_f = 0.0796(\varepsilon_t)^{-3.291}(E_1)^{-0.854} \tag{13.55}$$

Where:

N_f = number load applications to fatigue failure
ε_t = tensile strain at the bottom of asphalt layer
E_1 = elastic modulus of asphalt layer (psi)

(Assume that N_f is in ESALs.)

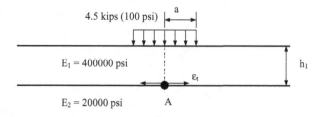

FIGURE 13.66 A full-depth asphalt pavement for Problem 13.29.

Solution:

Using the AI distress model for fatigue performance, the tensile strain at the bottom of the asphalt layer is calculated based on the design ESALs the pavement structure is designed for:

$$N_f = 0.0796(\varepsilon_t)^{-3.291}(E_1)^{-0.854}$$

$\Rightarrow$

$$4 \times 10^6 = 0.0796(\varepsilon_t)^{-3.291}(400000)^{-0.854}$$

Solving for ε_t provides the following solution:

$$\varepsilon_t = 160.8 \times 10^{-6} \cong 161 \ \mu\varepsilon$$

The tensile strain formula is used to determine the strain factor:

$$\varepsilon = \frac{q}{E_1}F_\varepsilon$$

$\Rightarrow$

$$161 \times 10^{-6} = \frac{100}{400000}F_\varepsilon$$

$\Rightarrow$

$$F_\varepsilon = 0.64$$

Using Figure 13.42 and the ratio $E_1/E_2 = 400000/20000 = 20$, the ratio h_1/a is determined as:

$$\frac{h_1}{a} = 1.6$$

But:

$$a = \sqrt{\frac{\left(\dfrac{P}{q}\right)}{\pi}}$$

$\Rightarrow$

$$a = \sqrt{\frac{\left(\dfrac{4500}{100}\right)}{\pi}} = 3.78 \ \text{in}$$

$\Rightarrow$

$$h_1 = 1.6(3.78) \cong 6.0 \ \text{in} \ (15 \ \text{cm})$$

The MS Excel worksheet used to perform the computations in this problem is shown in Figure 13.67.

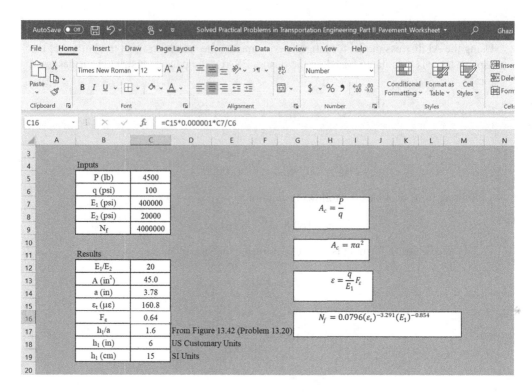

FIGURE 13.67 Image of the MS Excel worksheet used for estimating the tensile strain and the asphalt layer thickness for Problem 13.29.

13.30 Develop equivalent axle load factors (EALFs) for the pavement structure shown in Figure 13.68 using the following formula:

$$\text{EALF} = \frac{N_{f-18}}{N_{f-x}} \tag{13.56}$$

Where:
EALF = equivalent axle load factor for single axles
N_{f-18} = number of 18-kip single-axle load applications (repetitions) to failure (fatigue or rutting failure)
N_{f-x} = number of x-kip single-axle load applications (repetitions) to failure (fatigue or rutting failure)

The axle is a single axle with dual wheels.
Use the following Asphalt Institute (AI) distress models for fatigue and rutting performances:

$$N_f = 0.0796\left(\varepsilon_t\right)^{-3.291}\left(E_1\right)^{-0.854} \qquad \text{For fatigue performance} \tag{13.57}$$

$$N_f = 1.365 \times 10^{-9}\left(\varepsilon_c\right)^{-4.477} \qquad \text{For rutting performance} \tag{13.58}$$

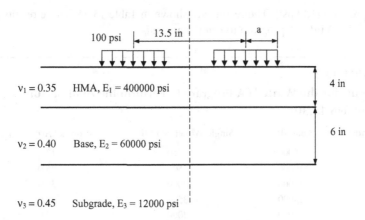

FIGURE 13.68 Dual wheel loading on a layered flexible pavement system for development of EALFs for Problem 13.30.

The following single-axle load values are used in the analysis (see Table 13.34).

TABLE 13.34
Single-Axle Load Values used to Develop EALFs for Problem 13.30

Load #	Load (lb)
1	5000
2	10000
3	12000
4	14000
5	16000
6	18000
7	20000
8	25000
9	30000
10	35000
11	40000
12	45000
13	50000

Determine the EALFs once using the AI fatigue distress model and another time using the AI rutting distress model. Compare the results obtained using the two models. Also compare the results with the AI's EALFs. The WinLEA program will be run 13 times to perform the pavement analysis for the 13 different loadings.

Solution:

The WinLEA program is used to perform the pavement analysis for critical tensile strain at the bottom of the asphalt layer and critical compressive strain on the top of the subgrade layer.

The single-axle load is applied on a set of dual wheels as shown in Table 13.35. Therefore, the wheel load is equal to the given load divided by four since the axle has four wheels. The contact area is determined by simply dividing the single wheel load by the applied pressure (100 psi). The results are shown in Table 13.35. These results are used as inputs for the WinLEA program to run the analysis.

TABLE 13.35
Inputs for the WinJULEA Program to Perform the Analysis for Problem 13.30

Load #	Load (lb)	Single Wheel Load (lb)	Contact Area (in²)
1	5000	1250	12.5
2	10000	2500	25.0
3	12000	3000	30.0
4	14000	3500	35.0
5	16000	4000	40.0
6	18000	4500	45.0
7	20000	5000	50.0
8	25000	6250	62.5
9	30000	7500	75.0
10	35000	8750	87.5
11	40000	10000	100.0
12	45000	11250	112.5
13	50000	12500	125.0

The WinLEA pavement response results are shown in Table 13.36.

TABLE 13.36
Pavement Critical Response (ε_t and ε_c) using WinJULEA for Problem 13.30

Load (lb)	ε_t ($\mu\varepsilon$)	ε_c ($\mu\varepsilon$)
5000	98	170
10000	163	335
12000	183	399
14000	202	461
16000	219	523
18000	234	583
20000	248	645
25000	279	802
30000	307	956
35000	335	1110
40000	373	1260
45000	408	1400
50000	439	1550

Based on the WinLEA pavement response results, the fatigue and rutting performances are determined using the AI distress models as shown below:

Sample Calculation:

For load = 5000 lb,

Based on the AI fatigue distress model, the EALF is calculated as follows:

$$EALF = \frac{N_{f-18}}{N_{f-x}}$$

$\Rightarrow$

$$EALF = \frac{0.0796(\varepsilon_{t-18})^{-3.291}(E_1)^{-0.854}}{0.0796(\varepsilon_{t-x})^{-3.291}(E_1)^{-0.854}}$$

$\Rightarrow$

$$EALF = \frac{(\varepsilon_{t-18})^{-3.291}}{(\varepsilon_{t-x})^{-3.291}}$$

Or:

$$EALF = \left(\frac{\varepsilon_{t-18}}{\varepsilon_{t-x}}\right)^{-3.291} \tag{13.59}$$

Where:

EALF = equivalent axle load factor for single axles

ε_{t-18} = maximum (critical) tensile strain at the bottom of the asphalt layer for the 18-kip single-axle load

ε_{t-x} = maximum (critical) tensile strain at the bottom of the asphalt layer for the x-kip single-axle load

Sulubstituting the tensile strains for x = 5000 lb and for the standard 18000 lb-load provides the following:

$$EALF = \left(\frac{\varepsilon_{t-18}}{\varepsilon_{t-5}}\right)^{-3.291}$$

$\Rightarrow$

$$EALF = \left(\frac{234}{98}\right)^{-3.291} = 0.05702$$

And based on the AI rutting distress model, the EALF is calculated as follows:

$$EALF = \frac{N_{f-18}}{N_{f-x}}$$

$\Rightarrow$

$$EALF = \frac{1.365 \times 10^{-9}(\varepsilon_{c-18})^{-4.477}}{1.365 \times 10^{-9}(\varepsilon_{c-x})^{-4.477}}$$

$\Rightarrow$

$$EALF = \frac{\left(\varepsilon_{c-18}\right)^{-4.477}}{\left(\varepsilon_{c-x}\right)^{-4.477}}$$

Or:

$$EALF = \left(\frac{\varepsilon_{c-18}}{\varepsilon_{c-x}}\right)^{-4.477} \tag{13.60}$$

Where:

EALF = equivalent axle load factor for single axles

ε_{c-18} = maximum (critical) vertical compressive strain on the top of the subgrade layer for the 18-kip single-axle load

ε_{c-x} = maximum (critical) vertical compressive strain on the top of the subgrade layer for the x-kip single-axle load

Substituting the vertical compressive strains for x = 5000 lb and for the standard 18000 lb-load provides the following:

$$EALF = \left(\frac{\varepsilon_{c-18}}{\varepsilon_{c-5}}\right)^{-4.477}$$

$\Rightarrow$

$$EALF = \left(\frac{583}{170}\right)^{-4.477} = 0.00402$$

The EALF results for all other loads based on fatigue performance and rutting performance are summarized in Tables 13.37 and 13.38, respectively.

TABLE 13.37
EALFs based on Fatigue Performance for Problem 13.30

Load (lb)	ε_t (με)	EALF	AI's EALF	% Difference
5000	98	0.05702	0.00500	91
10000	163	0.3042	0.0877	71
12000	183	0.445	0.189	58
14000	202	0.616	0.360	42
16000	219	0.804	0.623	23
18000	234	1.000	1.000	0
20000	248	1.21	1.51	25
25000	279	1.78	3.53	98
30000	307	2.44	6.97	186
35000	335	3.26	12.50	283
40000	373	4.64	21.08	354
45000	408	6.23	34.00	446
50000	439	7.93	52.88	567

TABLE 13.38

EALFs based on Rutting Performance for Problem 13.30

Load (lb)	ε_c ($\mu\varepsilon$)	EALF	AI's EALF	% Difference
5000	170	0.00402	0.00500	24
10000	335	0.0837	0.0877	5
12000	399	0.183	0.189	3
14000	461	0.350	0.360	3
16000	523	0.615	0.623	1
18000	583	1.000	1.000	0
20000	645	1.57	1.51	4
25000	802	4.17	3.53	15
30000	956	9.15	6.97	24
35000	1110	17.87	12.50	30
40000	1260	31.51	21.08	33
45000	1400	50.50	34.00	33
50000	1550	79.66	52.88	34

By comparing the results based on the analysis in this problem with the EALF values of the Asphalt Institute (AI), it is revealed that the results based on rutting performance provide more reasonable values compared to those based on fatigue performance.

The MS Excel worksheet that is used to perform the computations of this problem in an easy and efficient way is shown in Figure 13.69.

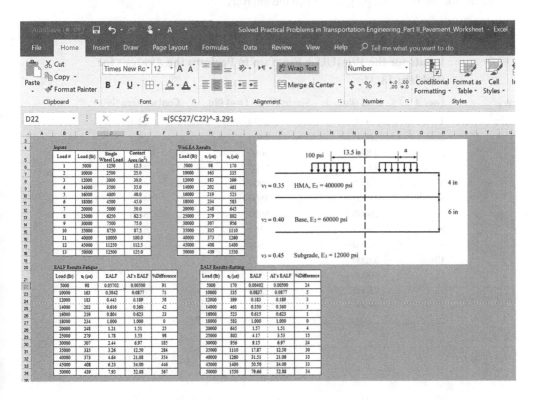

FIGURE 13.69 Image of the MS Excel worksheet used for the computations of Problem 13.30.

13.31 Repeat Problem 13.30 above and perform a similar analysis but for the pavement structure shown in Figure 13.70 where a full-depth asphalt pavement has been used instead of a conventional asphalt pavement.

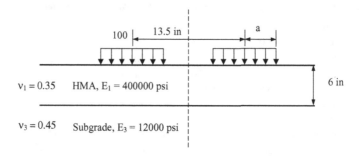

FIGURE 13.70   Dual wheel loading on a full-depth asphalt pavement system for development of EALFs for Problem 13.31.

Solution:

The WinLEA program is used to perform the pavement analysis for critical tensile strain at the bottom of the asphalt layer and critical compressive strain on the top of the subgrade layer.

The single-axle load is applied on a set of dual wheels as shown in the diagram. Therefore, the wheel load is equal to the given load divided by four since the axle has four wheels. The contact area is determined by simply dividing the single wheel load over the applied pressure (100 psi). The results are shown in Table 13.39. These results are used as inputs for the WinLEA program to run the analysis.

TABLE 13.39
Inputs for the WinJULEA Program to Perform the Analysis for Problem 13.31

Load #	Load (lb)	Single Wheel Load (lb)	Contact Area (in²)
1	5000	1250	12.5
2	10000	2500	25.0
3	12000	3000	30.0
4	14000	3500	35.0
5	16000	4000	40.0
6	18000	4500	45.0
7	20000	5000	50.0
8	25000	6250	62.5
9	30000	7500	75.0
10	35000	8750	87.5
11	40000	10000	100.0
12	45000	11250	112.5
13	50000	12500	125.0

The WinLEA pavement response results are shown in Table 13.40.

TABLE 13.40

Pavement Critical Response

(ε_t and ε_c) Using WinJULEA

for Problem 3.31

Load (lb)	ε_t ($\mu\varepsilon$)	ε_c ($\mu\varepsilon$)
5000	86	189
10000	161	355
12000	189	417
14000	215	475
16000	240	532
18000	264	587
20000	288	640
25000	343	765
30000	403	883
35000	460	994
40000	514	1090
45000	566	1180
50000	615	1300

Based on the WinLEA pavement response results, the fatigue and rutting performances are determined using the AI distress models as shown below:

Sample Calculation:

For load $= 5000$ lb,

Based on the AI fatigue distress model, the EALF is calculated as follows:

$$EALF = \left(\frac{\varepsilon_{t-18}}{\varepsilon_{t-x}} \right)^{-3.291}$$

Substituting the tensile strains for $x = 5000$ lb and for the standard 18000 lb-load provides the following:

$$EALF = \left(\frac{\varepsilon_{t-18}}{\varepsilon_{t-5}} \right)^{-3.291}$$

$\Rightarrow$

$$EALF = \left(\frac{264}{86} \right)^{-3.291} = 0.02494$$

And based on the AI rutting distress model, the EALF is calculated as follows:

$$EALF = \left(\frac{\varepsilon_{c-18}}{\varepsilon_{c-x}} \right)^{-4.477}$$

Substituting the vertical compressive strains for x = 5000 lb and for the standard 18000 lb-load provides the following:

$$\text{EALF} = \left(\frac{\varepsilon_{c-18}}{\varepsilon_{c-5}}\right)^{-4.477}$$

⇒

$$\text{EALF} = \left(\frac{587}{189}\right)^{-4.477} = 0.00626$$

The EALF results for all other loads based on fatigue performance and rutting performance are summarized in Tables 13.41 and 13.42, respectively.

TABLE 13.41

EALFs Based on Fatigue Performance for Problem 13.31

Load (lb)	ε_t (µε)	EALF	Al's EALF	% Difference
5000	98	0.02494	0.00500	80
10000	163	0.1964	0.0877	55
12000	183	0.333	0.189	43
14000	202	0.509	0.36	29
16000	219	0.731	0.623	15
18000	234	1.000	1.000	0
20000	248	1.33	1.51	13
25000	279	2.37	3.53	49
30000	307	4.02	6.97	73
35000	335	6.22	12.50	101
40000	373	8.96	21.08	135
45000	408	12.30	34.00	176
50000	439	16.17	52.88	227

TABLE 13.42

EALFs based on Rutting Performance for Problem 13.31

Load (lb)	ε_c (µε)	EALF	Al's EALF	% Difference
5000	170	0.00626	0.00500	20
10000	335	0.1052	0.0877	17
12000	399	0.216	0.189	13
14000	461	0.388	0.360	7
16000	523	0.644	0.623	3
18000	583	1.000	1.000	0
20000	645	1.47	1.51	3
25000	802	3.27	3.53	8
30000	956	6.22	6.97	12
35000	1110	10.57	12.50	18
40000	1260	15.97	21.08	32
45000	1400	22.78	34.00	49
50000	1550	35.15	52.88	50

By comparing the results based on the analysis in this problem with the EALF values of the Asphalt Institute (AI), it can be seen that the results based on rutting performance provide again more reasonable values compared to those based on fatigue performance. However, the EALF values show some improvement when compared with the results for conventional asphalt pavement in the previous problem.

The MS Excel worksheet that is used to perform all the computations in this problem efficiently is shown in Figure 13.71.

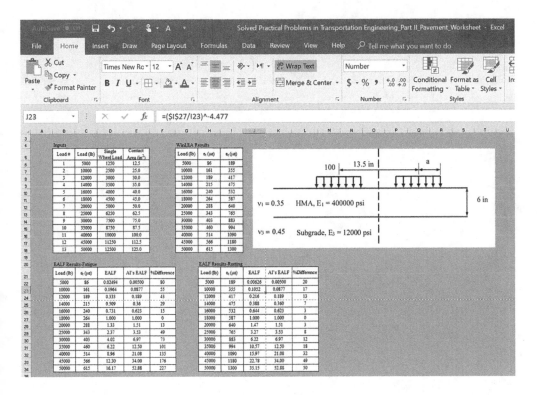

FIGURE 13.71 Image of the MS Excel worksheet used for the computations of Problem 13.31.

14 Design and Analysis of Rigid Pavements

Chapter 14 discusses an important subject in pavement engineering which focuses on the design and analysis of rigid pavements. The behavior of concrete material is different than the behavior of asphalt material in the pavement structure. The main inputs for empirical design of rigid pavements include traffic loading, material properties, reliability, performance indicators, pavement drainage quality, and load transfer coefficient. The properties of the concrete slab considered in the design include the modulus of elasticity and the modulus of rupture; whereas, the property of the subgrade layer under the rigid slab which is considered in the design is the effective modulus of subgrade reaction. The estimation of the effective modulus of subgrade reaction takes into account the monthly values of the soil resilient modulus, the subbase elastic (resilient) modulus and thickness, the depth of the rock foundation beneath the subgrade surface, the subbase loss of support due to erosion, and the projected rigid slab thickness. All these factors are actually taken into consideration in the design procedure of rigid pavements directly or indirectly. Reliability and therefore cost are a crucial element in the design process. The load transfer coefficient is one of the inputs in the design of rigid pavements that accounts for the ability of the concrete slab to transfer the load across joints and cracks. Therefore, the quality of the joint between rigid pavement slabs and the use of load transfer devices and tied pavement shoulders improve the load transfer across the joints for rigid pavements. On the other hand, the analysis of rigid pavements emphasizes the response and the behavior of rigid slabs under traffic loading and temperature changes. Consequently, curling stresses due to temperature and moisture changes will be highlighted. In addition, total bending (curling) stresses in rigid slabs are introduced on the basis of infinite plate assumption. The analysis of stresses and deflections due to traffic loading at critical locations on the concrete slab (interior, corner, and edge) are presented in this section. Single-wheel versus dual-wheel loading is considered as well in the analysis of stresses and deflections due to traffic loading.

14.1 The 20-year design slab thickness for a rigid pavement of a rural highway is 10.0 inches. Determine the ESALs that this rigid pavement can carry over the design life given the following design input data:
- The effective modulus of subgrade reaction = 100 pci
- The mean concrete modulus of rupture = 600 psi
- The concrete modulus of elasticity = 5000000 psi
- The load transfer coefficient = 3.2
- The drainage coefficient = 1.0
- The initial serviceability index = 4.5
- The terminal serviceability index = 2.5
- The reliability value = 95%
- The overall standard deviation = 0.29

Solution:

The following formula is used for the design of rigid pavements following the AASHTO design method:

$$\log_{10} W_{18}$$

$$= Z_R S_0 + 7.35 \log_{10}(D+1) - 0.06 + \frac{\log_{10}\left[\dfrac{\Delta PSI}{(4.5-1.5)}\right]}{1+\left[\dfrac{1.624\times10^7}{(D+1)^{8.46}}\right]}$$

$$+ \left(4.22 - 0.32\, p_t\right) \log_{10}\left\{\frac{S_c' C_D}{215.63 J}\left[\frac{D^{0.75} - 1.132}{\left(D^{0.75} - \dfrac{18.42}{\left[\dfrac{E_c}{k}\right]^{0.25}}\right)}\right]\right\}$$
(14.1)

Where:

W$_{18}$ = predicted number of 18,000-lb (80-kN) single-axle load applications

Z_R = standard normal variant for a given reliability

S_0 = overall standard deviation

D = thickness of rigid pavement (concrete slab) to the nearest half inch

$\Delta PSI = p_i - p_t$

p_i = initial serviceability index

p_t = terminal serviceability index

E_c = elastic modulus of the concrete to be used in construction

k = effective modulus of subgrade reaction

S_c' = modulus of rupture of concrete to be used in construction

J = load transfer coefficient (typical value = 3.2)

C_d = drainage coefficient

For a reliability level (R) of 95%, the Z_R value is −1.645. The S_0 value for rigid pavements is 0.29.

p_i = 4.5 (AASHTO recommended value for rigid pavements)

p_t = 2.5 (AASHTO recommended value for major highways)

$$\Delta PSI = 4.5 - 2.5 = 2.0$$

$\Rightarrow$

$$\log_{10} W_{18}$$

$$= -1.645(0.29) + 7.35\log_{10}(10+1) - 0.06 + \frac{\log_{10}\left[\dfrac{2.0}{(4.5-1.5)}\right]}{1 + \left[\dfrac{1.624 \times 10^7}{(10+1)^{8.46}}\right]}$$

$$+ \left(4.22 - 0.32(2.5)\right)\log_{10}\left\{\frac{600(1)}{215.63(3.2)} \left[\frac{10^{0.75} - 1.132}{\left(10^{0.75} - \dfrac{18.42}{\left[\dfrac{5\times10^6}{100}\right]^{0.25}}\right)}\right]\right\}$$

(14.2)

$$\Rightarrow$$

$$W_{18} = 5904387$$

In other words, the 10-in rigid pavement with the design data given in this problem can carry a traffic level of approximately 5.9 million ESALs over the design life of the pavement. The MS Excel worksheet as shown in Figure 14.1 is used to solve for D in the above equation.

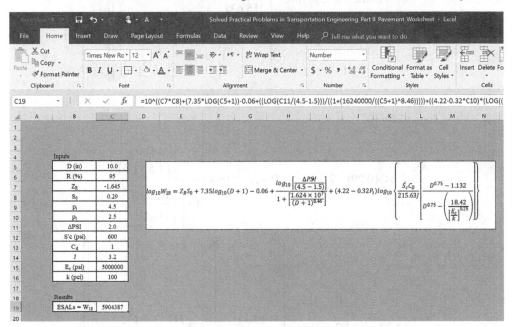

FIGURE 14.1 MS Excel worksheet image for the computations of Problem 14.1.

14.2 Design a rigid pavement for an urban highway to carry a total number of equivalent single-axle loads (ESALs) of 3.6 million over a design life of 20 years given the following input data:

- The effective modulus of subgrade reaction $= 75$ pci
- The mean concrete modulus of rupture $= 650$ psi
- The concrete modulus of elasticity $= 5000000$ psi
- The load transfer coefficient $= 3.2$
- The drainage coefficient $= 1.0$
- The initial serviceability index $= 4.5$
- The terminal serviceability index $= 2.5$
- The reliability value $= 95\%$
- The overall standard deviation $= 0.29$

Solution:

The same formula is used for the design of rigid pavements:

$$\log_{10} W_{18}$$

$$= Z_R S_0 + 7.35 \log_{10}(D+1) - 0.06 + \frac{\log_{10}\left[\dfrac{\Delta PSI}{(4.5-1.5)}\right]}{1+\left[\dfrac{1.624 \times 10^7}{(D+1)^{8.46}}\right]}$$

$$+ \left(4.22 - 0.32 p_t\right) \log_{10} \left\{ \left| \frac{S_c' C_D}{215.63 J} \right| \left[\frac{D^{0.75} - 1.132}{\left(D^{0.75} - \dfrac{18.42}{\left[\dfrac{E_c}{k}\right]^{0.25}}\right)} \right] \right\}$$

$\Rightarrow$

$$\log_{10}\left(3.6 \times 10^6\right)$$

$$= -1.645(0.29) + 7.35 \log_{10}(D+1) - 0.06 + \frac{\log_{10}\left[\dfrac{2.0}{(4.5-1.5)}\right]}{1+\left[\dfrac{1.624 \times 10^7}{(D+1)^{8.46}}\right]}$$

$$+ \left(4.22 - 0.32(2.5)\right) \log_{10} \left\{ \left| \frac{650(1)}{215.63(3.2)} \right| \left[\frac{D^{0.75} - 1.132}{\left(D^{0.75} - \dfrac{18.42}{\left[\dfrac{5 \times 10^6}{75}\right]^{0.25}}\right)} \right] \right\}$$

The MS Excel Solver is used to solve for the thickness of the concrete slab (D) in the above equation as shown in the procedure below:

The left side of the equation is set to:

$$\text{Left Side} = \text{Design ESALs}$$

In this case,

$$\text{Left Side} = \text{Design ESALs} = 3.5 \times 10^6$$

And the right side of the equation is set to:

Right side

$$= 10^{\left(-1.645(0.29)+7.35\log_{10}(D+1)-0.06+\dfrac{\log_{10}\left[\dfrac{2.0}{(4.5-1.5)}\right]}{1+\dfrac{1.624\times10^7}{(D+1)^{8.46}}}+(4.22-0.32(2.5))\log_{10}\left\{\dfrac{650(1)}{215.63(3.2)}\left[\dfrac{D^{0.75}-1.132}{D^{0.75}-\dfrac{18.42}{\left[\dfrac{5\times10^6}{75}\right]^{0.25}}}\right]\right\} \right)}$$

An error is defined as the difference between the left side and the right side, and in this case, the error is defined as follows:

$$\text{Error} = \left[\log(\text{Left Side}) - \log(\text{right side})\right]^2 \qquad (14.3)$$

The objective in Excel Solver is set to the "error" cell. The cell to be changed is the "D" cell. The Excel Solver keeps changing the value of D through an iterative approach until the error value is zero or minimum. That is the numerical solution for the thickness of the pavement (D). Solving the above equation provides the following solution:

$$D = 9.0 \text{ in } (\cong 23 \text{ cm})$$

The MS Excel worksheet used to solve for D is shown in Figure 14.2.

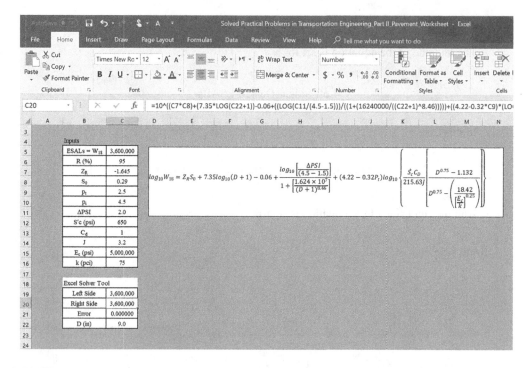

FIGURE 14.2   MS Excel worksheet image for the computations of Problem 14.2.

14.3 Design a rigid pavement (see Figure 14.3) for a new four-lane rural highway that connects two major cities using the AASHTO design procedure to carry the traffic mix shown in Figure 14.4. The elastic modulus (E_c) of the Portland cement concrete (PCC) used for the rigid slab is 5000000 psi and the effective modulus of subgrade reaction (k) of the slab over the subgrade layer and a granular base layer with 6-in thickness is 150 pci.

  The annual average daily traffic (AADT) that is expected to use the highway in both directions is 2500 veh/day in the first year, the traffic growth rate is 2%, and the design life of the pavement is 20 years. Assume that the traffic growth rate will remain the same during the design life of the pavement.

Given Data:

- The mean concrete modulus of rupture = 650 psi
- The load transfer coefficient = 3.2
- The initial serviceability index = 4.5
- The terminal serviceability index = 2.5
- The overall standard deviation = 0.29

  Drainage data: it is expected that it will take one day for water to be removed from the pavement, and the pavement structure will be exposed to moisture levels approaching saturation 30% of the time.

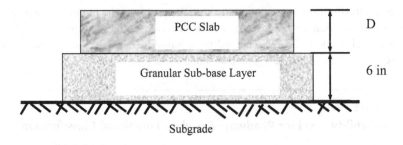

FIGURE 14.3   Designing a rigid pavement over a 6-in granular subbase layer.

Vehicle Type	Percentage, %
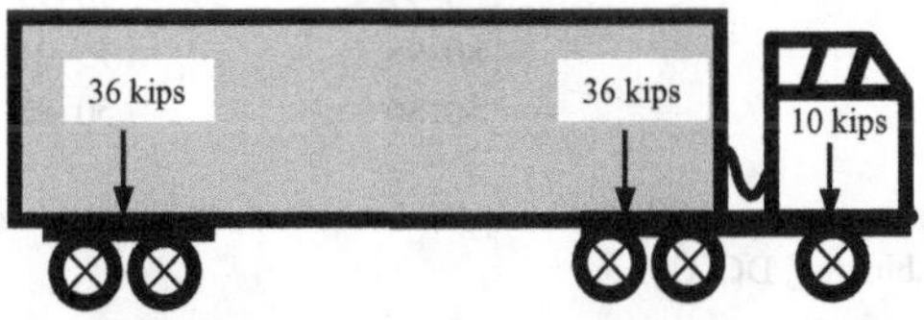 36 kips 36 kips 10 kips	10
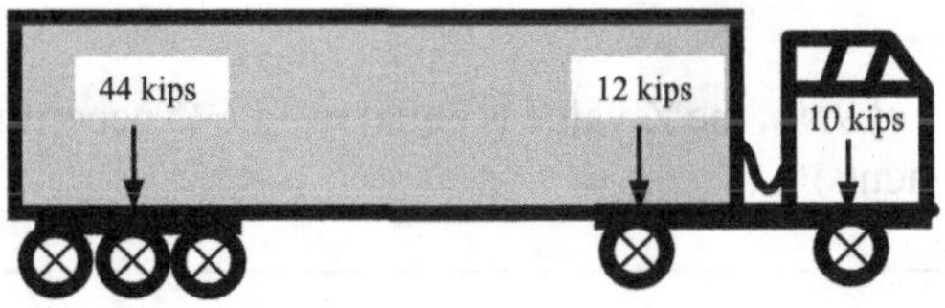 44 kips 12 kips 10 kips	10

FIGURE 14.4   Traffic mix expected to use the given 4-lane rural highway.

Solution:

The following formula is used for the design of rigid pavements following the AASHTO design method:

$$\log_{10} W_{18}$$

$$= Z_R S_0 + 7.35 \log_{10}(D+1) - 0.06 + \frac{\log_{10}\left[\dfrac{\Delta PSI}{(4.5-1.5)}\right]}{1 + \left[\dfrac{1.624 \times 10^7}{(D+1)^{8.46}}\right]}$$

$$+ (4.22 - 0.32 p_t) \log_{10}\left\{ \frac{S'_c C_D}{215.63 J}\left[\frac{D^{0.75} - 1.132}{D^{0.75} - \left(\dfrac{18.42}{\left[\dfrac{E_c}{k}\right]^{0.25}}\right)} \right] \right\}$$

Based on the highway information given in the problem, the highway is a major principal rural highway that serves two main cities. Therefore, the recommended value for the reliability level would be 75–95% according to Table 14.1. The highest value is selected in this case (95%) to be on the safe side (see Table 14.1).

TABLE 14.1
Reliability Level for Roadways Based on Functional Classification

Functional Classification	Recommended Level of Reliability (%)	
	Urban	Rural
Interstate and Other Freeways	85–99.9	80–99.9
Principal Arterials	80–99	75–95
Collectors	80–95	75–95
Local	50–80	50–80

Reproduced with permission from *AASHTO Guide for Design of Pavement Structures*, 1993, by AASHTO, Washington, DC, USA.

For a reliability (R) level of 95%, the Z-value is equal to −1.645 according to Table 14.2. $S_0 = 0.29$ (for rigid pavements)

TABLE 14.2
Standard Normal Deviates (Z- Value) at Various Levels of Reliability, R

Reliability, R (%)	Z_R
50	0.000
60	−0.253
70	−0.524
75	−0.674
80	−0.841
85	−1.037
90	−1.282
95	−1.645
97	−1.881
99	−2.327
99.9	−3.090

Reproduced with permission from *AASHTO Guide for Design of Pavement Structures*, 1993, by AASHTO, Washington, DC, USA.

$$\Delta PSI = p_i - p_t \qquad (14.4)$$

$\Rightarrow$

$$\Delta PSI = 4.5 - 2.5 = 2.0$$

The drainage coefficient is determined from Table 14.3 based on the given drainage data. For the condition "water removed within one day", the quality of drainage is "Good", and for the condition "30% of the time the pavement structure is exposed to moisture levels approaching saturation", the drainage coefficient (C_d) value is 1.00 (Table 14.3). Therefore, $C_d = 1.00$.

TABLE 14.3
Recommended Values of Drainage Coefficients (C_d) for Rigid Pavements

Quality of Drainage		Percentage of time pavement structure is exposed to moisture levels approaching saturation			
Rating	Water removed within	< 1%	1–5%	5–25%	> 25%
Excellent	2 hours	1.25–1.20	1.20–1.15	1.15–1.10	1.10
Good	1 day	1.20–1.15	1.15–1.10	1.10–1.00	1.00
Fair	1 week	1.15–1.10	1.10–1.00	1.00–0.90	0.90
Poor	1 month	1.10–1.00	1.00–0.90	0.90–0.80	0.80
Very Poor	Never drain	1.00–0.90	0.90–0.80	0.80–0.70	0.70

Reproduced with permission from *AASHTO Guide for Design of Pavement Structures*, 1993, by AASHTO, Washington, DC, USA.

Now the total ESALs should be determined as shown in the following procedure:

$$\text{ESAL}_i = (\text{AADT}_i)(365)(f_d)(G)(N_i)(f_{\text{LE}-i}) \tag{14.5}$$

The growth factor is calculated using the formula below:
$r = 2\%$
$n = 20$ years

$$G = \frac{(1+r)^n - 1}{r} \tag{14.6}$$

⇒

$$G = \frac{(1+0.02)^{20} - 1}{0.02} = 24.30$$

The load equivalency factor is determined using the power model developed in this book for single-axle loads for rigid pavements based on numbers in the table of equivalent axle load factors from the *AASHTO Guide for Design of Pavement Structures*, 1993, by AASHTO, Washington, DC, USA, as shown in the following relationship (see Figure 14.5).

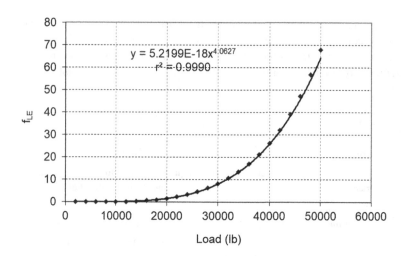

FIGURE 14.5 Load versus load equivalency factor for rigid pavements (Single-axle load, D=9 in, pt=2.5). Based on numbers in the table of equivalent axle load factors from *AASHTO Guide for Design of Pavement Structures*, 1993, by AASHTO, Washington, DC, USA.

The best-fit model that describes the relationship between the load and the load equivalency factor for single-axle loads is a power model with high coefficient of determination (r²) as shown below:

$$f_{LE} = 5.2199 \times 10^{-18} (P)^{4.0627} \tag{14.7}$$

Where:
f_{LE} = load equivalency factor for single-axle loads
P = single-axle load in (lb)

For tandem-axle loads, the load equivalency factor is determined using the power model developed in this book for tandem-axle loads based on numbers in the table of equivalent axle load factors from the *AASHTO Guide for Design of Pavement Structures*, 1993, by AASHTO, Washington, DC, USA, as shown in the following relationships (see Figures 14.6 and 14.7).

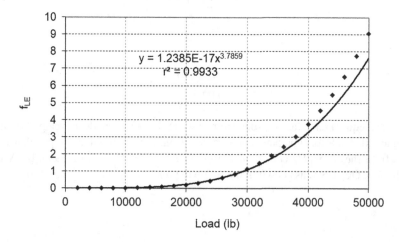

FIGURE 14.6 Load versus load equivalency factor for rigid pavements (Tandem-axle load ≤ 50000 lb, D=9 in, pt=2.5). Based on numbers in the table of equivalent axle load factors from *AASHTO Guide for Design of Pavement Structures*, 1993, by AASHTO, Washington, DC, USA.

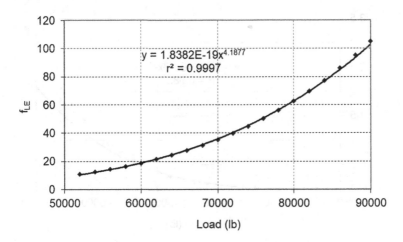

FIGURE 14.7 Load versus load equivalency factor for rigid pavements (Tandem-axle load > 50000 lb, D = 9 in, pt = 2.5). Based on numbers in the table of equivalent axle load factors from *AASHTO Guide for Design of Pavement Structures*, 1993, by AASHTO, Washington, DC, USA.

The best-fit model that describes the relationship between the load and the load equivalency factor for tandem-axle loads is a power model with high coefficient of determination (r^2) as shown below:

For load ≤ 50000 lb:

$$f_{LE} = 1.2385 \times 10^{-17}(P)^{3.7859} \tag{14.8}$$

For load > 50000 lb:

$$f_{LE} = 1.8382 \times 10^{-19}(P)^{4.1877} \tag{14.9}$$

Where:
f_{LE} = load equivalency factor for single-axle loads
P = tandem-axle load in (lb)

For tridem-axle loads, the load equivalency factor is determined using the power model developed in this booluk based on numbers in the table of equivalent axle load factors from the *AASHTO Guide for Design of Pavement Structures*, 1993, by AASHTO, Washington, DC, USA, as shown in the following relationship (see Figures 14.8 and 14.9).

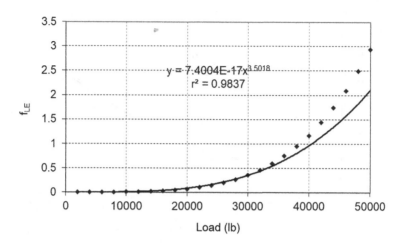

FIGURE 14.8 Load versus load equivalency factor for rigid pavements (Tridem-axle load ≤ 50000 lb, D=9 in, pt=2.5). Based on numbers in the table of equivalent axle load factors from *AASHTO Guide for Design of Pavement Structures*, 1993, by AASHTO, Washington, DC, USA.

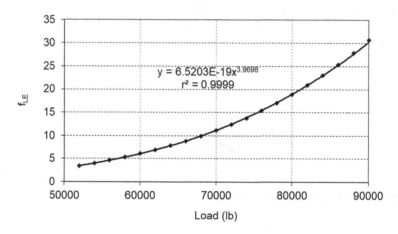

FIGURE 14.9 Load versus load equivalency factor for rigid pavements (Tridem-axle load > 50000, D=9 in, pt=2.5). Based on numbers in the table of equivalent axle load factors from *AASHTO Guide for Design of Pavement Structures*, 1993, by AASHTO, Washington, DC, USA.

The best-fit model that describes the relationship between the load and the load equivalency factor for tridem-axle loads is a power model with high coefficient of determination (r^2) as shown below:

For load ≤ 50000 lb:

$$f_{LE} = 7.4004 \times 10^{-17}(P)^{3.5018} \tag{14.10}$$

For load > 50000 lb:

$$f_{LE} = 6.5203 \times 10^{-19}(P)^{3.9698} \tag{14.11}$$

Where:
f_{LE} = load equivalency factor for single-axle loads
P = tridem-axle load in (lb)

Sample Calculation for the ESALs:
For the 36-kip tandem axle in truck 1:

$$AADT = 0.10\,(2500) = 250 \text{ truck/day}$$

Since the highway is a four-lane highway, the lane distribution factor is 45% (0.45) according to the Asphalt Institute (AI) values in Table 14.4. And based on the AASHTO recommended values in Table 14.5, for two lanes in each direction, $f_d = 80\text{–}100\%$ in one direction. The traffic in this problem is given for two directions; therefore, a value of 0.45 is used for f_d.

TABLE 14.4

Percentage of Total Truck Traffic in Design Lane (f_d)

Number of traffic lanes in two directions	% of Trucks in Design Lane
2	50
4	45 (35–48)
≥ 6	40 (25–48)

Reproduced with permission from *Thickness Design-Asphalt Pavements for Highways and Streets MS-1*, 1999, by Asphalt Institute, Lexington, Kentucky, USA.

TABLE 14.5

Lane Distribution Factor (f_d)

Number of lanes in each direction	% of 18-kip ESAL in design lane
1	100
2	80–100
3	60–80
4	50–75

Reproduced with permission from *AASHTO Guide for Design of Pavement Structures*, 1993, by AASHTO, Washington, DC, USA.

Or:

$f_d = 0.45$

$N = 2$ (since there are two axles with the same type and load; two 36-kip tandem axles)

$f_{LE} = 2.212$ (calculated below)

The power model (for load ≤ 50000 lb) for the load equivalency factor of tandem-axle loads developed above will be used:

$$f_{LE} = 1.2385 \times 10^{-17}(P)^{3.7859}$$

$\Rightarrow$

$$f_{LE} = 1.2385 \times 10^{-17}(36000)^{3.7859} = 2.212$$

$$ESAL_i = (AADT_i)(365)(f_d)(G)(N_i)(f_{LE-i})$$

$\Rightarrow$

$$\text{ESAL}_{36\text{-kip Tandem}} = (250)(365)(0.45)(24.3)(2)(2.212) = 4391737$$

The results of the other axle types are summarized in Table 14.6 based on the same formula. The modulus of subgrade reaction is determined using the following relationship:

TABLE 14.6

ESALs Computations for Problem 14.3

Truck Type	Axle Type	AADT	365	f_d	G	N	f_{LE}	ESALs
	Single	250	365	0.45	24.30	1	0.093	92,783
	Tandem					2	2.201	4,391,737
	Single	250				1	0.093	92,783
	Single					1	0.195	194,606
	Tridem					1	1.348	1,344,939
Total ESALs								6,116,847

$$k = \frac{q}{w_0} \tag{14.12}$$

This relationship is based on an analysis of a plate-bearing test. The ratio of the applied pressure (q) in the plate test and the resulting deflection (w_0) of the plate on a solid foundation is equal to the modulus of subgrade reaction. Therefore:

$$k = \frac{2M_R}{\pi\left(1-v^2\right)a} \tag{14.13}$$

Where:
k = modulus of subgrade reaction
q = applied pressure
w_0 = plate deflection
M_R = resilient modulus of subgrade
v = Poisson ratio of foundation (assumed as 0.45)
a = radius of plate (15 in)

Therefore, by substituting the above values, the formula becomes:

$$k = \frac{M_R}{18.8} \tag{14.14}$$

In this problem, the effective modulus of subgrade reaction that is used in the design formula is provided as 150 pci.

Now using the design formula for rigid pavements shown below:

$$\log_{10} W_{18}$$

$$= Z_R S_0 + 7.35 \log_{10}(D+1) - 0.06 + \frac{\log_{10}\left[\dfrac{\Delta PSI}{(4.5-1.5)}\right]}{1+\left[\dfrac{1.624\times10^7}{(D+1)^{8.46}}\right]}$$

$$+\left(4.22-0.32p_t\right)\log_{10}\left\{\dfrac{S_c'C_D}{215.63J}\left[\dfrac{D^{0.75}-1.132}{\left(D^{0.75}-\dfrac{18.42}{\left[\dfrac{E_c}{k}\right]^{0.25}}\right)}\right]\right\}$$

$$\Rightarrow$$

$$\log_{10}(6116847)$$

$$= (-1.645)(0.29) + 7.35\log_{10}(D+1) - 0.06 + \frac{\log_{10}\left[\dfrac{2.0}{(4.5-1.5)}\right]}{1+\left[\dfrac{1.624\times10^7}{(D+1)^{8.46}}\right]}$$

$$+\left(4.22-0.32\times2.5\right)\log_{10}\left\{\dfrac{650(1.00)}{215.63(3.2)}\left[\dfrac{D^{0.75}-1.132}{\left(D^{0.75}-\dfrac{18.42}{\left[\dfrac{5\times10^6}{150}\right]^{0.25}}\right)}\right]\right\}$$

The MS Excel Solver is used to solve for D in the above equation as shown in the procedure below:

The left side of the equation is set to:

$$\text{Left Side} = \text{Design ESALs}$$

In this case,

$$\text{Left Side} = \text{Design ESALs} = 6116847$$

And the right side of the equation is set to:

Right side

$$= 10^{\left\{(-1.645)(0.29)+7.35\log_{10}(D+1)-0.06+\dfrac{\log_{10}\left[\dfrac{2.0}{(4.5-1.5)}\right]}{1+\dfrac{1.624\times10^7}{(D+1)^{8.46}}}+(4.22-0.32\times2.5)\log_{10}\left[\dfrac{650(1.00)}{215.63(3.2)}\right]\dfrac{D^{0.75}-1.132}{D^{0.75}-\dfrac{18.42}{\left[\dfrac{5\times10^6}{150}\right]^{0.25}}}\right\}}$$

An error is defined as the difference between the left side and the right side, and in this case, the error is defined as follows:

$$\text{Error} = \left[\log\left(\text{Left Side}\right)-\log\left(\text{right side}\right)\right]^2$$

The MS Excel Solver performs an iterative approach to find the numerical solution of the rigid pavement thickness (D). The Solver keeps doing iteration by changing the value of the cell that corresponds to D and targeting the cell that corresponds to the error defined above. When the error is close to zero or reaches a minimum value, the solution converges to the objective D value.

$$D = 9.5 \text{ in } \left(\cong 24 \text{ cm}\right)$$

The MS Excel worksheet used to perform the computations and the solver solution is shown in Figure 14.10.

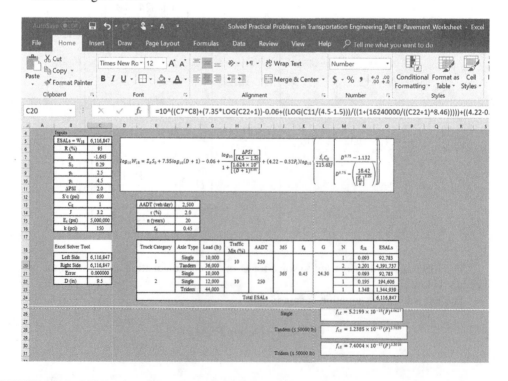

FIGURE 14.10   MS Excel worksheet image for the computations of Problem 14.3.

14.4 The 20-year design slab thickness for a rural highway rigid pavement is 8.0 inches. The truck traffic that is expected to use the highway pavement is of the type shown in Figure 14.11. Determine the ESALs that this rigid pavement can carry and the maximum allowable load that can be carried by the tandem axle shown below given the following design input data (note: ignore the effect of passenger cars):

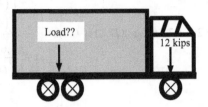

FIGURE 14.11 Type of truck traffic expected to use the highway for Problem 14.4.

- The traffic growth factor = 29.8
- The design lane factor = 0.45
- The 1st year AADT = 200 truck/day
- The effective modulus of subgrade reaction = 180 pci
- The mean concrete modulus of rupture = 620 psi
- The concrete modulus of elasticity = 5,000,000 psi
- The load transfer coefficient = 3.2
- The drainage coefficient = 1.0
- The initial serviceability index = 4.5
- The terminal serviceability index = 2.5
- The reliability value = 95%
- The overall standard deviation = 0.29

Solution:

For a reliability (R) level of 95%, the Z-value is equal to −1.645.

Using the design formula shown below, the ESALs that the rigid pavement can carry can be determined:

$$\log_{10} W_{18}$$

$$= Z_R S_0 + 7.35 \log_{10}(D+1) - 0.06 + \frac{\log_{10}\left[\dfrac{\Delta PSI}{(4.5-1.5)}\right]}{1+\left[\dfrac{1.624\times10^7}{(D+1)^{8.46}}\right]}$$

$$+\left(4.22-0.32 p_t\right)\log_{10}\left\{\frac{S'_c C_D}{215.63J}\left[\frac{D^{0.75}-1.132}{\left(D^{0.75}-\dfrac{18.42}{\left[\dfrac{E_c}{k}\right]^{0.25}}\right)}\right]\right\}$$

Substituting all design inputs into the above equation

$\Rightarrow$

$\log_{10} W_{18}$

$$= (-1.645)(0.29) + 7.35\log_{10}(D+1) - 0.06 + \dfrac{\log_{10}\left[\dfrac{(4.5-2.5)}{(4.5-1.5)}\right]}{1+\left[\dfrac{1.624\times10^7}{(D+1)^{8.46}}\right]}$$

$$+\left(4.22 - 0.32\times2.5\right)\log_{10}\left\{\left[\dfrac{620(1.0)}{215.63(3.2)}\right]\left[\dfrac{D^{0.75} - 1.132}{\left(D^{0.75} - \dfrac{18.42}{\left[\dfrac{5\times10^6}{180}\right]^{0.25}}\right)}\right]\right\}$$

Solving for W_{18} provides the following solution:

$$W_{18} = 5494255 \text{ ESALs}$$

But the estimated number of ESALs based on the AADT is calculated as below:

$$ESAL_i = (AADT_i)(365)(f_d)(G)(N_i)(f_{LE-i})$$

For the 12-kip single axle:
$f_d = 0.45$ (given)
$N = 1$ (since there is only one axle with the same type and load; one 12-kip single axle)
$f_{LE} = 0.195$ (computed below)

The power model developed earlier for the load equivalency factor of single-axle loads will be used to compute f_{LE} for the 12-kip single-axle load as follows:

$$f_{LE} = 5.2199\times10^{-18}(P)^{4.0627}$$

$\Rightarrow$

$$f_{LE} = 2.1159\times10^{-17}(12000)^{3.9120} = 0.195$$

$$ESAL_i = (AADT_i)(365)(f_d)(G)(N_i)(f_{LE-i})$$

$\Rightarrow$

$$ESAL_{12\text{-kip Single}} = (200)(365)(0.45)(29.8)(1)(0.195) = 190943$$

For the other single axle:
$f_d = 0.45$ (given)
$N = 1$
f_{LE} is unknown.

$$ESAL_i = (AADT_i)(365)(f_d)(G)(N_i)(f_{LE-i})$$

$\Rightarrow$

$$ESAL_{x\text{-Single}} = (200)(365)(0.45)(29.8)(1)(f_{LE-x}) = 978930\ f_{LE-x}$$

The predicted number of ESALs based on AADT should be equal to the total number of design ESALs the pavement structure can carry in the design life; therefore:

$$W_{18} = 5494255 = 190943 + 978930\ f_{LE-x}$$

$\Rightarrow$

$$f_{LE-x} = 5.417$$

Using the power model developed earlier for the load equivalency factor of tandem-axle loads for rigid pavements, the tandem-axle load that corresponds to a f_{LE} of 5.417 is determined as shown below:
For load ≤ 50000 lb:

$$f_{LE} = 1.2385 \times 10^{-17}(P)^{3.7859}$$

$\Rightarrow$

$$5.417 = 1.2385 \times 10^{-17}(P)^{3.7859}$$

$\Rightarrow$

$P = 45670\ \text{lb}(\cong 203\ \text{kN}) \leq 50000\ \text{lb}$, which means the power model used above is the correct one.

Therefore, the maximum allowable load that can be carried on the tandem-axle of the truck using the 8-in rigid pavement under the existing conditions is about 46 kips.

The MS Excel worksheet used to determine the ESALs that the rigid pavement can carry and the maximum allowable tandem-axle load of the truck given in this problem is shown in Figure 14.12.

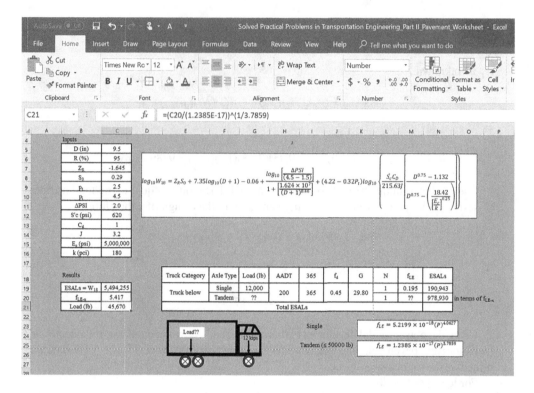

FIGURE 14.12 MS Excel worksheet image for the computations of Problem 14.4.

14.5 A rigid concrete pavement is designed using the AASHTO design procedure to carry a predicted number of 18-kip ESALs equal to 2.6 million for 20 years. The following input values are given:

- The effective modulus of subgrade reaction (k) = 120 pci
- The mean concrete modulus of rupture $(S_c') = 700$ psi
- The concrete modulus of elasticity $(E_c) = 4000000$ psi
- The load transfer coefficient (J) = 3.0
- The drainage coefficient $(C_d) = 0.7$
- The initial serviceability index $(p_i) = 4.5$
- The terminal serviceability index $(p_t) = 2.5$
- Reliability (R) = 90 percent
- Overall standard deviation $(S_0) = 0.30$

Determine the design thickness (D). If the reliability level is 60% instead of 90%, what would be the design thickness of the rigid pavement? What is the reduction in thickness in this case?

Solution:

The following formula is used for the design of rigid pavements following the AASHTO design method:

$$\log_{10} W_{18}$$

$$= Z_R S_0 + 7.35 \log_{10}(D+1) - 0.06 + \dfrac{\log_{10}\left[\dfrac{\Delta PSI}{(4.5-1.5)}\right]}{1+\left[\dfrac{1.624 \times 10^7}{(D+1)^{8.46}}\right]}$$

$$+ (4.22 - 0.32 p_t) \log_{10}\left\{ \dfrac{S'_c C_D}{215.63 J} \left[\dfrac{D^{0.75} - 1.132}{\left(D^{0.75} - \dfrac{18.42}{\left[\dfrac{E_c}{k}\right]^{0.25}} \right)} \right] \right\}$$

For a reliability (R) level of 90%, the Z-value is equal to -1.282.

$$\Delta PSI = p_i - p_t$$

$\Rightarrow$

$$\Delta PSI = 4.5 - 2.5 = 2.0$$

The following design inputs are given:
$S_0 = 0.30$
$C_d = 0.7$
$J = 3.0$
$E_c = 4000000$ psi
$S'_c = 700$ psi
$k = 120$ pci
Design ESALs $= 2.6 \times 10^6$

$\Rightarrow$

$$\log_{10}\left(2.6\times10^6\right)$$

$$= (-1.282)(0.30) + 7.35\log_{10}(D+1) - 0.06 + \frac{\log_{10}\left[\dfrac{2.0}{(4.5-1.5)}\right]}{1+\left[\dfrac{1.624\times10^7}{(D+1)^{8.46}}\right]}$$

$$+\left(4.22-0.32\times2.5\right)\log_{10}\left\{\frac{700(0.7)}{215.63(3.0)}\left[\frac{D^{0.75}-1.132}{\left(D^{0.75}-\dfrac{18.42}{\left[\dfrac{4\times10^6}{120}\right]^{0.25}}\right)}\right]\right\}$$

The MS Excel Solver is used to solve for D in the above equation; the solution obtained using the Excel Solver tool is:

$$D = 9.0 \text{ in } (\cong 23 \text{ cm})$$

If the reliability level is changed from 90% to 60%, only the Z_R used in the design formula will be different (in this case, $Z_R = -0.253$). Substituting the design inputs into the formula provides:

$$\log_{10}\left(2.6\times10^6\right)$$

$$= (-0.253)(0.30) + 7.35\log_{10}(D+1) - 0.06 + \frac{\log_{10}\left[\dfrac{2.0}{(4.5-1.5)}\right]}{1+\left[\dfrac{1.624\times10^7}{(D+1)^{8.46}}\right]}$$

$$+\left(4.22-0.32\times2.5\right)\log_{10}\left\{\frac{700(0.7)}{215.63(3.0)}\left[\frac{D^{0.75}-1.132}{\left(D^{0.75}-\dfrac{18.42}{\left[\dfrac{4\times10^6}{120}\right]^{0.25}}\right)}\right]\right\}$$

Using the MS Excel Solver tool, this provides a solution for D:

$$D = 8.0 \text{ in } (\cong 20.5 \text{ cm})$$

The reduction in the design thickness is 1.0 inch and the percentage of the reduction is estimated as:

$$\text{Reduction in } D = 100 \times \frac{(9-8)}{9} \cong 11\%$$

The MS Excel worksheet used to perform the computations and the solver solution is shown in Figure 14.13.

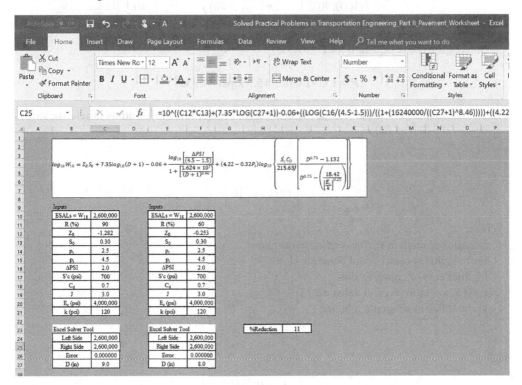

FIGURE 14.13 MS Excel worksheet image for the computations of Problem 14.5.

14.6 An interior circular loading with a radius of 6 inches is applied on a concrete slab with a radius of relative stiffness of 40 inches. The modulus of subgrade reaction (k) of the slab is 100 pci, the modulus of concrete is 4000000 psi, and the Poisson ratio is 0.15. If the maximum deflection due to this loading is 0.012 inch, determine the interior loading and the thickness of the slab.

Solution:

Westergaard formula for the deflection due to interior loading is used to solve for the interior loading in this problem as shown below:

$$\Delta_i = \frac{P}{8kl^2}\left\{1+\frac{1}{2\pi}\left[\ln\left(\frac{a}{2l}\right)-0.673\right]\left[\frac{a}{l}\right]^2\right\} \qquad (14.15)$$

Where:

Δ_i = deflection in the interior of the slab
P = applied load
k = modulus of subgrade reaction
l = radius of relative stiffness of concrete slab
a = radius of contact area

$\Rightarrow$

$$0.012 = \frac{P}{8(100)(40)^2}\left\{1+\frac{1}{2\pi}\left[\ln\left(\frac{6}{2(40)}\right)-0.673\right]\left[\frac{6}{40}\right]^2\right\}$$

Solving the above equation for P provides:

$$P = 15541.6 \text{ lb} \cong 15.5 \text{ kips} \left(\cong 69 \text{ kN}\right)$$

The thickness of the slab is determined from the formula of the radius of relative stiffness of the slab as follows:

$$l = \left[\frac{Eh^3}{12\left(1-v^2\right)k}\right]^{0.25} \tag{14.16}$$

Where:

l = radius of relative stiffness of concrete slab
E = modulus of elasticity of concrete slab
h = thickness of concrete slab
v = Poisson ratio (typical value = 0.15)
k = modulus of subgrade reaction

$\Rightarrow$

$$40 = \left[\frac{\left(4\times10^6\right)h^3}{12\left(1-(0.15)^2\right)100}\right]^{0.25}$$

Solving the above equation for h $\Rightarrow$

$$h = 9.1 \text{ in} \left(\cong 23 \text{ cm}\right)$$

The MS Excel worksheet is used to perform the computations of this problem in a rapid and efficient way as shown in Figure 14.14.

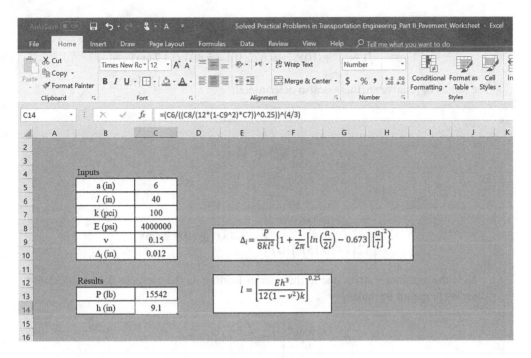

FIGURE 14.14   MS Excel worksheet image for the computations of Problem 14.6.

14.7   An interior circular loading of 9 kips with a contact radius of 6 inches is applied on a
10-inch concrete slab with a modulus of subgrade reaction (k) of 100 pci. The modulus of
concrete (E) is 4000000 psi, and the Poisson ratio is 0.15. Determine the maximum stress
and deflection due to this loading.

Solution:

Westergaard formula for the stress due to interior loading is used as shown below:

$$\sigma_i = \frac{0.316P}{h^2}\left[4\log\left(\frac{l}{b}\right)+1.069\right] \tag{14.17}$$

Where:
σ_i = tensile stress in the interior of the slab
P = applied load
h = thickness of the slab
l = radius of relative stiffness of concrete slab

$$b = a \quad \text{if} \quad a \geq 1.724\,h \tag{14.18}$$

$$b = \left[\sqrt{1.6a^2 + h^2}\right] - 0.675\,h \quad \text{if} \quad a < 1.724\,h \tag{14.19}$$

Where:
a = radius of contact area
Since $a = 6$ in $< 1.724\,h = 17.24$ in, therefore, the value of b is determined as below:

$$b = \left[\sqrt{1.6a^2 + h^2}\right] - 0.675\,h$$

$\Rightarrow$

$$b = \left[\sqrt{1.6(6)^2 + (10)^2} \right] - 0.675(10) = 5.804 \ (14.7 \text{ cm})$$

The radius of relative stiffness is also determined:

$$l = \left[\frac{Eh^3}{12\left(1 - v^2\right)k} \right]^{0.25}$$

$\Rightarrow$

$$l = \left[\frac{\left(4 \times 10^6\right)\left(10\right)^3}{12\left(1 - (0.15)^2\right)100} \right]^{0.25} = 43.0 \text{ in } (109.2 \text{ cm})$$

Now the maximum stress due to the interior loading on the concrete slab is calculated using Westergaard formula:

$$\sigma_i = \frac{0.316P}{h^2}\left[4\log\left(\frac{l}{b}\right) + 1.069 \right]$$

$\Rightarrow$

$$\sigma_i = \frac{0.316(9000)}{(10)^2}\left[4\log\left(\frac{43.0}{5.804}\right) + 1.069 \right] = 129.3 \text{ psi } (892 \text{ kPa})$$

The deflection due to the interior loading on the concrete slab is computed using Westergaard formula for deflection as follows:

$$\Delta_i = \frac{P}{8kl^2}\left\{ 1 + \frac{1}{2\pi}\left[\ln\left(\frac{a}{2l}\right) - 0.673 \right]\left[\frac{a}{l}\right]^2 \right\}$$

$\Rightarrow$

$$\Delta_i = \frac{9000}{8(100)(43.0)^2}\left\{ 1 + \frac{1}{2\pi}\left[\ln\left(\frac{6}{2(43.0)}\right) - 0.673 \right]\left[\frac{6}{43.0}\right]^2 \right\}$$

$$= 0.006 \text{ in } (0.015 \text{ cm})$$

The MS Excel worksheet used to perform the computations in this problem to determine the maximum stress and deflection due to the interior loading on the concrete slab is shown in Figure 14.15.

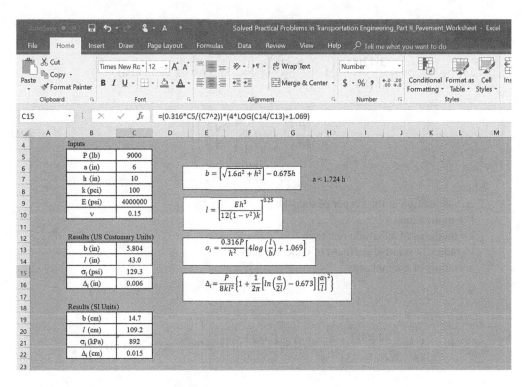

FIGURE 14.15 MS Excel worksheet image for the computations of Problem 14.7.

14.8 A concrete slab is subjected to a corner circular loading of 12 kips. The thickness of the slab is 8 inches and the modulus of subgrade reaction (k) = 100 pci. If the maximum stress due to this loading is 220 psi, the radius of the circular area is 6 inches, determine the maximum stress and deflection due to the corner loading (assume the elastic modulus of concrete is 4000000 psi and the Poisson ratio is 0.15).

Solution:

$$l = \left[\frac{Eh^3}{12\left(1 - v^2\right)k} \right]^{0.25}$$

⇒

$$l = \left[\frac{\left(4 \times 10^6\right)\left(8\right)^3}{12\left(1 - \left(0.15\right)^2\right)100} \right]^{0.25} = 36.35 \text{ in } (92.3 \text{ cm})$$

Westergaard formula for the maximum stress due to corner loading is used as shown below:

$$\sigma_c = \frac{3P}{h^2} \left[1 - \left(\frac{a\sqrt{2}}{l} \right)^{0.6} \right] \tag{14.20}$$

Where:
σ_c = stress at the corner of the slab
P = applied load
h = thickness of concrete slab
a = radius of contact area
l = radius of relative stiffness of concrete slab

$\Rightarrow$

$$\sigma_c = \frac{3(12000)}{(8)^2}\left[1-\left(\frac{6\sqrt{2}}{36.35}\right)^{0.6}\right] = 327.5 \text{ psi } (\cong 2258 \text{ kPa})$$

$$\Delta_c = \frac{P}{kl^2}\left[1.1-0.88\left(\frac{a\sqrt{2}}{l}\right)\right] \qquad (14.21)$$

Where:

Δ_c = deflection at the corner of the slab
P = applied load
k = modulus of subgrade reaction
l = radius of relative stiffness of concrete slab
a = radius of contact area

$\Rightarrow$

$$\Delta_c = \frac{12000}{100(36.35)^2}\left[1.1-0.88\left(\frac{6\sqrt{2}}{36.35}\right)\right] = 0.0812 \text{ in } (0.206 \text{ cm})$$

Using MS Excel worksheets makes the solution always easy and fast. The MS Excel worksheet used to perform the computations in this problem to determine the maximum stress and deflection due to the corner loading on the concrete slab is shown in Figure 14.16.

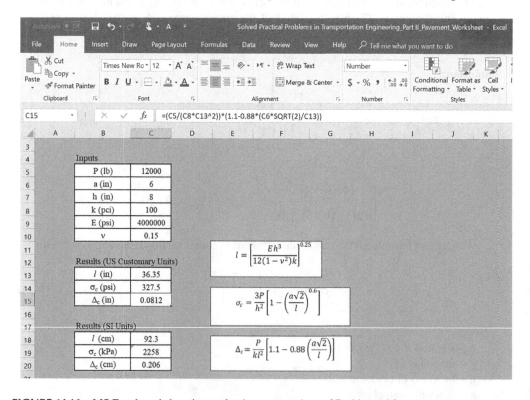

FIGURE 14.16   MS Excel worksheet image for the computations of Problem 14.8.

14.9 A corner circular loading with a contact radius of 4 inches is applied on a concrete slab with a modulus of subgrade reaction (k) of 150 pci creating a corner deflection of 0.0155 inch. The modulus of concrete is 4000000 psi, and the Poisson ratio is 0.15. If the radius of relative stiffness (*l*) for the concrete slab is 44.5 inches, determine the thickness of the slab, the corner loading, and the corner stress due to this loading.

Solution:

$$l = \left[\frac{Eh^3}{12(1-v^2)k} \right]^{0.25}$$

⇒

$$44.5 = \left[\frac{(4\times10^6)h^3}{12(1-(0.15)^2)150} \right]^{0.25}$$

Solving the above equation for h provides the following solution:

$$h = 12.0 \text{ in } (\cong 30 \text{ cm})$$

Westergaard formula for corner deflection is used to determine the applied load:

$$\Delta_c = \frac{P}{kl^2} \left[1.1 - 0.88 \left(\frac{a\sqrt{2}}{l} \right) \right]$$

⇒

$$0.0155 = \frac{P}{150(44.5)^2} \left[1.1 - 0.88 \left(\frac{4\sqrt{2}}{44.5} \right) \right]$$

⇒

$$P = 4659 \text{ lb } (\cong 20.7 \text{ kN})$$

The corner stress is calculated using Westergaard formula for corner stress:

$$\sigma_c = \frac{3P}{h^2} \left[1 - \left(\frac{a\sqrt{2}}{l} \right)^{0.6} \right]$$

⇒

$$\sigma_c = \frac{3(4659)}{(12)^2} \left[1 - \left(\frac{4\sqrt{2}}{44.5} \right)^{0.6} \right] = 69.0 \text{ psi } (\cong 476 \text{ kPa})$$

The MS Excel worksheet used to perform the computations in this problem makes the solution rapid and efficient as shown in Figure 14.17.

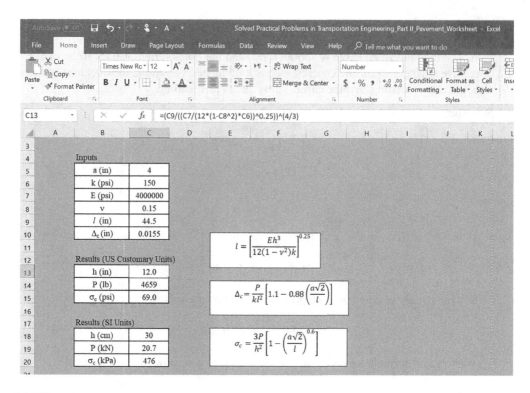

FIGURE 14.17 MS Excel worksheet image for the computations of Problem 14.9.

14.10 A concrete slab with a modulus of subgrade reaction (k) of 120 pci is subjected to an edge circular loading with a radius of 6 inches resulting in a deflection of 0.012 inch. If the modulus of elasticity of the concrete is 4000000 psi, the Poisson ratio is 0.15, and the radius of relative stiffness of the concrete slab is 40 inches, determine the applied edge loading, the thickness of the slab, and the stress due to the edge loading.

Solution:

Westergaard formula for deflection due to edge circular loading is used:

$$\Delta_{\text{e-circle}} = \frac{0.431P}{kl^2}\left[1 - 0.82\left(\frac{a}{l}\right)\right]$$ (14.22)

Where:
   $\Delta_{\text{e-circle}}$ = deflection due to edge circular loading
   P = applied load
   k = modulus of subgrade reaction
   l = radius of relative stiffness of concrete slab
   a = radius of contact area

$\Rightarrow$

$$0.012 = \frac{0.431P}{120(40)^2}\left[1 - 0.82\left(\frac{6}{40}\right)\right]$$

$\Rightarrow$

$$P = 6095 \text{ lb } (\cong 27 \text{ kN})$$

$$l = \left[\frac{Eh^3}{12(1-v^2)k} \right]^{0.25}$$

$$\Rightarrow$$

$$40 = \left[\frac{(4 \times 10^6)h^3}{12(1-(0.15)^2)120} \right]^{0.25}$$

Solving the above equation for h provides the following solution:

$$h = 9.7 \text{ in } (\cong 25 \text{ cm})$$

Westergaard formula for stress due to edge circular loading is used to calculate the edge stress on the concrete slab:

$$\sigma_{\text{e-circle}} = \frac{0.803P}{h^2} \left[4\log\left(\frac{l}{a}\right) + 0.666\left(\frac{a}{l}\right) - 0.034 \right] \tag{14.23}$$

Where:
$\sigma_{\text{e-circle}}$ = deflection due to edge circular loading
P = applied load
h = thickness of concrete slab
l = radius of relative stiffness of concrete slab
a = radius of contact area

$$\Rightarrow$$

$$\sigma_{\text{e-circle}} = \frac{0.803(6095)}{(9.7)^2} \left[4\log\left(\frac{40}{6}\right) + 0.666\left(\frac{6}{40}\right) - 0.034 \right]$$

$$= 176.4 \text{ psi}$$

$$(\cong 1216 \text{ kPa})$$

The MS Excel worksheet used to perform the computations related to this problem to provide a fast and efficient solution is shown in Figure 14.18.

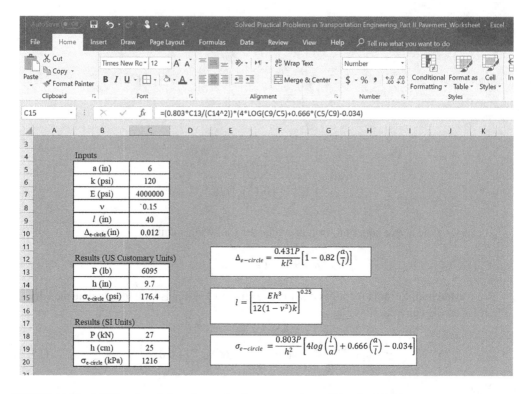

FIGURE 14.18   MS Excel worksheet image for the computations of Problem 14.10.

14.11 If the same loading in Problem 14.10 is applied at the corner of the concrete slab, determine the resulting stress in this case. What is the reduction in the stress between edge and corner?

Solution:

$$\sigma_c = \frac{3P}{h^2}\left[1 - \left(\frac{a\sqrt{2}}{l}\right)^{0.6}\right]$$

$\Rightarrow$

$$\sigma_c = \frac{3(6095)}{(9.7)^2}\left[1 - \left(\frac{6\sqrt{2}}{40}\right)^{0.6}\right] = 117.7 \text{ psi}$$

The reduction in the stress between the edge and the corner is equal to:

Reduction in Stress = 176.4 − 117.7 = 58.7 psi

The percentage of reduction in the stress between the edge and the corner is:

$$\%\text{Reduction in Stress} = 100 \times \frac{58.7}{176.4} \cong 33\%$$

The MS Excel worksheet shown in Figure 14.19 is used to perform the computations in this problem.

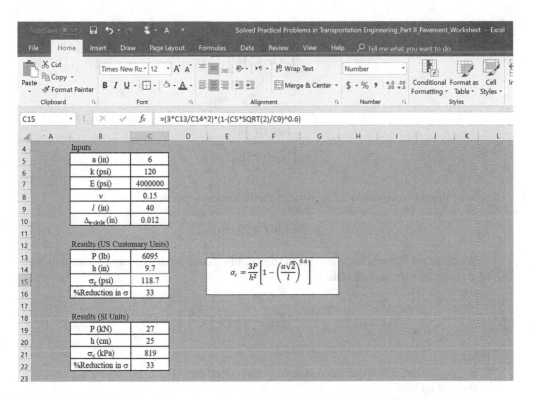

FIGURE 14.19 MS Excel worksheet image for the computations of Problem 14.11.

14.12 In Problem 14.10, if the edge loading is a semi-circular loading, determine the applied edge loading, the thickness of the slab, and the stress due to the edge loading.

Solution:

Westergaard formula for the deflection due to edge semi-circular loading is used:

$$\Delta_{\text{e-semicircle}} = \frac{0.431P}{kl^2}\left[1 - 0.349\left(\frac{a}{l}\right)\right] \tag{14.24}$$

Where:

$\Delta_{\text{e-semicircle}}$ = deflection due to edge semi-circular loading
P = applied load
k = modulus of subgrade reaction
l = radius of relative stiffness of concrete slab
a = radius of contact area

$\Rightarrow$

$$0.012 = \frac{0.431P}{120(40)^2}\left[1 - 0.349\left(\frac{6}{40}\right)\right]$$

$\Rightarrow$

$$P = 5641 \text{ lb } (\cong 25 \text{ kN})$$

$$l = \left[\frac{Eh^3}{12\left(1-v^2\right)k} \right]^{0.25}$$

$\Rightarrow$

$$40 = \left[\frac{\left(4\times10^6\right)h^3}{12\left(1-(0.15)^2\right)120} \right]^{0.25}$$

Solving the above equation for h provides the following solution:

$$h = 9.7 \text{ in } (\cong 25 \text{ cm})$$

Westergaard formula for stress due to edge semi-circular loading is used to calculate the edge stress on the concrete slab:

$$\sigma_{\text{e-semicircle}} = \frac{0.803P}{h^2}\left[4\log\left(\frac{l}{a}\right) + 0.282\left(\frac{a}{l}\right) + 0.650 \right] \tag{14.25}$$

Where:

$\sigma_{\text{e-semicircle}}$ = deflection due to edge semi-circular loading
P = applied load
h = thickness of concrete slab
l = radius of relative stiffness of concrete slab
a = radius of contact area

$\Rightarrow$

$$\sigma_{\text{e-circle}} = \frac{0.803(6095)}{(9.7)^2}\left[4\log\left(\frac{40}{6}\right) + 0.282\left(\frac{6}{40}\right) + 0.650 \right] = 193.7 \text{ psi}$$

$$\left(\cong 1335 \text{ kPa}\right)$$

The MS Excel worksheet shown in Figure 14.20 is used to perform the computations related to this problem to provide a fast and efficient solution.

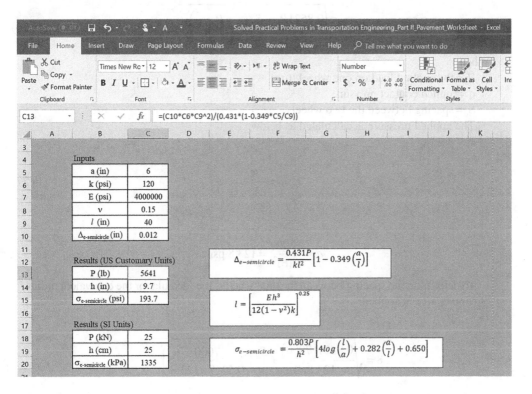

FIGURE 14.20 MS Excel worksheet image for the computations of Problem 14.12.

14.13 A concrete slab is subjected to a corner loading of 14 kips on a set of dual wheels as shown in Figure 14.21. The thickness of the slab is 8 inches. If the stress due to this loading is 338.4 psi, the equivalent radius of the contact area for the dual wheels is 8 inches, and the radius of a single circular contact area is 6 inches, determine the spacing between the dual wheels from center to center, the radius of relative stiffness of the concrete slab, and the modulus of subgrade reaction.

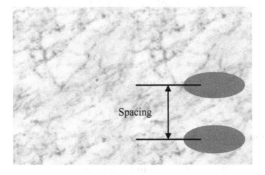

FIGURE 14.21 A concrete slab subjected to 14-kip corner loading on dual wheels.

Solution:

When the load is applied on a set of dual wheels, an equivalent radius of contact area is determined using the formula below:

$$a = \sqrt{\frac{0.8521 P_d}{q\pi} + \frac{S_d}{\pi}\left(\frac{P_d}{0.5227q}\right)^{0.5}} \qquad (14.26)$$

Where:

P_d = the applied load on one wheel (tire)

q = applied contact pressure

S_d = spacing between the two wheels

The applied pressure is calculated as below:

$$q = \frac{P}{A_c} \qquad (14.27)$$

$\Rightarrow$

$$q = \frac{14000}{\pi(6)^2} = 123.8 \text{ psi}$$

To determine the spacing between the dual wheels, the formula for the equivalent radius of contact area is applied:

$\Rightarrow$

$$8 = \sqrt{\frac{0.8521(7000)}{123.8\pi} + \frac{S_d}{\pi}\left(\frac{7000}{0.5227 \times 123.8}\right)^{0.5}}$$

Solving for S_d provides the following solution:

$\Rightarrow$

$$S_d = 14.7 \text{ in } (\cong 37 \text{ cm})$$

Since the stress at the corner of the slab is given, the radius of relative stiffness is determined using the stress formula as shown below:

$$\sigma_c = \frac{3P}{h^2}\left[1 - \left(\frac{a\sqrt{2}}{l}\right)^{0.6}\right]$$

$\Rightarrow$

$$338.4 = \frac{3(14000)}{(8)^2}\left[1 - \left(\frac{8\sqrt{2}}{l}\right)^{0.6}\right]$$

Solving this equation for l provides the following solution:

$$l = 37.9 \text{ in } (\cong 96 \text{ cm})$$

To determine the modulus of subgrade reaction, the formula for the radius of relative stiffness is applied:

$$l = \left[\frac{Eh^3}{12(1-v^2)k}\right]^{0.25}$$

$\Rightarrow$

$$37.9 = \left[\frac{4 \times 10^6 (8)^3}{12\left(1-(0.15)^2\right)k}\right]^{0.25}$$

Solving the above equation for k provides the solution below:

$$k \cong 85 \text{ pci } (\cong 23 \text{ MPa/m})$$

The MS Excel worksheet (see Figure 14.22) is used to perform the computations related to this problem to provide a fast and efficient solution particularly back-calculations are needed to determine the required parameters in this problem.

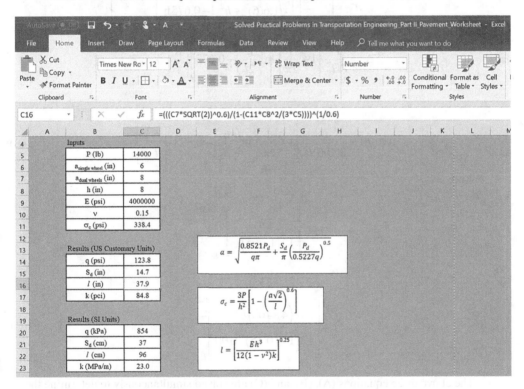

FIGURE 14.22 MS Excel worksheet image for the computations of Problem 14.13.

14.14 An interior circular loading with a contact radius of 6 inches is applied on a concrete slab with a modulus of subgrade reaction (k) of 100 pci. The modulus of concrete is 4000000 psi, and the Poisson ratio is 0.15. If the maximum stress and deflection due to this loading are 259.0 psi and 0.0120 inch, respectively, determine the thickness and the radius of relative stiffness of the concrete slab, and the interior loading.

Solution:

Westergaard formulas for stress and deflection due to interior loading are applied:

$$\sigma_i = \frac{0.316P}{h^2}\left[4\log\left(\frac{l}{b}\right)+1.069\right]$$

a=6 in, assume that a=6 in<1.724 h, therefore, the value of b is determined as below:

$$b = \left[\sqrt{1.6a^2 + h^2} \right] - 0.675h$$

$\Rightarrow$

$$b = \left[\sqrt{1.6(6)^2 + h^2} \right] - 0.675h$$

$\Rightarrow$

$$259.0 = \frac{0.316P}{h^2} \left[4\log \left(\frac{l}{\left[\sqrt{1.6(6)^2 + h^2} \right] - 0.675h} \right) + 1.069 \right] \quad \text{(A)}$$

$$\Delta_i = \frac{P}{8kl^2} \left\{ 1 + \frac{1}{2\pi} \left[\ln \left(\frac{a}{2l} \right) - 0.673 \right] \left[\frac{a}{l} \right]^2 \right\}$$

$\Rightarrow$

$$0.0120 = \frac{P}{8(100)l^2} \left\{ 1 + \frac{1}{2\pi} \left[\ln \left(\frac{6}{2l} \right) - 0.673 \right] \left[\frac{6}{l} \right]^2 \right\} \quad \text{(B)}$$

And:

$$l = \left[\frac{Eh^3}{12\left(1-v^2\right)k} \right]^{0.25}$$

$\Rightarrow$

$$l = \left[\frac{\left(4 \times 10^6\right)h^3}{12\left(1-(0.15)^2\right)100} \right]^{0.25} \quad \text{(C)}$$

The above three equations (A), (B), and (C) are solved simultaneously to determine the values of h, l, and P. This task is very difficult (or impossible) to do manually. Therefore, the MS Excel Solver tool is used to obtain the solution using iterative approach as shown in Figure 14.23.

$$h = 9.6 \text{ in } (\cong 24 \text{ cm})$$

$$l = 41.7 \text{ in } (\cong 106 \text{ cm})$$

$$P = 16838 \text{ lb } (\cong 75 \text{ kN})$$

The MS Excel worksheet with the Solver tool set up are shown in Figure 14.23.

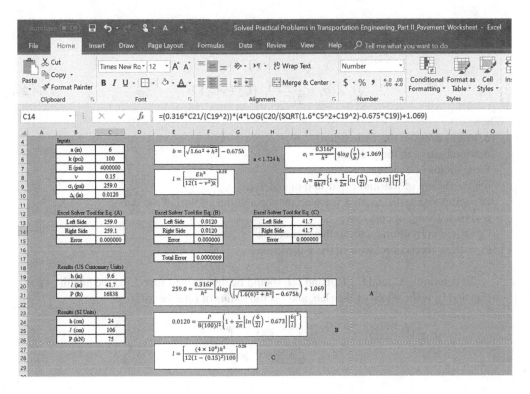

FIGURE 14.23 MS Excel worksheet image for the computations of Problem 14.14.

14.15 A 20 ft × 10 ft × 8 in-concrete slab with a modulus of subgrade reaction (k) of 100 pci is subjected to a temperature differential of 25°F. Determine the maximum curling stress in the interior and at the edge of the slab if the coefficient of thermal expansion of concrete (α_t) is 5×10^{-6} in/in/°F, the elastic modulus of concrete is 4×10^6 psi, and the Poisson ratio is 0.15 (see Figure 14.24).

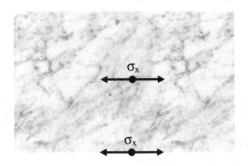

FIGURE 14.24 A concrete slab subjected to a temperature differential of 25°F.

Solution:

The curling stresses are explained by theory of plate on a Winkler or liquid foundation expressed by a series of springs under the rigid slab.

The curling stresses in the interior of a finite slab are determined using the following formulas:

$$\sigma_x = \frac{E\alpha_t \Delta t}{2(1-v^2)}\left(C_x + vC_y\right) \tag{14.28}$$

$$\sigma_y = \frac{E\alpha_t \Delta t}{2(1-v^2)}\left(C_y + vC_x\right) \tag{14.29}$$

Where:

σ_x = total stress in the x-direction
σ_y = total stress in the y-direction
E = modulus of concrete
α_t = coefficient of thermal expansion of concrete
Δt = temperature differential between the top and the bottom of the slab
v = Poisson ratio of concrete
C_x = stress correction factor in the x-direction
C_y = stress correction factor in the y-direction

To determine C_x and C_y, L_x/l and L_y/l should be determined first. L_x is the length of the slab and L_y is the width of the slab, and l is the radius of relative stiffness.

$$l = \left[\frac{Eh^3}{12(1-v^2)k}\right]^{0.25}$$

$\Rightarrow$

$$l = \left[\frac{4 \times 10^6 (8)^3}{12\left(1-(0.15)^2\right)100}\right]^{0.25} = 36.7 \text{ in}$$

$\Rightarrow$

$$\frac{L_x}{l} = \frac{20 \times 12}{36.7} = 6.5$$

$$\frac{L_y}{l} = \frac{10 \times 12}{36.7} = 3.3$$

The following figure is reproduced based on numbers from the chart of the stress correction factors in *Pavement Analysis and Design* by Yang H. Huang, 2004. It provides the stress corrections factors for finite slabs. The fifth-order polynomial is developed in this book based on numbers from the stress correction factors chart shown below with a high coefficient of determination (see Figure 14.25).

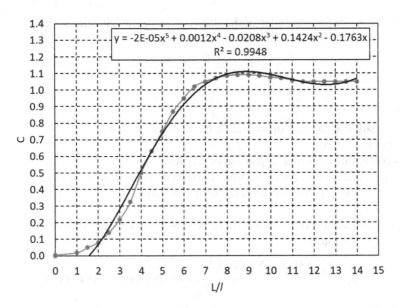

FIGURE 14.25 Stress correction factors for concrete slab. Huang, Yang H., *Pavement Analysis and Design*, 2nd Ed., ©2004. Reprinted by permission of Pearson Education, Inc., New York, New York, USA

From the figure or from the developed polynomial model, C_x and C_y are determined using the values L_x/l and L_y/l, respectively:

$$C_x = 1.07$$

$$C_y = 0.36$$

$$\sigma_x = \frac{E\alpha_t \Delta t}{2(1-v^2)}\left(C_x + vC_y\right)$$

$\Rightarrow$

$$\sigma_x = \frac{\left(4\times10^6\right)\left(5\times10^{-6}\right)(25)}{2\left(1-(0.15)^2\right)}\left(1.07 + 0.15(0.36)\right) = 287.5 \text{ psi } (\cong 1982 \text{ kPa})$$

$$\sigma_y = \frac{E\alpha_t \Delta t}{2(1-v^2)}\left(C_y + vC_x\right)$$

$\Rightarrow$

$$\sigma_y = \frac{\left(4\times10^6\right)\left(5\times10^{-6}\right)(25)}{2\left(1-(0.15)^2\right)}\left(0.36 + 0.15(1.07)\right) = 133.1 \text{ psi } (\cong 918 \text{ kPa})$$

Therefore, the maximum curling stress in the interior of the slab is equal to 287.5 psi ($\cong$ 1982 kPa).

The curling stress at the edge of the slab is given by the following formula:

$$\sigma_{x \text{ or } y} = \frac{C_{x \text{ or } y} E \alpha_t \Delta t}{2}$$

(14.30)

Where:

$\sigma_{x \text{ or } y}$ = stress at the edge of the slab in the x- or y-direction

E = modulus of concrete

α_t = coefficient of thermal expansion of concrete

Δt = temperature differential between the top and the bottom of the slab

$C_{x \text{ or } y}$ = stress correction factor in the x- or y-direction

In this case, the maximum stress at the edge is σ_x since C_x is higher than C_y. Therefore, the edge stress is equal to:

$$\sigma_x = \frac{(1.07)(4 \times 10^6)(5 \times 10^{-6})(25)}{2} = 267.5 \text{ psi } (\cong 1844 \text{ kPa})$$

14.16 Determine the curling (bending) stresses in the interior of the slab in Problem 14.15 if the concrete slab is assumed to be an infinite plate.

Solution:

In case of an infinite plate, the total stress in the x-direction or y-direction in the interior of the slab is given by the following formula:

$$\sigma_{x \text{ or } y} = \frac{E \alpha_t \Delta t}{2(1-v)}$$

(14.31)

Where:

$\sigma_{x \text{ or } y}$ = stress in the interior of infinite slab in the x- or y-direction

E = modulus of concrete

α_t = coefficient of thermal expansion of concrete

Δt = temperature differential between the top and the bottom of the slab

v = Poisson ratio of concrete

$\Rightarrow$

$$\sigma_{x \text{ or } y} = \frac{(4 \times 10^6)(5 \times 10^{-6})(25)}{2(1-0.15)} = 294.1 \text{ psi } (\cong 2028 \text{ kPa})$$

Multiple Choice Questions and Answers for Chapter 6
Terminology, Concepts, and Theory

1. The R-value in the Hveem stabilometer test of a soil sample is defined as:
   a. The ratio of lateral pressure to the vertical pressure
   b. **The ratio of the applied vertical pressure to the resulting lateral pressure**
   c. The ratio of the applied vertical pressure to the developed lateral deformation
   d. The ratio of the lateral deformation to the vertical deformation
   e. The ratio of the vertical deformation to the developed lateral pressure
2. The R-value in the Hveem stabilometer test of a soil sample is defined as:
   a. The ratio of lateral pressure to the vertical pressure
   b. **The ratio of the applied vertical pressure to the resulting lateral pressure**
   c. The ratio of the applied vertical pressure to the developed lateral deformation
   d. The ratio of the lateral deformation to the vertical deformation
   e. The ratio of the vertical deformation to the developed lateral pressure
3. One of the following statements is correct:
   a. Granular materials are called stress-softening materials due to the proportional relationship between the applied stress and the resulting resilient modulus
   b. **Granular materials are called stress-hardening materials due to the proportional relationship between the applied stress and the resulting resilient modulus**
   c. Fine-grained materials are stress-hardening materials due to the fact that the applied deviator stress is inversely related to the resulting resilient modulus
   d. Fine-grained materials are stress-hardening materials due to the fact that the applied deviator stress is proportionally related to the resulting resilient modulus
   e. Granular materials and fine-grained materials are both stress-hardening materials
4. For fine-grained soil materials, as the deviator stress increases:
   a. The resilient modulus increases
   b. The resilient modulus keeps constant
   c. The resilient modulus increases for a while and then decreases
   d. The resilient modulus decreases for a while and then increases
   e. **The resilient modulus decreases to a relatively low value and then stabilizes at that value**
5. In calculating the resilient modulus for unbound granular materials:
   a. **The recoverable part of the material strain is considered**
   b. The plastic part of the material strain is considered
   c. The total strain of the materials is considered
   d. None of the above
   e. The recoverable part or the plastic part of the material strain is considered
6. The relationship between the applied stress and the resilient modulus for unbound granular materials is:
   a. Linear
   b. Nonlinear

   c. Inverse

   d. Proportional

   e. **b and d**

7. The standard penetration loading rate in the California Bearing Ratio (CBR) test is 0.05 in/min (1.3 mm/min). If this rate is to be changed to a higher value, what would the new CBR value be:

   a. Same as old one

   b. lower

   c. **Higher**

   d. Unknown

   e. Dependent on the new rate value

8. The following pavement structure is called:

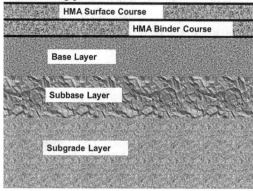

   a. Composite pavement

   b. **Conventional asphalt pavement**

   c. Full-depth asphalt pavement

   d. Rigid pavement

   e. Contained-rock asphalt mat (CRAM) pavement

9. The pavement structure shown below represents a:

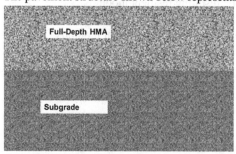

   a. Composite pavement

   b. Conventional asphalt pavement

   c. **Full-depth asphalt pavement**

   d. Rigid pavement

   e. Contained-rock asphalt mat (CRAM) pavement

10. The structure shown below is called:

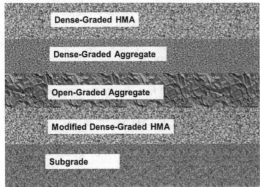

a. Composite pavement
b. Conventional asphalt pavement
c. Full-depth asphalt pavement
d. Rigid pavement
e. **Contained-rock asphalt mat (CRAM) pavement**

11. One of the following is <u>not</u> among the types of flexible pavements:
a. Conventional asphalt pavement
b. Full-depth asphalt pavement
c. Contained-rock asphalt mat (CRAM) pavement
d. **Jointed-plain concrete pavement (JPCP)**
e. None of the above

12. One of the following is not among the types of rigid pavements:
a. Jointed-plain concrete pavement (JPCP)
b. Jointed reinforced concrete pavement (JRCP)
c. Continuous reinforced concrete pavement (CRCP)
d. Prestressed concrete pavement (PCP)
e. **Full-depth asphalt pavement**

13. The thin asphalt surface treatment that is applied on a flexible asphalt pavement surface layer is called:
a. Tack coat
b. Prime coat
c. **Seal coat**
d. Surface course
e. Binder course

14. The very light asphalt application that is applied between the binder course and the surface course of a flexible pavement is called:
a. **Tack coat**
b. Prime coat
c. Seal coat
d. Surface course
e. Binder course

15. The application of a low-viscosity cutback asphalt between the granular base and the asphalt layer of a flexible pavement is called:
a. Tack coat
b. **Prime coat**
c. Seal coat
d. Surface course
e. Binder course

16. One of the following is <u>not</u> among the advantages of using full-depth asphalt pavements:
    a. No entrapped water into the pavement structure due to the absence of permeable granular layers
    b. Construction time required is reduced
    c. **Less cost compared to conventional asphalt pavement**
    d. They are less affected by moisture or frost
    e. Little or no reduction in subgrade strength due to the fact that moisture is not built up in subgrade underneath this type of pavement
17. One of the advantages of contained-rock asphalt mat (CRAM) pavements is:
    a. Construction time required is reduced
    b. They have extended construction seasons
    c. They are less affected by moisture
    d. No water is entrapped into the pavement
    e. **They are capable of controlling surface water by the open-graded aggregate layer**
18. Steel reinforcement in jointed reinforced concrete pavements (JRCP) is used to:
    a. Increase load capacity
    b. Increase joint spacing
    c. Reduce spalling
    d. Keep cracks tied together
    e. **b and d**
19. Steel reinforcement in jointed reinforced concrete pavements (JRCP) is used to:
    a. Increase load capacity
    b. Increase joint spacing
    c. Reduce spalling
    d. Keep cracks tied together
    e. **b and d**
20. Joints are used in jointed concrete pavements to:
    a. **Minimize cracks in concrete slabs**
    b. Reduce spalling and faulting
    c. Increase load capacity
    d. Keep cracks tied together
    e. None of the above
21. For prestressed concrete pavements (PCP), one of the following statements is valid:
    a. The joint spacing is increased significantly compared to typical concrete pavements
    b. The required thickness is reduced due to the increase in load capacity of the pavement
    c. The required thickness is the same as typical concrete pavements
    d. All of the above
    e. **a and b**
22. A composite pavement consists of:
    a. Portland cement concrete (PCC) slab over hot-mix asphalt (HMA) pavement
    b. PCC slab over another PCC slab
    c. HMA layer directly over PCC pavement
    d. HMA layer over thick granular materials over PCC pavement
    e. **c or d**
23. Granular materials used for pavement design are defined as those containing:
    a. **35% or less material passing No. 200 (75 μm) sieve**
    b. More than 35% material passing No. 200 (75 μm) sieve
    c. 50% or less material passing No. 200 (75 μm) sieve
    d. More than 50% material passing No. 200 (75 μm) sieve
    e. 35% or less material passing No. 4 (4.75 mm) sieve

24. Clay and silt-clay materials used for pavement design are those containing:
    a. 35% or less material passing No. 200 (75 µm) sieve
    b. **More than 35% material passing No. 200 (75 µm) sieve**
    c. 50% or less material passing No. 200 (75 µm) sieve
    d. More than 50% material passing No. 200 (75 µm) sieve
    e. 35% or less material passing No. 4 (4.75 mm) sieve
25. The ratio of the pressure for a soil material to that of a standard crushed rock (typically 1,000 and 1,500 psi at 0.1 and 0.2 in-penetration, respectively) in a standard penetration test is called:
    a. Resilient modulus
    b. R-value
    c. **California bearing ratio (CBR)**
    d. Shear strength
    e. Cohesion
26. The R-value in the Hveem stabilometer test of a soil sample is defined as:
    a. The ratio of lateral pressure to the vertical pressure
    b. **The ratio of the applied vertical pressure to the resulting lateral pressure**
    c. The ratio of the applied vertical pressure to the developed lateral deformation
    d. The ratio of the lateral deformation to the vertical deformation
    e. The ratio of the vertical deformation to the developed lateral pressure
27. One of the following statements is correct:
    a. Granular materials are called stress-softening materials due to the proportional relationship between the applied stress and the resulting resilient modulus
    b. **Granular materials are called stress-hardening materials due to the proportional relationship between the applied stress and the resulting resilient modulus**
    c. Fine-grained materials are stress-hardening materials due to the fact that the applied deviator stress is inversely related to the resulting resilient modulus
    d. Fine-grained materials are stress-hardening materials due to the fact that the applied deviator stress is proportionally related to the resulting resilient modulus
    e. Granular materials and fine-grained materials are both stress-hardening materials
28. For fine-grained soil materials, as the deviator stress increases:
    a. The resilient modulus increases
    b. The resilient modulus keeps constant
    c. The resilient modulus increases for a while and then decreases
    d. The resilient modulus decreases for a while and then increases
    e. **The resilient modulus decreases to a relatively low value and then stabilizes at that value**
29. For fine-grained soil materials, as the deviator stress increases:
    a. The resilient modulus increases
    b. The resilient modulus keeps constant
    c. The resilient modulus increases for a while and then decreases
    d. The resilient modulus decreases for a while and then increases
    e. **The resilient modulus decreases to a relatively low value and then stabilizes at that value**
30. In calculating the resilient modulus for unbound granular materials:
    a. **The recoverable part of the material strain is considered**
    b. The plastic part of the material strain is considered
    c. The total strain of the materials is considered
    d. None of the above
    e. The recoverable part or the plastic part of the material strain is considered

31. The relationship between the applied stress and the resilient modulus for unbound granular materials is:
    a. Linear
    b. Nonlinear
    c. Inverse
    d. Proportional
    e. **b and d**
32. The standard penetration loading rate in the California Bearing Ratio (CBR) test is 0.05 in/min (1.3 mm/min). If this rate is to be changed to a higher value, what would the new CBR value be:
    a. Same as old one
    b. Lower
    c. **Higher**
    d. Unknown
    e. Dependent on the new rate value
33. The relationship between resilient modulus and CBR is:
    a. **Linear**
    b. Nonlinear
    c. Inverse
    d. Exponential
    e. b and d
34. In a modified Proctor test compared to a standard Proctor test for soil materials:
    a. The compactive effort is higher
    b. The falling height is higher
    c. The number of blows applied is more
    d. The number of layers is five compared to three
    e. **a, b, and d**
35. The moisture content at the maximum dry density in a modified Proctor test compared to a standard Proctor test is:
    a. Higher
    b. **Lower**
    c. Similar
    d. Dependent on material properties
    e. None of the above
36. Granular bases are used in pavements to:
    a. Increase structural load capacity of pavements
    b. Prevent pumping in rigid pavements
    c. Prevent volume changes in subgrade
    d. Protect against frost action
    e. **All of the above**
37. The ratio of width to thickness of an aggregate particle is called:
    a. Elongation
    b. **Flatness**
    c. Angularity
    d. Fractured face
    e. None of the above
38. The ratio of length to width of an aggregate particle is called:
    a. **Elongation**
    b. Flatness
    c. Angularity
    d. Fractured face
    e. None of the above

39. Flat and elongated (F&E) coarse aggregate particles in asphalt mixtures can lead to:
    a. Reduction in voids in mineral aggregate (VMA)
    b. Mixture instability
    c. Rutting
    d. Shoving
    e. **All of the above**
40. In Superpave, the amount of coarse aggregate particles having fractured (rough) faces is called:
    a. Fine aggregate angularity (FAA)
    b. Flat and elongated (F&E) particles
    c. **Coarse aggregate angularity (CAA)**
    d. Sand equivalent
    e. Coarse aggregate toughness
41. Aggregate angularity is desired in asphalt paving mixtures to:
    a. Reduce fatigue cracking
    b. Minimize thermal cracking
    c. Ensure adequate aggregate interlock
    d. Prevent excessive permanent deformation
    e. **c and d**
42. One of the following aggregate properties does <u>not</u> affect the uncompacted void content of fine aggregate:
    a. Shape
    b. Surface texture
    c. Angularity
    d. Gradation
    e. **Color**
43. The property that measures the amount of clay material contained in fine aggregate is:
    a. Flatness
    b. **Sand equivalent**
    c. Angularity
    d. Texture
    e. Toughness
44. Clay material is <u>not</u> desired in fine aggregate used for asphalt paving mixtures because it leads to:
    a. Fatigue cracking
    b. Bleeding
    c. Thermal cracking
    d. High stiffness
    e. **Stripping and moisture damage**
45. One of the following statements is a valid statement about the two asphalt binders: AC-30 and AC-10:
    a. AC-30 has higher penetration than AC-10
    b. **AC-30 has higher viscosity than AC-10**
    c. AC-30 has lower high-performance grade (PG) than AC-10
    d. AC-30 has equal viscosity as AC-10
    e. None of the above
46. One of the following statements is a valid statement about the two asphalt binders: AC-30 and AC-10:
    a. AC-30 has higher penetration than AC-10
    b. **AC-30 has higher viscosity than AC-10**

   c. AC-30 has lower high-performance grade (PG) than AC-10

   d. AC-30 has equal viscosity as AC-10

   e. None of the above

47. Absolute viscosity of asphalt binder in Poises is converted to kinematic viscosity of asphalt binder in Stokes by dividing it by:

   a. A factor of 1000

   b. A factor of 10

   c. **The density of asphalt binder in g/cm³**

   d. The density of water in g/cm³

   e. None of the above

48. The following are common grades of asphalt residue from the Rotating or Rolling Thin-Film Oven (RTFO) test:

   a. AR5, AR10, AR20, AR40

   b. AR1, AR2, AR3, AR4

   c. 40/50, 60/70, 85/100, 120/150

   d. AC5, AC10, AC20, AC40

   e. **AR1000, AR2000, AR4000, AR8000, AR16000**

49. The standard test temperature and time for the Rolling Thin-Film Oven (RTFO) test are:

   a. **163°C and 75 minutes**

   b. 135°C and 75 minutes

   c. 163°C and 60 minutes

   d. 135°C and 60 minutes

   e. 60°C and 75 minutes

50. The RTFO test simulates:

   a. The long-term aging that happens during the service life of the asphalt pavement

   b. **The short-term aging that occurs in asphalt binders during mixing and lay-down conditions**

   c. Aging that occurs during the first two years of the asphalt pavement's service life

   d. Aging that occurs during the first five years of the asphalt pavement's service life

   e. The long-term aging that happens during 10 years of the asphalt pavement's service life

51. The Pressure Aging Vessel (PAV) test simulates:

   a. The long-term aging that happens after 20 years of the asphalt pavement's service life

   b. The short-term aging that occurs in asphalt binders during mixing and lay-down conditions

   c. Aging that occurs during the first two years of the asphalt pavement's service life

   d. Aging that occurs during the first year of the asphalt pavement's service life

   e. **The long-term aging that happens after 5 to 10 years of the asphalt pavement's service life**

52. The standard testing conditions of the PAV test are:

   a. 300 psi air pressure, 90, 100, or 110°C temperature, and 75-min. time

   b. 100 psi air pressure, 90, 100, or 110°C temperature, and 20-hour time

   c. 300 psi air pressure, 163°C temperature, and 20-hour time

   d. **300 psi air pressure, 90, 100, or 110°C temperature, and 20-hour time**

   e. 300 psi air pressure, 135°C temperature, and 20-hour time

53. The rotational viscosity is determined for asphalt binders at standard conditions that include:

   a. Temperature = 135°C, rotational speed = 5 rpm

   b. **Temperature = 135°C, rotational speed = 20 rpm**

   c. Temperature = 25°C, rotational speed = 20 rpm

   d. Temperature = 60°C, rotational speed = 10 rpm

   e. Temperature = 60°C, rotational speed = 20 rpm

54. The rotational viscosity for asphalt binders simulates:
    a. Workability of asphalt binders at service temperatures
    b. Behavior of asphalt binders at high temperatures for rutting
    c. **Workability of asphalt binders at mixing and laydown temperatures**
    d. Performance of asphalt binders at warm temperatures for fatigue cracking
    e. Behavior of asphalt binders at low temperatures for thermal cracking
55. According to the Superpave system, the maximum limit for the rotational viscosity of asphalt binders is:
    a. 3000 Pa.s
    b. **3 Pa.s**
    c. 3000 Poise
    d. 1 Pa.s
    e. 1000 cPoise
56. The standard loading frequency that the Dynamic Shear Rheometer (DSR) applies on asphalt binders in a standard Superpave DSR test is:
    a. 0.1 Hz
    b. 1 Hz
    c. 2 Hz
    d. **1.59 Hz**
    e. 10 Hz
57. The radius of a standard DSR asphalt binder sample is:
    a. 5.0 mm
    b. **12.5 mm or 4.0 mm based on aging level and temperature**
    c. 12.5 mm
    d. 4.0 mm
    e. 10.0 mm
58. The height (thickness) of a standard DSR asphalt binder sample is:
    a. 1.0 mm
    b. **1.0 mm or 2.0 mm based on aging level and temperature**
    c. 2.0 mm
    d. 0.5 mm
    e. 3.0 mm
59. The time lag between the maximum (amplitude) applied shear stress and the maximum (amplitude) resulting shear strain in a DSR test is called:
    a. Shift factor
    b. Maximum applied torque
    c. Rotational angle
    d. Complex shear modulus
    e. **Phase angle**
60. The elastic part of the DSR complex shear modulus (G*) value is given by:
    a. **|G*| cos δ**
    b. |G*| sin δ
    c. |G*| tan δ
    d. |G*|/cos δ
    e. |G*|/sin δ
61. The viscous part of the DSR complex shear modulus (G*) value is given by:
    a. |G*| cos δ
    b. **|G*| sin δ**
    c. |G*| tan δ
    d. |G*|/cos δ
    e. |G*|/sin δ

62. The rutting parameter according to the Superpave system is known to be:
    a. $|G^*|\cos\delta$
    b. $|G^*|\sin\delta$
    c. $|G^*|\tan\delta$
    d. $|G^*|/\cos\delta$
    e. **$|G^*|/\sin\delta$**

63. The fatigue parameter according to the Superpave system is known to be:
    a. $|G^*|\cos\delta$
    b. **$|G^*|\sin\delta$**
    c. $|G^*|\tan\delta$
    d. $|G^*|/\cos\delta$
    e. $|G^*|/\sin\delta$

64. To verify the value of the rutting parameter for asphalt binders, the standard Superpave DSR test is applied on the asphalt binder sample under the following aging conditions:
    a. Nonaged (fresh) asphalt binder
    b. Rolling Thin-Film Oven (RTFO)-aged asphalt binder
    c. Pressure Aging Vessel (PAV)-aged asphalt binder
    d. b or c
    e. **a or b**

65. To verify the value of the fatigue parameter for asphalt binders, the standard Superpave DSR test is applied on the asphalt binder sample under the following aging conditions:
    a. None-aged (fresh) asphalt binder
    b. Rolling Thin-Film Oven (RTFO)-aged asphalt binder
    c. **Pressure Aging Vessel (PAV)-aged asphalt binder**
    d. b or c
    e. a or b

66. The Superpave system specifies the following value for $|G^*|/\sin\delta$ of asphalt binders:
    a. **Minimum of 1.0 kPa for fresh asphalt binders and minimum of 2.2 kPa for RTFO-aged asphalt binders**
    b. Maximum of 5000 kPa
    c. Minimum of 5 kPa
    d. Maximum of 300 MPa
    e. Minimum of 1000 kPa

67. The Superpave system specifies the following value for $|G^*|\sin\delta$ of asphalt binders:
    a. Minimum of 1.0 kPa for fresh asphalt binders and minimum of 2.2 kPa for RTFO-aged asphalt binders
    b. **Maximum of 5000 kPa**
    c. Minimum of 5 kPa
    d. Maximum of 300 MPa
    e. Minimum of 1000 kPa

68. The Superpave Bending Beam Rheometer (BBR) test is conducted on asphalt binders under the following aging conditions:
    a. Original (unaged) asphalt binders
    b. RTFO-aged asphalt binders
    c. **PAV-aged asphalt binders**
    d. 2-hour RTFO-aged asphalt binders
    e. 5-hour RTFO-aged asphalt binders

69. In a standard Superpave BBR test, the asphalt binder beam is subjected to a small creep load over 240 seconds; the amount of the load is:
    a. **0.981 N**
    b. 100 N

    c. 2 N
    d. 10 N
    e. 5 N

70. The creep stiffness [S(t)] of the asphalt binder beam in a standard Superpave BBR test as a function of time (t) is best fitted by the following relationship:

    a. $S(t) = A + Bt + Ct^2$
    b. $S(t) = A\log(t)$
    c. $S(t) = A + B\log(t) + C\big[\log(t)\big]^2$
    d. $S(t) = A + B\log(t)$
    e. $S(t) = At^B$

71. The BBR test simulates asphalt binder stiffness under the following conditions:

    a. After 10 hours of loading at the minimum asphalt pavement design temperature
    b. **After 2 hours of loading at the minimum asphalt pavement design temperature**
    c. After 20 hours of loading at the minimum asphalt pavement design temperature
    d. After 1.25 hours of loading at the minimum asphalt pavement design temperature
    e. After 5 hours of loading at the minimum asphalt pavement design temperature

72. The rate of change of BBR creep stiffness with time (m-value) is related to:

    a. The ability of the asphalt pavement to shrink
    b. The ability of the asphalt pavement to relieve tensile stresses at high temperatures
    c. The ability of the asphalt pavement to relieve tensile stresses at intermediate temperatures
    d. The ability of the asphalt pavement to relieve compressive stresses
    e. **The ability of the asphalt pavement to relieve thermal stresses (tensile stresses at low temperatures)**

73. The Superpave asphalt binder specifications require the following values for creep stiffness (S) and the m-value at 60 seconds:

    a. $S \leq 300$ MPa, m-value ≤ 0.3
    b. $S \leq 600$ MPa, m-value ≤ 0.3
    c. $S \geq 300$ MPa, m-value ≤ 0.3
    d. $S \geq 600$ MPa, m-value ≥ 0.3
    e. **$S \leq 300$ MPa, m-value ≥ 0.3**

74. The Direct Tension (DT) test is used in the Superpave system because:

    a. Creep stiffness as measured by the BBR test does not predict thermal cracking
    b. m-value as measured by the BBR test is not enough to predict thermal cracking
    c. It is a replacement test for the BBR test
    d. **Creep stiffness as measured by the BBR test is not sufficient to predict thermal cracking in some asphalt binders with high creep stiffness**
    e. m-value as measured by the BBR test is not related to thermal cracking

75. The Superpave DT test is used for asphalt binders having m-value of at least 0.3 and creep stiffness in the range of:

    a. ≤ 600 MPa
    b. ≥ 300 MPa
    c. **300–600 MPa**
    d. $S \leq 300$ MPa
    e. $S \leq 500$ MPa

76. In the Superpave DT test, the strain at failure is defined as the elongation of the asphalt binder sample at failure divided by:

    a. Length of the sample
    b. **Effective gage length**
    c. Gage length

    d. Original cross-sectional area

    e. 100

77. The specifications of the Superpave system require a minimum value for the failure strain from the DT test of:

    a. **1%**

    b. 2%

    c. 3%

    d. 4%

    e. 5%

78. The Superpave DT test is conducted on asphalt binders under the following aging conditions:

    a. Original (unaged) asphalt binders

    b. RTFO-aged asphalt binders

    c. **PAV-aged asphalt binders**

    d. 2-hour RTFO-aged asphalt binders

    e. 5-hour RTFO-aged asphalt binders

79. The standard Superpave DT test is performed using a loading rate of:

    a. 5 cm/minute

    b. 2.5 cm/minute

    c. 5 mm/minute

    d. 2 mm/minute

    e. **1 mm/minute**

80. The Superpave Gyratory Compactor (SGC) is believed to:

    a. Simulate Marshall compaction

    b. Simulate compaction under heavy traffic

    c. **Simulate compaction that takes place in field**

    d. Simulate compaction under rubber rollers

    e. Simulate compaction under steel rollers

81. The SGC initial number of gyrations ($N_{initial}$) is used to measure:

    a. **Mixture compactability during construction**

    b. Mixture density similar to that in the field after the design number of single-axle loads

    c. Mixture density that should never be exceeded in the field

    d. Mixture density after five years of the asphalt pavement's service life

    e. Mixture density after two years of the asphalt pavement's service life

82. The SGC design number of gyrations (N_{design}) is used to measure:

    a. Mixture compactability during construction

    b. **Mixture density similar to that in the field after the design traffic loadings**

    c. Mixture density that should never be exceeded in the field

    d. Mixture density after five years of the asphalt pavement's service life

    e. Mixture density after two years of the asphalt pavement's service life

83. The SGC maximum number of gyrations ($N_{maximum}$) is used to measure:

    a. Mixture compactability during construction

    b. Mixture density similar to that in the field after the design number of single-axle loads

    c. **Mixture density that should never be exceeded in the field**

    d. Mixture density after five years of the asphalt pavement's service life

    e. Mixture density after two years of the asphalt pavement's service life

84. The SGC is done to:

    a. Simulate field densification under traffic and climatic conditions

    b. Measure compactability

    c. Be favorable to quality control (QC)

    d. Accommodate large aggregates

    e. **All of the above**

85. The standard dimensions of the SGC specimens for mix design are:
    a. 150 mm in diameter and 100 mm in height
    b. 100 mm in diameter and 115 mm in height
    c. **150 mm in diameter and 115 mm in height**
    d. 100 in diameter and 50 mm in height
    e. 150 mm in diameter and 50 mm in height
86. The standard values of the SGC inputs: the vertical pressure, the angle of gyration, and the rate of gyration are, respectively:
    a. 300 kPa, 1.15°, and 30 gyrations/minute
    b. 300 kPa, 1.25°, and 30 gyrations/minute
    c. 100 kPa, 1.25°, and 60 gyrations/minute
    d. **600 kPa, 1.25°, and 30 gyrations/minute**
    e. 600 kPa, 1.00°, and 60 gyrations/minute
87. The appropriate SGC operation mode that should be used to prepare specimens for mix design purposes is:
    a. **"number of gyrations" mode**
    b. "height" mode
    c. "angle of gyration" mode
    d. "rate of gyration" mode
    e. Any of the above
88. The appropriate SGC operation mode that should be used to prepare specimens for research purposes or further performance testing is:
    a. "number of gyrations" mode
    b. **"height" mode**
    c. "angle of gyration" mode
    d. "rate of gyration" mode
    e. Any of the above
89. The design number of gyrations (N_{design}) in the SGC depends on:
    a. The minimum air temperature and the design traffic equivalent single-axle loads (ESALs)
    b. The average design high air temperature
    c. The minimum air temperature
    d. The design traffic ESALs
    e. **The average design high air temperature and the design ESALs**
90. When the SGC initial number of gyrations ($N_{initial}$) is below a specific level, the asphalt mixture tends to be:
    a. **Tender**
    b. Stiff
    c. Rutting resistant
    d. Fatigue resistant
    e. Properly designed
91. The SGC maximum number of gyrations ($N_{maximum}$) is given as a function of N_{design} by the following formula:
    a. $\log N_{maximum} = 1.10 \log N_{design}$
    b. $\log N_{maximum} = 0.45 \log N_{design}$
    c. $\log N_{maximum} = 0.80 \log N_{design}$
    d. $\log N_{maximum} = 1.30 \log N_{design}$
    e. $\log N_{maximum} = 2.20 \log N_{design}$
92. The SGC initial number of gyrations ($N_{initial}$) is given as a function of N_{design} by the following formula:
    a. $\log N_{maximum} = 1.10 \log N_{design}$
    b. $\log N_{maximum} = 0.45 \log N_{design}$

    c. $\log N_{maximum} = 0.80 \log N_{design}$
    d. $\log N_{maximum} = 1.30 \log N_{design}$
    e. $\log N_{maximum} = 2.20 \log N_{design}$

93. The aging requirements for the SGC loose asphalt mixtures prior to mixing and compaction are:
    a. Short-term aging of samples is done for 4 hours at 135°C in a forced-draft oven
    b. Short-term aging is done for samples for 2 hours at 135°C in a forced-draft oven
    c. Long-term aging is done for samples for 10 hours in a forced-draft oven
    d. **Short-term aging is done for samples for 2 hours at the equiviscous compaction temperature in a forced-draft oven**
    e. Long-term aging is done for samples for 20 hours in a forced-draft oven

94. The Superpave system specifies the following values for the % G_{mm} (percent maximum specific gravity or mixture density) at $N_{initial}$ and $N_{maximum}$ when the traffic level ≥ 3 million ESALs:
    a. **≤ 89% and ≤ 98%, respectively**
    b. ≤ 89% and ≤ 100%, respectively
    c. ≤ 85% and ≤ 98%, respectively
    d. ≤ 89% and ≤ 96%, respectively
    e. ≤ 90% and ≤ 96%, respectively

95. The Superpave system specifies the following value for the % G_{mm} (percent maximum specific gravity or mixture density) at N_{design}:
    a. 100%
    b. 97%
    c. 95%
    d. 89%
    e. **96%**

96. One of the following mix properties is not controlled by the Superpave mix design criteria:
    a. Voids in mineral aggregate (VMA)
    b. Voids filled with asphalt (VFA)
    c. Dust proportion (DP)
    d. Air voids (Va)
    e. **Fatigue resistance**

97. The Superpave mix design method evaluates one of the following performances for the mixture through the design process:
    a. Fatigue performance
    b. Low-temperature performance
    c. **Moisture susceptibility**
    d. Roughness
    e. None of the above

98. In the Superpave mix design method, the aggregate is evaluated and selected based on:
    a. Only consensus properties
    b. Only source properties
    c. Only chemical properties
    d. **Both consensus and source properties**
    e. Only mechanical properties

99. One of the following aggregate gradation charts is used to define permissible aggregate gradations in the Superpave mix design method:
    a. The Fuller and Thompson 0.5-power chart
    b. A logarithmic chart
    c. A linear chart
    d. A semi-logarithmic chart

   e. **The FHWA 0.45 power chart**

100. The definition "one standard sieve size larger than the first sieve to retain more than 10%" refers to:
   a. Restricted zone
   b. Critical sieve size
   c. Maximum aggregate size
   d. **Nominal maximum aggregate size (NMAS)**
   e. Control sieve size

101. The definition "one standard sieve size larger than the nominal maximum aggregate size" refers to:
   a. Restricted zone
   b. Critical sieve size
   c. **Maximum aggregate size (MAS)**
   d. Nominal maximum aggregate size (NMAS)
   e. Control sieve size

102. Superpave specifies gradation control points: maximum size, nominal maximum size, and key sieves for aggregate blends, which are:
   a. **No. 8 and No. 200**
   b. No. 4 and No. 8
   c. No. 8 and No. 16
   d. No. 16 and No. 30
   e. No. 100 and No. 200

103. In the original Superpave mix design method, one of the following statements is correct with regard to aggregate gradation criteria:
   a. Same gradation criteria are used for all gradations
   b. **Unique gradation criteria are used for each gradation with a specific NMAS**
   c. Only the Superpave restricted zone boundary is different from a gradation to another
   d. Only the criteria for the maximum size and the nominal maximum size are different from a gradation to another
   e. One band with a maximum limit and minimum limit is used for all gradations

104. In the Superpave mix design method, the criteria for aggregate consensus properties are set based upon:
   a. Traffic level (ESALs) only
   b. **Both traffic level and layer position**
   c. Layer position
   d. NMAS
   e. MAS

105. The Voids in Mineral Aggregate (VMA) criteria in the Superpave mix design method are based on:
   a. Traffic level (ESALs) only
   b. Both traffic level and layer position
   c. Layer position only
   d. **NMAS**
   e. MAS

106. The Voids Filled with Asphalt (VFA) criteria in the Superpave mix design method are based on:
   a. **Traffic level (ESALs) only**
   b. Both traffic level and layer position
   c. Layer position only
   d. NMAS
   e. MAS

107. The criteria for the percent maximum specific gravity ($\%G_{mm}$) at the initial number of gyrations ($N_{initial}$) in the Superpave mix design method are based on:
   a. **Traffic level (ESALs)**
   b. Both traffic level and layer position
   c. Layer position
   d. NMAS
   e. MAS

108. The criteria for the percent maximum specific gravity ($\%G_{mm}$) at the design number of gyrations (N_{design}) in the Superpave mix design method are:
   a. 98.0%
   b. 5.0%
   c. 95.0%
   d. **96.0%**
   e. 100.0%

109. The criteria for the percent maximum specific gravity ($\%G_{mm}$) at the maximum number of gyrations ($N_{maximum}$) in the Superpave mix design method are:
   a. **$\leq 98.0\%$**
   b. $\leq 5.0\%$
   c. $\leq 95.0\%$
   d. $\leq 96.0\%$
   e. $\leq 89.0\%$

110. The dust proportion (DP) requirement in the Superpave mix design method is:
   a. $1.0 - 2.0$
   b. $0.0 - 1.0$
   c. **$0.6 - 1.2$**
   d. $0.0 - 4.0$
   e. $3.0 - 5.0$

111. The design mixture is evaluated for moisture sensitivity (susceptibility) in the Superpave mix design method using:
   a. The French rutting test
   b. The Hamburg wheel-tracking test
   c. The asphalt pavement analyzer
   d. The asphalt mixture performance tester (AMPT)
   e. **The indirect tensile (IDT) strength test**

112. In the Superpave mix design method, the moisture susceptibility test is conducted on six Superpave gyratory specimens cut to 150 mm (diameter) $\times$ 50 mm (thickness) dimensions to determine the:
   a. Flow number ratio
   b. Dynamic modulus ratio
   c. **Tensile strength ratio**
   d. Resilient modulus ratio
   e. Creep stiffness ratio

113. Three of the six specimens in the moisture susceptibility test of the Superpave mix design method are conditioned using:
   a. Freezing at $-10°C$
   b. **Freeze and thaw cycles**
   c. Submerging in a 25°C bath
   d. Submerging in a 90°C bath
   e. Freezing at $-30°C$

114. The indirect tensile (IDT) strength test is conducted on fabricated Superpave gyratory specimens at a temperature and loading rate equal to:
    a. 60°C and 2 in/minute, respectively
    b. 60°C and 1 in/minute, respectively
    c. 25°C and 1 in/minute, respectively
    d. 25°C and 0.5 in/minute, respectively
    e. **25°C and 2 in/minute, respectively**

115. The moisture sensitivity in the Superpave mix design method is measured by the ratio of the average tensile strength of the conditioned (saturated) specimens to that of the dry (controlled) specimens known as the Tensile Strength Ratio (TSR). The criteria for the TSR in the Superpave is:
    a. ≥ 0.90
    b. ≥ 0.70
    c. ≥ 0.60
    d. $\mathbf{\geq 0.80}$
    e. ≥ 0.50

116. The voids in mineral aggregate as a volume phase in the asphalt mixture includes the following two volume phases:
    a. **The air voids and the voids filled with asphalt**
    b. The air voids and the absorbed asphalt volume
    c. The air voids and the bulk volume of the aggregate
    d. The apparent volume of the aggregate and the asphalt binder volume
    e. The air voids and the asphalt binder volume

117. The total (bulk) volume of the asphalt mixture minus the volume of the air voids provides the void-less volume of the mixtures that is used to determine:
    a. The bulk specific gravity of the mixture (G_{mb})
    b. The apparent specific gravity of the aggregate in the mixture (G_{sa})
    c. **The theoretical maximum specific gravity of the mixture (G_{mm})**
    d. The bulk specific gravity of the aggregate in the mixture (G_{sb})
    e. The effective specific gravity of the aggregate in the mixture (G_{se})

118. The asphalt binder volume minus the volume of the voids filled with asphalt represents the volume of the following phase in the asphalt mixture:
    a. The effective asphalt binder
    b. The bulk aggregates
    c. The voids in the asphalt mixture
    d. **The absorbed asphalt binder**
    e. The voids in the aggregate

119. The total (bulk) of the asphalt mixture minus the bulk volume of the aggregate in the mixture represents the volume of the following phase:
    a. The effective asphalt binder
    b. The voids in mineral aggregate
    c. **The voids in the asphalt mixture**
    d. The absorbed asphalt binder
    e. The asphalt binder

120. The bulk volume of the aggregate in the asphalt mixture minus the volume of the absorbed asphalt represents:
    a. The volume of the effective asphalt binder
    b. The volume of the voids in mineral aggregate
    c. The volume of the voids in the asphalt mixture

    d.   **The effective volume of the aggregate**
    e.   The apparent volume of the aggregate
121. The total (bulk) volume of the asphalt mixture consists of the volume of air voids plus the volume of asphalt binder plus:
    a.   The volume of the voids in mineral aggregate
    b.   The bulk volume of the aggregate
    c.   **The effective volume of the aggregate**
    d.   The volume of the effective asphalt binder
    e.   The apparent volume of the aggregate
122. The ability of the pavement to satisfy the needs of the road users, which is also related to skid resistance, safety, and the visual appearance of the road pavement is called:
    a.   Structural behavior
    b.   **Functional behavior**
    c.   Empirical behavior
    d.   Mechanistic behavior
    e.   Mechanistic-empirical behavior
123. The ability of the pavement to resist the traffic loadings under the existing environmental conditions is called:
    a.   **Structural behavior**
    b.   Functional behavior
    c.   Empirical behavior
    d.   Mechanistic behavior
    e.   Mechanistic-empirical behavior
124. The pavement index that is based on mathematical calculations of physical measurements of pavement distresses to predict the serviceability of the pavement is called:
    a.   Structural index
    b.   Functional index
    c.   Present Serviceability Rating (PSR)
    d.   **Present Serviceability Index (PSI)**
    e.   Pavement condition
125. The initial serviceability index (p_i) value recommended by the AASHTO road test in the design of new flexible pavements is:
    a.   4.0
    b.   **4.2**
    c.   5.0
    d.   3.0
    e.   2.5
126. The terminal serviceability index (p_t) value recommended by the AASHTO road test in the design of new flexible pavements for major highways is:
    a.   **2.5 or 3.0**
    b.   2.0
    c.   1.5
    d.   4.0
    e.   1.0
127. The terminal serviceability index (p_t) value recommended by the AASHTO road test in the design of new flexible pavements for minor (low class) highways is:
    a.   2.5 or 3.0
    b.   **2.0**
    c.   1.5
    d.   4.0
    e.   1.0

128. The standard axle used to compute the number of equivalent single-axle loads (ESALs) for traffic load applications on flexible pavements is:
    a. 12000 lb (53 kN)
    b. 10000 lb (44 kN)
    c. 9000 lb (40 kN)
    d. 20000 lb (89 kN)
    e. **18000 lb (80 kN)**

129. The "N" in the formula used to determine the ESALs for traffic load applications on flexible pavements refers to:
    a. The total number of lanes in the highway
    b. The number of trucks considered for design
    c. The design life (number of years)
    d. **The number of axles with the same type and load for a specific vehicle category**
    e. The number of layers

130. The traffic growth factor (G) used in the ESALs formula takes into consideration:
    a. Traffic growth rate only
    b. Design period only
    c. Traffic growth rate and traffic distribution
    d. Traffic distribution and design period
    e. **Traffic growth rate and design period**

131. The quality of the material used in each pavement layer is determined in the AASHTO empirical pavement design method in terms of:
    a. The California bearing ratio (CBR)
    b. The resilient modulus (M_R)
    c. **Layer coefficients (a_1, a_2, and a_3)**
    d. PSI
    e. Drainage coefficients

132. The value of drainage coefficient (m) for untreated base and subbase layers of flexible pavements is controlled by two factors, which are:
    a. **The time required to remove water from the pavement layer and the percentage of time the pavement is exposed to moisture levels approaching saturation**
    b. The time required to remove water from the pavement layer and the location of the pavement layer
    c. The percentage of time the pavement is exposed to moisture levels approaching saturation and the thickness of the layer
    d. The thickness of the layer and the material used in the layer
    e. The time required to remove water from the pavement layer and the material type used in the layer

133. The major distresses that are of significance in flexible pavement design are:
    a. Fatigue cracking, block cracking, and bleeding
    b. **Fatigue cracking, rutting, and low-temperature cracking**
    c. Edge cracking, corrugation, and slippage
    d. Rutting, depressions, and bleeding
    e. Fatigue cracking, raveling, and bleeding

134. The critical pavement response in a conventional flexible pavement for fatigue cracking is:
    a. The horizontal tensile strain at the bottom of the base layer
    b. The vertical compressive strain at the top of the subbase layer
    c. The horizontal strain at the bottom of the subbase layer
    d. The vertical compressive strain at the top of the subgrade layer
    e. **The horizontal tensile strain at the bottom of the asphalt layer**

135. The most critical pavement response in a conventional flexible pavement for rutting is:
    a. The horizontal tensile strain at the bottom of the base layer
    b. The vertical compressive strain at the top of the subbase layer
    c. The horizontal strain at the bottom of the subbase layer
    d. **The vertical compressive strain at the top of the subgrade layer**
    e. The horizontal tensile strain at the bottom of the asphalt layer
136. A distress model in a mechanistic-empirical pavement design method is a function that:
    a. **Determines the pavement distress from the pavement response**
    b. Determines the structural pavement design from the pavement response
    c. Determines the final pavement design from the pavement response
    d. Determines the critical pavement response
    e. All of the above
137. The pavement response in a flexible pavement is based on many factors including:
    a. Wheel load
    b. Depth in pavement
    c. Modulus of layers
    d. Thickness of pavement layers
    e. **All of the above**
138. In the flexible plate analysis for one-layered systems:
    a. The strain distribution is uniform
    b. The pressure is different from point to point under the plate
    c. The deflection is different from point to point under the plate
    d. The pressure distribution is uniform
    e. **c and d**
139. In the rigid plate analysis for one-layered systems:
    a. The pressure distribution is uniform
    b. The pressure distribution is not uniform
    c. The deflection is different from point to point under the plate
    d. The deflection is the same at all points on the plate
    e. **b and d**
140. The subgrade strength considered in the AASHTO design method for rigid pavements is given in terms of:
    a. The resilient modulus (M_R)
    b. The California bearing ratio (CBR)
    c. The layer coefficient
    d. **The Westergaard modulus of subgrade reaction (k)**
    e. The maximum dry density
141. The effective modulus of subgrade reaction for the subgrade layer in rigid pavements is used as an input in the AASHTO design procedure after modifying the composite modulus of subgrade reaction for:
    a. Subbase thickness
    b. Subbase elastic modulus
    c. Soil resilient modulus
    d. Faulting and pumping
    e. **The potential loss of subbase support due to erosion and the effect of rigid foundation near subgrade surface**
142. The property of concrete that is used in the AASHTO design procedure for rigid pavements is:
    a. Strength
    b. Modulus of elasticity
    c. a and b

    d. Modulus of rupture

    e. **b and d**

143. The major distresses that are of significance in rigid pavement design are:

    a. **Joint faulting, joint spalling, linear cracking, and pumping**

    b. Transverse cracking, pumping, lane/shoulder drop off, and divided slab

    c. Joint faulting, lane/shoulder drop off, divided slab, and D-cracking

    d. Corner spalling, polished aggregate, scaling, and blow-up

    e. Transverse cracking, pumping, punchouts, and pop-outs

144. One of the following is <u>not</u> among the main causes of joint faulting in rigid pavements:

    a. Settlement due to weak foundation

    b. Curling of the slab

    c. Erosion in the subgrade soil

    d. Sever pumping

    e. **Concrete shrinkage**

145. One of the following is <u>not</u> among the main causes of joint spalling in rigid pavements:

    a. Lack of joint sealant and incompressible materials in the joint

    b. High stresses at the joint due to high traffic loading

    c. **Erosion in the subgrade soil**

    d. Improper dowel alignment

    e. Freeze–thaw cycles at the joint

146. The major causes of linear cracking in rigid pavements are:

    a. Repeated traffic loading

    b. Repeated moisture loading

    c. Curling stresses

    d. Improper dowel alignment

    e. **a, b, and c**

147. The major causes of pumping in rigid pavements are:

    a. Water and fine materials

    b. Lack or improper load transfer at the joint

    c. Curling stresses

    d. Repeated traffic loading

    e. **a, b, and d**

148. The present serviceability index (PSI) for rigid pavements is a function of:

    a. Slope variance, spalling, and pumping

    b. Cracking, faulting, and pumping

    c. **Slope variance, cracking, and patching**

    d. Faulting, spalling, and pumping

    e. Slope variance, faulting, and pumping

149. Curling stresses in rigid pavements occur during the day in the form of:

    a. **Compressive stresses at the top of the slab and tensile stresses at the bottom of the slab**

    b. Tensile stresses at the top of the slab and compressive stresses at the bottom of the slab

    c. Only tensile stresses at the top of the slab

    d. Only compressive stresses at the bottom of the slab

    e. None of the above

150. Curling stresses in rigid pavements occur during the night in the form of:

    a. Compressive stresses at the top of the slab and tensile stresses at the bottom of the slab

    b. **Tensile stresses at the top of the slab and compressive stresses at the bottom of the slab**

    c. Only compressive stresses at the top of the slab

    d. Only tensile stresses at the bottom of the slab

    e. None of the above

151. Stresses due to traffic loading in rigid pavements are ordered (in value) relative to the location of the load as below:
    a. **Edge of the slab > corner of the slab > interior of the slab**
    b. Corner of the slab > edge of the slab > interior of the slab
    c. Corner of the slab > interior of the slab > edge of the slab
    d. Interior of the slab > corner of the slab > edge of the slab
    e. Interior of the slab > edge of the slab > corner of the slab

152. Joints are provided in rigid pavements to:
    a. Transfer load
    b. Prevent pumping
    c. Prevent faulting
    d. **Prevent premature cracks due to temperature and moisture changes from occurring**
    e. Prevent blow-up

153. One of the following is <u>not</u> among the types of joints in rigid pavements:
    a. Contraction joints
    b. Expansion joints
    c. Construction joints
    d. Longitudinal joints
    e. **None of the above**

154. The type of joint that is used to relieve tensile stresses in rigid pavement is:
    a. **Contraction joint**
    b. Expansion joint
    c. Construction joint
    d. Longitudinal joint
    e. Tension joint

155. The type of joint that is used in the transverse direction for the relief of compressive stresses in rigid pavement is:
    a. Contraction joint
    b. **Expansion joint**
    c. Construction joint
    d. Longitudinal joint
    e. Tension joint

156. The type of joint that is placed at the location of contraction joint is called:
    a. Contraction joint
    b. Expansion joint
    c. **Construction butt joint**
    d. Longitudinal joint
    e. Tension joint

157. The type of joint that is placed at the location of contraction joint is called:
    a. Contraction dummy groove joint
    b. Expansion joint
    c. **Construction butt joint**
    d. Longitudinal joint
    e. Tension joint

158. The type of joint that is used in case if work must stop due to an emergency situation is:
    a. Contraction dummy groove joint
    b. Contraction pre-molded strip joint
    c. Construction butt joint
    d. **Construction key joint**
    e. Longitudinal joint

159. The type of joint that is used to relieve curling and warping stresses in rigid pavement is:
    a. Contraction joint
    b. Expansion joint
    c. Construction joint
    d. **Longitudinal joint**
    e. Tension joint
160. For lane-at-a time construction of rigid pavements, the proper type of longitudinal joint that should be used is:
    a. Butt joint
    b. key joint
    c. Dummy groove joint
    d. Ribbon joint
    e. **a or b**
161. For full-width construction of rigid pavements, the proper type of longitudinal joint that should be used is:
    a. Butt joint or key joint
    b. Dummy groove joint
    c. Ribbon (or pre-molded strip) joint
    d. Deformed plate joint
    e. **b, c, or d**

References

American Association of State Highway and Transportation Officials (AASHTO), AASHTO Guide for Design of Pavement Structures, American Association of State Highway and Transportation Officials (AASHTO), Washington, DC, 1993.

Asphalt Institute, *Asphalt Mix Design Methods, Asphalt Institute Manual Series No. 2 (MS-2)*, 7th Edition, Asphalt Institute, Lexington, KY, 2014.

Asphalt Institute, Asphalt Pavement Thickness Design: A Simplified Version of the 1981 Edition of the Asphalt Institute's Thickness Design Manual Series No. 1 (MS-1), Asphalt Institute, Lexington, KY, 2001.

Asphalt Institute, Research and Development of the Asphalt Institute's Thickness Design Manual (MS-1), 9th Edition, Research Report No. 82-2 (RR-82-2), Asphalt Institute Building, College Park, MD, 1982.

Asphalt Institute, *Superpave Mix Design, Asphalt Institute Superpave Series No. 2 (SP-2)*, 3rd Edition, Asphalt Institute, Lexington, KY, 2001.

Asphalt Institute, Superpave Performance Graded Asphalt Binder Specifications and Testing, Asphalt Institute Superpave Series No. 1 (SP-1), Asphalt Institute, Lexington, KY, 2003.

Asphalt Institute, Thickness Design-Asphalt Pavements for Highways and Streets, Asphalt Institute Manual Series No. 1 (MS-1), Asphalt Institute, Lexington, KY, 1999.

Highway Capacity Manual (HCM), Transportation Research Board, National Research Council, Washington, DC, 2000.

Manual on Uniform Traffic Control Devices (MUTCD), The Federal Highway Administration (FHWA), US Department of Transportation, 2010.

Microsoft Excel, Office 365, 2016.

Microsoft Excel Solver Tool, Office 365, 2016.

Nicholas J. Garber and Lester A. Hoel, *Traffic and Highway Engineering*, 4th Edition, Cengage Learning, Inc., 2009.

Steven C. Chapra and Raymond P. Canale, *Numerical Methods for Engineers*, McGraw Hill Education, New York, 2015.

WinJULE Program, Windows-Based Layered Elastic Analysis Program, Engineering Research and Development Center (ERDC), Airfields and Pavements Branch, Vicksburg, MS, 2003.

Yang H. Huang, *Pavement Analysis and Design*, 2nd Edition, Pearson Education, Inc., New York, NY, 2004.

Index

Printed in the United States
by Baker & Taylor Publisher Services